Mathematical Engineering

Series editors

Claus Hillermeier, Neubiberg, Germany
Jörg Schröder, Essen, Germany
Bernhard Weigand, Stuttgart, Germany

For further volumes:
http://www.springer.com/series/8445

Joël Chaskalovic

Mathematical and Numerical Methods for Partial Differential Equations

Applications for Engineering Sciences

 Springer

Joël Chaskalovic
Institut Jean le Rond d'Alembert
University Pierre and Marie Curie
Paris
France

ISSN 2192-4732   ISSN 2192-4740   (electronic)
ISBN 978-3-319-37556-4   ISBN 978-3-319-03563-5   (eBook)
DOI 10.1007/978-3-319-03563-5
Springer Cham Heidelberg New York Dordrecht London

Printed on acid-free paper

Springer is part of Springer Science+Business Media (www.springer.com)

To Anate,
For these years of patience
and cognitive orthogonality,
so complementary and constructive,
which keep us moving forward!

To Eva and Arielle,
who have endured the torment
of mathematical formulas
and metaphysical discourse.
Anyway, highly formative!

Preface

This book is an introduction to the mathematical analysis and approximation of partial differential equations using the finite-element method. It arises in part from third-year undergraduate and graduate lectures given by the author at the University of Pierre and Marie Curie (Paris VI) over the past 15 years, under the headings "Partial Differential Equations" and "Numerical Methods for Mechanics," but also from two books published by the author in 2004 and 2008 [1, 2].

Over the past 60 years, numerical analysis has witnessed a spectacular development. Of course, this has much to do with the unprecedented explosion of computer technology, which has literally taken over the world, generating hitherto undreamt of computational capacity. The whole field of mathematical activity that constitutes numerical analysis might well be described as the mathematics of mathematics, in the same sense in which one speaks of a police of police. For whenever a mathematical technique cannot be implemented in the context of industrial applications for reasons of operational inadequacy, numerical analysis will step into solve the problem by identifying the most suitable approximation process. From this moment on, all other branches of mathematics can be called into "force a way through," estimating a solution, combining tricks of the trade with lucidity, and all this within as rigorous a mathematical framework as possible.

One of the favored areas of application of numerical analysis is the approximation of partial differential equations, a major factor in the modeling of real systems. Indeed, whether we consider applications in physics or mechanics, in economics, the media, marketing, or the field of finance, the mathematical formulation of the system under investigation leads frequently to ordinary or partial differential equations that require solution. For this reason, many methods have been devised to solve such equations. Among the best known are finite-difference methods, finite-volume methods, singularity or integral methods, spectral methods, and variational finite-difference methods. However, there can be no doubt whatever that it is the finite-element method that has literally revolutionized the numerical approximation of partial differential equations.

With its unequaled flexibility, the finite-element method surely constitutes the most widely used method for approximate solution of mathematical models in the engineering sciences. Given the high level of mathematical know-how required to implement this method, several specialists in numerical analysis, such as P. A. Raviart [4], have presented the subject in advanced-level university courses, reserved for students with the necessary mathematical prerequisites, particularly in functional analysis, which is key to a theoretical understanding of the finite-element method. For other students, those not specializing in mathematics, such as undergraduate and graduate students in physics or mechanics, but also engineering students, namely those who consume mathematics at a range of different levels, the book [3] published by D. Euvrard in the 1990s provided a course for those not so familiar with the tools of functional analysis.

The outstanding success of these two books merely reflects the high level of teaching of numerical analysis in the mechanics department at the University of Pierre and Marie Curie, and later in the engineering faculty, by P. A. Raviart and subsequently by D. Euvrard. Concerning the present work, the idea itself and the core of the book result from the significant interaction between the various aspects of my teaching activities, and in particular, undergraduate- and master's-level mechanics courses at the University of Pierre and Marie Curie, including methods of functional analysis and numerical methods applied to mechanics and the mechanics of deformable solids. Indeed, the aim was to pursue the enterprise of the previously cited authors while supplementing the course in such a way as to achieve a new balance between a work devoted to specialists and one "dealing with operational aspects, while merely touching upon the mathematical under-pinnings," to quote Euvrard [3 p. 198], who had considerable influence on my thinking during those years of teaching.

My training and my appreciation of the subjects underlying mathematical and numerical analysis, and in particular those relating to partial differential equations, can be largely attributed to Prof. Gérard Tronel, a notably active and devoted member of the numerical analysis teaching group at the Jacques Louis Lions research center of the University of Pierre and Marie Curie. The new balance presented in the present work owes much to the relevance of Prof. Tronel's pedagogical approach, from which I benefited enormously, first as a student, then as a colleague. I extend my warmest thanks to him.

This new form of presentation of 1-dimensional applications to the resistance of materials has been followed by third year undergraduate and master's students in mechanics at UPMC. The present book discusses these examples and extends to other applications. The first step was to identify certain key tools of functional analysis, which are presented without proof unless they exhibit some specific didactic relevance or a potential for implementation in the context of specific applications. The second was to deploy and exploit these results in the examination of problems relating to the existence, uniqueness, and regularity of the weak solutions and their equivalence with the strong solutions.

In this respect, the Part I of present book contains, on the one hand, an intro-duction to the techniques of functional analysis insofar as they are relevant to partial differential equations, and on the other, a discussion of the finite-element method complementary to that given by Euvrard [3]. The Part II presents and solves several problems that exploit these techniques of functional analysis while constructing the kind of node equations that characterize every numerical imple-mentation of the finite-element method. There is also a detailed comparison of finite-element methods and finite-difference schemes, which correspond to the node equations when the mesh is regular and for certain types of finite-element analysis. Particular attention has been paid to the assembly process as illustrated in problems of resistance of materials.

In this context, I would like to pay homage to the memory of Claude Kammoun, who introduced me to these techniques in the context of the resistance of materials. I would also like to thank my colleagues at the UPMC—Benoit Goyeau, Cédric Croizet, Diana Baltéan, Catherine Weisman, and Hélène Dumontet—and those at the Ecole Nationale Supérieure in Cachan—Ludovic Chamoin and Florent Pled—and also Professor Franck Assous, for many discussions that have done much to extend and deepen my understanding and given me the impetus to write the present book. I extend my warmest thanks to all of them.

Paris, France, May 2013 Joël Chaskalovic

References

1. J. Chaskalovic, *Méthodes des éléments finis pour les sciences de l'ingénieur*, ed. by Lavoisier (2004)
2. J. Chaskalovic, *Finite Element Methods for Engineering Sciences*, (Springer-Verlag, New York, 2008)
3. D. Euvrard, *Résolution des équations aux dérivées partielles de la physique, de la mécanique et des sciences de l'ingénieur*, (Masson, Paris, 1994)
4. P.A. Raviart, J.M. Thomas, *Introduction à l'analyse numérique des équations aux dérivées partielles*, (Masson, Paris, 1988)

Contents

Part I
Theory

Chapter 1
Applications of Functional Analysis to Partial Differential Equations

1.1 Basic Issues in the Mathematical Analysis of Partial Differential Equations

1.1.1 The Standard Problem

The classic example used to illustrate the basic problems raised by the mathematical analysis of partial differential equations (PDEs) concerns the mechanics of deformable solids, and in particular a homogeneous and isotropic elastic membrane occupying a region Ω in the plane $(O, \mathbf{x}, \mathbf{y})$. This membrane is subjected to a surface density of transverse forces $\mathbf{f} = f(x, y)\mathbf{z}$, where $\mathbf{z}$ is a unit vector in the orthonormal basis $(\mathbf{x}, \mathbf{y}, \mathbf{z})$ of $\mathbb{R}^3$.

If $\mathbf{u} = u(x, y)\mathbf{z}$ represents the field of displacements of the membrane and if we apply the theory of linear elasticity under the assumption of small perturbations, then u will be the solution to the following problem:

Find real-valued u defined on Ω such that

$$(\textbf{CP}) \quad \begin{cases} -\Delta u(x, y) = f(x, y), & \forall (x, y) \in \Omega, \quad (1.1) \\ u = 0, & \text{on } \partial\Omega, \quad (1.2) \end{cases}$$

where $\Delta \equiv \partial^2/\partial x^2 + \partial^2/\partial y^2$ is the Laplacian operator and $\partial\Omega$ is the boundary of the membrane Ω. The condition (1.2) says that the membrane is fixed along its boundary $\partial\Omega$. This is known as the strong formulation of the boundary value problem, or the continuous problem (**CP**).

The continuous problem raises several questions, applicable to any problem formulated in terms of differential or partial differential equations:

1. **Existence of Solutions.** Insofar as (**CP**) is a mathematical model for some aspect of physical reality, there is no immediate guarantee that it will have solutions.

J. Chaskalovic, *Mathematical and Numerical Methods for Partial Differential Equations*, Mathematical Engineering, DOI: 10.1007/978-3-319-03563-5_1, © Springer International Publishing Switzerland 2014

2. **Uniqueness of the Solution.** If solutions do exist, how many are there?

3. **Dependence of the Solution on the Input Data.** In the present case, we should ask how sensitive the solution u is to the input data for **(CP)**, i.e., the force field **f** and the domain Ω. More precisely, how does the solution u change when these data are adjusted?

These three properties, which characterize solutions to **(CP)**, lead naturally to the notion of a well-posed problem in the sense of Hadamard:

Definition 1.1 The continuous problem **(CP)** is Hadamard well posed if for any data f, there exists one and only one solution u of **(CP)** that depends continuously on the data f in the sense that

$$(f_n \longrightarrow f) \implies (u_n \longrightarrow u). \tag{1.3}$$

As it stands, this definition is not complete, because it depends on how we define the convergence of the function sequences, i.e., the forces f_n and the displacements u_n. From a mathematical standpoint, convergence will be defined in terms of the norms that characterize the regularity of the given functions.

So we come naturally to the need to define convergence, and this will depend on the nature of the functions f and u. For example, we will have to consider the following situations:

- If f is continuous on Ω, convergence is measured relative to the uniform convergence norm $\|f\|_\infty$ defined by

$$\|f\|_\infty \equiv \sup_{x \in \Omega} |f(x)|. \tag{1.4}$$

- If f is a square integrable function on Ω, we prefer the L^2 norm defined by

$$\|f\|_{L^2} \equiv \sqrt{\int_\Omega |f(x)|^2 dx}. \tag{1.5}$$

4. **Regularity of the Solution.** This aspect of the problem has led to many developments and given rise to its own field of mathematical analysis. It involves *a priori* estimation of solutions in order to describe their regularity, independently of whether they exist.

More precisely, the aim is to characterize solutions in terms of the problem data. This is crucial, because it fixes the functional framework that will be used to search for solutions to **(CP)**.

5. **Methods of Resolution.** This last point concerns solution of the mathematical formulation of **(CP)** in order to arrive at a suitable result u. There are two main lines of attack. One involves analytic solution, where u is calculated exactly,

while the other proceeds by approximation, adapting numerical methods to the mathematical formulation of the continuous problem (**CP**). The second of these will be discussed in detail in the second part of the book.

1.1.2 Implications of the Mathematical Analysis

From a mathematical standpoint, the continuous problem (**CP**) belongs to a broader class of problems that can be described formally in the following way:

$$(\textbf{CP}) \quad \begin{cases} \text{Find } u \in V \text{ a solution to} \\ A(u) = f, \end{cases} \tag{1.6}$$

where

- V is a suitably chosen function space, and

- A is a specific operator on V with values in a second function space W, which characterizes the regularity of the external effects f acting on the system.

This formalism raises several fundamental questions:

- What is a suitable choice for the function space V?

- What about the space W?

- What are the most suitable norms or topologies, considering the aims of the mathematical analysis?

- The notion of Hadamard well-posedness is not fully determined without first specifying the function spaces and the norms. Indeed, a different choice here can lead to quite contradictory results regarding the nature of the solutions and their properties for a given continuous problem (**CP**) as specified by (1.6).

Investigation of the existence, uniqueness, and regularity of the solution to such a problem will be carried out using a new formulation, the variational formulation, which is better suited both to the mathematical analysis and to the methods of numerical approximation to be developed later in the book.

1.1.3 Strong and Weak Formulations

To present the transformation of the strong (or continuous) formulation to the weak (or variational) formulation, rather than considering the membrane introduced at the beginning of the chapter, we return to our original model and simplify by rewriting it in a one-dimensional space of $x \in \mathbb{R}$ such that $0 \leq x \leq L$, where L stands for the arbitrary length of a beam.

Then the model **(CP)** given in (1.1)–(1.2) takes the following form:

Find a real-valued function u on $[0, L]$ that solves

(CP) $\quad \begin{cases} -u''(x) = f(x), & 0 \le x \le L, \\ u(0) = u(L) = 0. \end{cases}$ $\qquad\qquad$ (1.7)

(1.8)

A variational formulation **(VP)** is associated with the continuous problem **(CP)** by considering the solution u as a particular element to be found in a function space V.

If v denotes the generic element of this space V, we multiply the differential equation (1.7) by v and integrate from 0 to L to obtain

$$-\int_0^L u'' v \, dx = \int_0^L f v \, dx.$$ $\qquad\qquad$ (1.9)

We then integrate by parts to bring in, as far as possible, the boundary conditions (1.8) of **(CP)**:

$$\int_0^L u'v' dx - u'(L)v(L) + u'(0)v(0) = \int_0^L f v \, dx.$$ $\qquad\qquad$ (1.10)

Noting that the boundary conditions on u do not appear in (1.10), we require all elements v of V to satisfy

$$v(0) = v(L) = 0.$$

Then, if we find a solution u in V, it will automatically satisfy the conditions $u(0) = u(L) = 0$, simply because it will be one of the functions v in V. We then obtain the following variational formulation **(VP)**:

Find $u \in V \equiv \{v : [0, L] \longrightarrow \mathbb{R} \text{ such that } v(0) = v(L) = 0\}$ that is a solution of

(VP) $\quad \int_0^L u'v' dx = \int_0^L f v \, dx, \quad \forall v \in V.$ $\qquad\qquad$ (1.11)

We have thus formally established that if u is a solution of the continuous problem **(CP)** given by (1.7)–(1.8), then u is also a solution of the variational problem **(VP)** given by (1.11).

However, this is only a formal result. One must also undertake more careful analysis of the following fundamental questions:

1. What is the function space in which **(CP)** is posed? The answer to this question depends directly on the regularity of the input data f.

2. In the same vein, what function space V authorizes the variational formulation **(VP)**? The main point here is to guarantee convergence of the integrals appearing in **(VP)**. This, in turn, depends directly on both the data f and the differential operator appearing in **(CP)** and transformed by integration by parts in **(VP)**.

3. Since the numerical approximation by finite elements will be developed and carried out on the basis of the variational formulation **(VP)**, it will be essential to check the equivalence of the two formulations **(CP)** and **(VP)**. Indeed, without this equivalence, one could not justify considering approximations to the solution u of **(VP)** as approximations to the solution u of **(CP)**.

   In other words, although we know that a solution to **(CP)** will always be a solution to **(VP)** by the very construction of the variational formulation **(VP)** described above, what can we say about the converse, i.e., **(VP)** $\Rightarrow$ **(CP)**?

4. Once we have established the equivalence of the two formulations, should we consider the existence and uniqueness of the solution u of **(CP)**, or would it be easier to address this question in the context of the variational formulation **(VP)**?

5. Finally, the problem of *a priori* estimation of the solution u of **(CP)** should also be considered from the point of view of the variational formulation **(VP)**.

6. In the case of the variational formulation **(VP)** we are considering, there is another equivalent formulation that takes the form of the following minimization problem:

$$
\textbf{(MP)} \quad
\begin{cases}
\text{Find } u \in V \text{ a solution to} \\[4pt]
J(u) \equiv \min_{v \in V} J(v), \\[4pt]
\text{where } J(v) = \dfrac{1}{2} \int_0^L v'^2 \, dx - \int_0^L f v \, dx.
\end{cases}
\tag{1.12}
$$

The equivalence of the formulations **(VP)** and **(MP)** will be established in a more general context in Sect. 1.6 as part of Theorem 1.13.

These are the questions that motivate the discussion in the following chapters and the presentation of suitable mathematical tools to answer them for a range of problems involving partial differential equations.

1.2 Banach Spaces

In this section, we recall the main definitions and properties required for the rest of the book. We shall limit our discussion to illustrating results rather than trying to provide systematic proofs, since the reader can find these in any standard textbook on Hilbert spaces (see, for example, [1]).

We begin by defining Cauchy sequences in a normed vector space $(E, \| \cdot \|_E)$, assuming that the reader is already familiar with the basic notions of vector spaces and norms.

Definition 1.2 A Cauchy sequence $(u_n)_{n\in\mathbb{N}}$ in a normed vector space $(E, \|\cdot\|_E)$ has the following property:

$$\forall \varepsilon > 0, \; \exists \, N(\varepsilon) > 0 \text{ such that: } \; n > m > N(\varepsilon) \Longrightarrow \|u_n - u_m\|_E < \varepsilon. \tag{1.13}$$

Convergent sequences are defined analogously in a normed vector space $(E, \|\cdot\|_E)$.

Definition 1.3 A sequence $(u_n)_{n\in\mathbb{N}}$ in a normed vector space $(E, \|\cdot\|_E)$ converges to a limit l belonging to E if

$$\forall \varepsilon > 0, \; \exists \, N(\varepsilon) > 0 \text{ such that: } \; n > N(\varepsilon) \Longrightarrow \|u_n - l\|_E < \varepsilon. \tag{1.14}$$

Note. Definitions 1.2 and 1.3 are not intrinsic, since they depend on the norm $\|\cdot\|_E$ on the space E. In other words, if a space E can be equipped with two norms $\|\cdot\|_1$ and $\|\cdot\|_2$, it may be that E has Cauchy sequences or convergent sequences for one of the two norms that are not Cauchy or convergent, respectively, for the other.

We gather the main properties of Cauchy sequences in Banach spaces in the following lemma.

Lemma 1.1 *Let $(u_n)_{n\in\mathbb{N}}$ be a sequence in a normed vector space $(E, \|\cdot\|_E)$. Then we have:*

1. *If $(u_n)_{n\in\mathbb{N}}$ is a Cauchy sequence for the norm $\|\cdot\|_E$, it is bounded for this norm.*

2. *If $(u_n)_{n\in\mathbb{N}}$ is a Cauchy sequence for the norm $\|\cdot\|_E$, then every subsequence $(u_{\alpha(n)})_{n\in\mathbb{N}}$ is also a Cauchy sequence in E for that norm.*

3. *If $(u_n)_{n\in\mathbb{N}}$ is a Cauchy sequence for the norm $\|\cdot\|_E$ that has a convergent subsequence, then $(u_n)_{n\in\mathbb{N}}$ is also convergent.*

4. *If $(u_n)_{n\in\mathbb{N}}$ is a convergent sequence for the norm $\|\cdot\|_E$, then $(u_n)_{n\in\mathbb{N}}$ is also a Cauchy sequence.*

Note that the converse of the last property is false. Indeed, there are Cauchy sequences that do not converge in their space. For example, the sequence of rational numbers $(u_n)_{n\in\mathbb{N}} \in \mathbb{Q}$ given by

$$u_n \equiv \frac{1}{0!} + \frac{1}{1!} + \frac{1}{2!} + \cdots + \frac{1}{n!}$$

is Cauchy in $\mathbb{Q}$ but converges to constant e, which belongs to $\mathbb{R}$ but not $\mathbb{Q}$. It was this observation that motivated the introduction of Banach spaces, defined as follows.

Definition 1.4 A normed vector space $(E, \|\cdot\|_E)$ is a Banach space if it is complete, i.e., if every Cauchy sequence $(u_n)_{n\in\mathbb{N}}$ in E converges in E.

Once again, we note that the characterization of a given space E as a Banach space depends directly on the choice of norm $\|\cdot\|_E$. It follows that a given space may be Banach for one norm but not for another.

1.2.1 Examples of Banach Spaces

Example 1

The space of real numbers $\mathbb{R}$ is complete. This property can be considered to result directly from the construction of $\mathbb{R}$, or it can be established in the following way. According to the Bolzano–Weierstrass theorem, every bounded sequence of real numbers has a convergent subsequence. Hence, if $(u_n)_{n\in\mathbb{N}}$ is a Cauchy sequence, it is bounded, by the first property of Lemma 1.1, and so contains a convergent subsequence. It follows that $(u_n)_{n\in\mathbb{N}}$ is itself a convergent sequence by the third property of Lemma 1.1.

Example 2

The space $C^0([-1, 1])$ of continuous real-valued functions on $[-1, 1]$ is a Banach space for the uniform convergence norm $\|\cdot\|_\infty$ defined by

$$\|f\|_\infty \equiv \sup_{-1\leq x\leq 1} |f(x)|. \tag{1.15}$$

Indeed, if $(f_n)_{n\in\mathbb{N}}$ is a Cauchy sequence of functions in $C^0([-1, 1])$ for the norm (1.15), then

$$\forall \varepsilon > 0, \ \exists N(\varepsilon) > 0 \text{ such that } \forall n, m > N(\varepsilon), \tag{1.16}$$

$$\|f_n - f_m\|_\infty \equiv \sup_{-1\leq x\leq 1} |f_n(x) - f_m(x)| < \varepsilon.$$

If we fix any x in $[-1, 1]$, the inequality (1.16) implies that

$$|f_n(x) - f_m(x)| < \varepsilon, \tag{1.17}$$

so the sequence of real numbers $\left\{f_n(x)\right\}_{n\in\mathbb{N}}$ is Cauchy in $\mathbb{R}$. But then, since $\mathbb{R}$ is a Banach space for the norm defined by the absolute value, we deduce that the sequence

$\{f_n(x)\}_{n \in \mathbb{N}}$ converges in $\mathbb{R}$ to some number, which we shall denote by $f(x)$. We then note that this works for all $x \in [-1, 1]$.

Now consider the function f defined by the above process, i.e.,

$$\forall x \in [-1, 1]: \quad f(x) \equiv \lim_{n \to +\infty} f_n(x). \tag{1.18}$$

We shall show that f_n converges to f for the uniform convergence norm (1.15). Now, we have already seen that if $(f_n)_{n \in \mathbb{N}}$ is Cauchy for the norm $\| \cdot \|_\infty$ defined by (1.15), then $\forall x \in [-1, 1]$, we have

$$\forall \varepsilon > 0, \ \exists N(\varepsilon) > 0 \text{ such that: } \forall n, m > N(\varepsilon), \ |f_n(x) - f_m(x)| < \varepsilon. \tag{1.19}$$

In the previous inequality, we fix n and note that the sequence $\{|f_m(x) - f_n(x)|\}_{m \in \mathbb{N}}$ converges to the number $|f(x) - f_n(x)|$ for each value of x. Since the terms of the sequence $\{|f_m(x) - f_n(x)|\}_{m \in \mathbb{N}}$ are bounded by ε, the limit of this sequence is also less than ε, i.e.,

$$\forall n > N(\varepsilon) \text{ and } \forall x \in [-1, 1], \ |f(x) - f_n(x)| < \varepsilon, \tag{1.20}$$

whence

$$\sup_{-1 \le x \le 1} |f(x) - f_n(x)| < \varepsilon. \tag{1.21}$$

In terms of the norm $\| \cdot \|_\infty$, this can be written

$$\lim_{n \to +\infty} \|f_n - f\|_\infty = 0. \tag{1.22}$$

To complete the proof, we must show that the limit function f is indeed in the set $C^0([-1, 1])$. But a function is continuous on a bounded interval if it is uniformly continuous on that interval, i.e., if

$$\forall \varepsilon > 0, \ \exists \eta > 0 \text{ such that } \forall (x, y) \in [-1, 1]^2, \ |x-y| < \eta \implies |f(x)-f(y)| < \varepsilon. \tag{1.23}$$

However, the difference $|f(x) - f(y)|$ can be rewritten as

$$|f(x) - f(y)| = |f(x) - f_n(x) + f_n(x) - f_n(y) + f_n(y) - f(y)| \tag{1.24}$$

$$\le |f(x) - f_n(x)| + |f_n(x) - f_n(y)| + |f_n(y) - f(y)|.$$

Now since the functions f_n are continuous, we have

$$\forall \varepsilon > 0, \ \exists \eta(x, \varepsilon) > 0 \text{ s.t. } \forall (x, y) \in [-1, 1]^2, \ |x - y| < \eta \implies |f_n(x) - f_n(y)| < \varepsilon/3. \tag{1.25}$$

Having shown earlier that the sequence f_n converges to f for the norm $\|\cdot\|_\infty$, we also have

$$\forall x \in [-1, 1], \ \forall \varepsilon > 0, \ \exists N(\varepsilon) > 0 \text{ s.t. } n \geq N(\varepsilon) \Longrightarrow |f_n(x) - f(x)| < \varepsilon/3. \tag{1.26}$$

Putting (1.25) and (1.26) in the inequality (1.24), we finally obtain

$$\forall \varepsilon > 0, \ \exists \eta(x, \varepsilon) > 0 \text{ s.t. } \forall (x, y) \in [-1, 1]^2, \ |x - y| < \eta \Longrightarrow |f(x) - f(y)| < \varepsilon. \tag{1.27}$$

This ends the proof that the space $C^0([-1, 1])$ of continuous real-valued functions on the interval $[-1, 1]$ is a Banach space for the norm $\|\cdot\|_\infty$.

Note, however, that $C^0([-1, 1])$ is not a Banach space if equipped with the L^2 norm defined by

$$\|f\|_{L^2} \equiv \sqrt{\int_{-1}^{1} |f(x)|^2 dx}.$$

To see this, consider the sequence of functions $(f_n)_{n\in\mathbb{N}}$ defined by

$$f_n(x) \equiv \begin{cases} 0 & \text{if } x \in [-1, 0], \\ nx & \text{if } x \in (0, 1/n], \\ 1 & \text{if } x \in (1/n, 1]. \end{cases} \tag{1.28}$$

Let us show that this sequence of functions is Cauchy for the L^2 norm, but that it does not converge to an element of $C^0([-1, 1])$.

- We begin by showing that $(f_n)_{n\in\mathbb{N}}$ is Cauchy for the L^2 norm.

  Let n and m be two natural numbers such that $n < m$. We find the L^2 norm of the difference $f_n - f_m$:

$$\|f_n - f_m\|_{L^2}^2 = \int_{-1}^{1} |f_n(x) - f_m(x)|^2 dx = \int_{0}^{1/n} |f_n(x) - f_m(x)|^2 dx. \tag{1.29}$$

  Since

$$|f_n(x) - f_m(x)| \leq |f_n(x)| + |f_m(x)| \leq 2 \implies \|f_n - f_m\|_{L^2}^2 \leq \int_{0}^{1/n} 4 \, dx = \frac{4}{n},$$

  if the difference $f_n - f_m$ is to be bounded by ε in the L^2 norm, we can just require

$$\frac{4}{n} < \varepsilon.$$

Hence,

$$\exists N(\varepsilon) > 0 \text{ such that } m > n > N(\varepsilon) \equiv \frac{4}{\varepsilon} \implies \|f_n - f_m\|_{L^2} \leq \varepsilon,$$

which shows that the sequence $(f_n)_{n \in \mathbb{N}}$ is Cauchy for the L^2 norm.

- Now assume that $C^0([-1, 1])$ is complete for the L^2 norm.

  If this were so, there would be a function f, continuous on $[-1, 1]$, such that

  $$\lim_{n \to +\infty} \|f_n - f\|_{L^2} = 0.$$

In particular, this would imply that

$$\forall \varepsilon > 0, \ \exists N(\varepsilon) \text{ s.t. } n > N(\varepsilon) \implies \int_{-1}^{0} |f_n(x) - f(x)|^2 dx < \varepsilon. \quad (1.30)$$

Now, $\forall x \in [-1, 0]$, we have $f_n(x) = 0$, so

$$\int_{-1}^{0} |f(x)|^2 dx < \varepsilon, \quad \forall \varepsilon > 0.$$

Finally, $f(x) = 0, \forall x \in [-1, 0]$.

Now let α be a strictly positive real number. By hypothesis, we would also have

$$\forall \varepsilon > 0, \ \exists N(\varepsilon) \text{ s.t. } n > N(\varepsilon) \implies \int_{\alpha}^{1} |f_n(x) - f(x)|^2 dx < \varepsilon. \quad (1.31)$$

But $\forall x \in [\alpha, 1]$, we have $f_n(x) = 1$ as soon as $n > 1/\alpha$.

Therefore, for $n > \max\left(1/\alpha, N(\varepsilon)\right)$, we would have

$$\int_{\alpha}^{1} |1 - f(x)|^2 dx < \varepsilon, \quad \forall \varepsilon > 0. \quad (1.32)$$

It follows that $\forall \alpha > 0$, we have $f(x) = 1, \forall x \in [\alpha, 1]$, whereupon f cannot be continuous at $x = 0$, the limit from the left being 0 and that from the right being 1.

The sequence $\{f_n\}_{n \in \mathbb{N}}$ is thus Cauchy for the L^2 norm but does not converge to a continuous function for this same norm. The space $C^0([-1, 1])$ is thus not complete for the L^2 norm.

1.2.2 Properties of Banach Spaces

The aim of the lemma 1.2 is to characterize closed subspaces in Banach spaces.

Lemma 1.2 *Let $(E, \| \cdot \|_E)$ be a Banach space and F a subset of E. Then F is closed in E if and only if F contains the limits of all its Cauchy sequences.*

Some Basic Topology

- A subset F of E is closed in E if and only if its complement in E is open.

- A subset O of E is open if it is a neighborhood of each of its points. This can be expressed in the following way:

$$\forall x \in O, \ \exists r > 0 \text{ such that } B(x, r) \subset O, \tag{1.33}$$

where $B(x, r)$ is an open ball of radius r centered at x defined by:

$$B(x, r) \equiv \{ y \in E : \| y - x \|_E < r \}. \tag{1.34}$$

Lemma 1.3 *Every finite-dimensional vector space E is complete for all norms defined on E.*

This is a consequence of the following two results, which fully characterize finite-dimensional vector spaces in this context:

1. Every vector space of dimension N is isomorphic to $\mathbb{R}^N$, which is complete in the same way as $\mathbb{R}$.

2. All norms on a finite-dimensional vector space are equivalent.

1.2.3 Continuous Linear and Bilinear Maps

Definition 1.5 Let $(E, \| \cdot \|_E)$ and $(F, \| \cdot \|_F)$ be two normed vector spaces. Let $\mathscr{L}(E, F)$ be the space of linear maps from E to F, i.e.,

$$\forall f \in \mathscr{L}(E, F), \ \forall \lambda \in \mathbb{R}, \ \forall (x, y) \in E^2: \ f(x + \lambda y) = f(x) + \lambda f(y).$$

The following characterization of continuous linear maps will be used systematically in our subsequent applications to partial differential equations.

Lemma 1.4 *Let f be an element of $\mathscr{L}(E, F)$. Then the following three properties are equivalent:*

1. *f is continuous on E.*

2. *f is continuous at 0.*

3. *f is Lipschitz continuous, i.e.,*

$$\exists M > 0 \text{ such that } \forall x \in E: \ \|f(x)\|_F \leq M\|x\|_E.$$

Remark: If E is a finite-dimensional space, then every linear map is continuous. However, this is not true in general in infinite dimensions.

The above three properties can be extended to bilinear maps as follows.

Definition 1.6 Let $(E_1, \|\cdot\|_{E_1})$ and $(E_2, \|\cdot\|_{E_2})$ be two normed vector spaces. Let $a(.,..)$ be a map from $E_1 \times E_2$ to another normed vector space $(F, \|\cdot\|_F)$:

$$\begin{aligned}
a : E_1 \times E_2 &\longrightarrow F, \\
(x, y) &\mapsto a(x, y).
\end{aligned} \tag{1.35}$$

Then $a(\cdot, \cdot)$ is a bilinear form if

$$\forall \lambda \in \mathbb{R}, \ \forall (x_1, x_2) \in E_1^2, \ \forall y \in E_2, \ a(\lambda x_1 + x_2, y) = \lambda a(x_1, y) + a(x_2, y),$$

$$\forall \lambda \in \mathbb{R}, \ \forall x \in E_1, \ \forall (y_1, y_2) \in E_2^2, \ a(x, \lambda y_1 + y_2) = \lambda a(x, y_1) + a(x, y_2).$$

The space of bilinear maps from $E_1 \times E_2$ to F is denoted by $\mathscr{L}[(E_1, E_2; F)]$.

We can now characterize the bilinear maps in $\mathscr{L}[(E_1, E_2; F)]$ by straightforward extension of the properties already listed for the linear maps in $\mathscr{L}(E, F)$.

Lemma 1.5 *Let $a(\cdot, \cdot)$ be a map in $\mathscr{L}[(E_1, E_2; F)]$. Then the following three properties are equivalent:*

1. *$a(\cdot, \cdot)$ is continuous on $E_1 \times E_2$ equipped with the product topology. In other words, we define the norm of an element $x = (x_1, x_2)$ in $E_1 \times E_2$ by $\|(x_1, x_2)\|_{E_1 \times E_2} \equiv \max\left(\|x_1\|_{E_1}, \|x_2\|_{E_2}\right).$*

2. *$a(\cdot, \cdot)$ is continuous at the origin $(0, 0)$.*

3. *$\exists M > 0$ such that $\forall (x_1, x_2) \in E_1 \times E_2, \ \left\|a(x_1, x_2)\right\|_F \leq M\|x_1\|_{E_1} \cdot \|x_2\|_{E_2}.$*

1.3 Hilbert Spaces

1.3.1 The Scalar Product and Its Properties

Definition 1.7 Let E be a real vector space. A scalar product is defined on E by any map φ from $E \times E$ to $\mathbb{R}$ satisfying the following conditions:

1. φ is a bilinear map.

2. φ is symmetric: $\forall (x, y) \in E^2, \varphi(x, y) = \varphi(y, x)$.

3. φ is positive definite. That is:

   - $\varphi(x, x) \geq 0, \ \forall x \in E$ (positive),

   - $\varphi(x, x) = 0 \implies x = 0$ (definite).

Remark:

- Since a scalar product φ is real-valued, it is a bilinear form.

- We shall use one of the following to denote the scalar product φ :

$$\forall (x, y) \in E^2, \quad \varphi(x, y) \equiv \langle x, y \rangle \equiv (x, y) \equiv ((x, y)) \equiv [x, y] \equiv \cdots .$$

- Definition 1.7 formulates the well-known properties of the Euclidean scalar product defined for vectors in $\mathbb{R}^n$ by

$$\varphi(x, y) \equiv (x, y) \equiv \sum_{i=1}^{n} x_i y_i .$$

- This definition can also be used to introduce a notion of projection for continuous functions on an interval $[a, b]$, i.e., for the function space $E \equiv C^0([a, b])$. We consider the scalar product φ defined by

$$\varphi : E \times E \longrightarrow \mathbb{R},$$
$$(f, g) \longmapsto \varphi(f, g) \equiv (f, g) \equiv \int_a^b fg \, \mathrm{d}x.$$

All the properties of this scalar product are immediate, and in particular the definiteness of φ :

$$\varphi(f, f) \equiv \int_a^b f^2 \mathrm{d}x = 0 \implies f = 0.$$

This follows because every continuous nonnegative function on an interval $[a, b]$, such as f^2 in the present case, with zero integral over that same interval must necessarily be the zero function on $[a, b]$.

Let us consider two of the main properties of the scalar product. The first is the Cauchy–Schwarz inequality:

Lemma 1.6 *If E is a real vector space and φ a scalar product on E, then*

$$\forall (x, y) \in E^2, \quad \|\varphi(x, y)\| \leq \sqrt{\varphi(x, x)}\sqrt{\varphi(y, y)}.$$

The second fundamental property of the scalar product concerns the norm it induces in the following way:

Lemma 1.7 *If E is a real vector space and φ a scalar product on E, then the map N from E to $\mathbb{R}$ defined by*

$$\begin{aligned} N : E &\longrightarrow \mathbb{R}, \\ x &\longmapsto N(x) \equiv \sqrt{\varphi(x, x)}, \end{aligned} \tag{1.36}$$

is a norm on E. We write $\|x\| \equiv \sqrt{\varphi(x, x)}$.

Remark: The converse is false here, i.e., not all norms can be associated with a scalar product. Indeed, this motivates the introduction of Hilbert spaces in the Sect. 1.3.2.

1.3.2 Definition and Properties of a Hilbert Space

Definition 1.8 A normed vector space $(E, \| \cdot \|_E)$ is a Hilbert space if it is complete and if its norm derives from a scalar product.

In other words, a Hilbert space is a Banach space $(E, \| \cdot \|_E)$ with a norm that derives from a scalar product in the following way: there is a positive definite symmetric bilinear form φ on $E \times E$ such that

$$\forall x \in E, \quad \|x\|_E \equiv \sqrt{\varphi(x, x)}. \tag{1.37}$$

Remark: As mentioned earlier, we can always associate a norm with a scalar product. In the present case, a Hilbert space characterizes the opposite property, which is not always possible.

A well-known example of a Hilbert space is the set of square-integrable functions on a given interval (a, b), that is,

$$E \equiv L^2(a, b) \equiv \left\{ f : (a, b) \longrightarrow \mathbb{R} \text{ such that } \int_a^b f^2 \, dx < +\infty \right\},$$

equipped with the scalar product

$$\langle f, g \rangle \equiv \int_a^b fg \, dx$$

and the associated norm

$$\| f \|^2 = \langle f, f \rangle = \int_a^b f^2 \, dx.$$

An important property with many applications to establish the existence and uniqueness of solutions to variational problems concerns the characterization of closed vector subspaces in a Hilbert space.

Lemma 1.8 *Let* $(H, \| \cdot \|_H)$ *be a Hilbert space and* F *a closed subspace of* H. *Then* F *is itself a Hilbert space for the induced norm* $\| \cdot \|_H$.

We shall now review some properties of the convex subsets of a Hilbert space.

1.3.3 Convex Subsets of a Hilbert Space

Definition 1.9 Let K be a subset of a Hilbert space V. Then K is convex if and only if

$$\forall (x, y) \in K^2, \quad \forall t \in [0, 1], \quad (1 - t)x + ty \in K.$$

In other words, a subset K is convex if for every pair of points (x, y) in K, the set of points on the "straight line" joining these points, described by $(1 - t)x + ty$ as t ranges over $[0, 1]$, is entirely contained within K.

This notion can be illustrated by the following examples:

- Every open or closed ball in a normed vector space H is convex. Geometrically, at least in 2 or 3 dimensions, this property seems rather obvious. It is proved in the following way in an arbitrary Hilbert space equipped with a norm $\| \cdot \|_V$. Since translation and homothety conserve convexity, we may consider without loss of generality the unit ball $B(0,1)$. In this case, if $(x,y) \in [B(0,1)]^2$, then

$$\forall t \in [0,1], \quad \|(1-t)x + ty\|_V \leq (1-t)\|x\|_V + t\|y\|_V \leq (1-t) + t = 1,$$

which proves that the general element of the "straight line" $(1-t)x + ty$ passing through the points x and y remains inside the unit ball $B(0,1)$.

- The unit sphere $S(0,1)$ is not convex. Once again, in 3-dimensional geometry, this result seems rather obvious, because apart from the endpoints (x,y) of a diameter of a given sphere, all the other points are located inside the sphere, which contradicts the requirements of convexity. This property can also be established without considering 3-dimensional geometry, since if $x \in S(0,1)$, then $-x$ also belongs to the unit sphere. However, the convex combination

$$\frac{x}{2} - \frac{x}{2} = 0$$

of the two points does not belong to $S(0,1)$.

- A vector subspace K of a Hilbert space V is convex. This follows directly from the definition of a vector subspace, which is stable under all linear combinations.

The last two results discussed here concern projection onto a closed convex subset.

Lemma 1.9 *If K is a nonempty closed convex subset of a normed vector space V that is not necessarily a Hilbert space, then for every x in V, there is a unique x_K in K such that*

$$\|x - x_K\| = \inf_{y \in K} \|x - y\|;$$

x_K is called the projection of x on K.

The property described in Lemma 1.9 shows that x_K achieves the minimum distance from x to the convex subset K. Note that this property will prove crucial when we come to consider inner approximation in the context of the finite-element method (see Sect. 2.6).

Lemma 1.10 *If V is a Hilbert space and K a closed vector subspace of H, and if x_K in K satisfies*

$$\langle x - x_K, y \rangle = 0, \quad \forall x \in V, \ \forall y \in K, \tag{1.38}$$

then x_K is called the orthogonal projection of x on K.

1.4 Basics of Distribution Theory

1.4.1 Motivation: Intuitive Definition of the Weak Derivative

The aim of distribution theory is to extend the notion of function to more general and less restrictive mathematical objects. A first step can be taken in this direction by considering the well-known formula for integration by parts.

Indeed, if f and φ are two sufficiently well behaved real-valued functions defined on (a,b), this formula is

$$\int_a^b f(x)\varphi'(x)\,\mathrm{d}x = -\int_a^b f'(x)\varphi(x)\,\mathrm{d}x + \left[f(x)\varphi(x) \right]_a^b. \qquad (1.39)$$

In order to simplify the analysis, if we assume that the product $f\varphi$ is zero at $x = a$ and $x = b$, or that $f\varphi$ is zero outside a closed, bounded subset K of the interval (a,b), the relation (1.39) becomes

$$\int_K f(x)\varphi'(x)\,\mathrm{d}x = -\int_K f'(x)\varphi(x)\,\mathrm{d}x. \qquad (1.40)$$

In order to be more precise about the statement that f and φ are sufficiently well behaved, note that in the relation (1.40), each of the integrals is well defined whenever f and φ are both differentiable, but also if both the products $f\varphi'$ and $f'\varphi$ are integrable over the whole of K. But if f and φ are two functions defined, continuous, and differentiable on K, the integrals of $f\varphi'$ and $f'\varphi$ over K will *ipso facto* be convergent.

These considerations raise the following question: what are sufficient conditions on f and φ to ensure that $f'\varphi$ is integrable? One possible answer is provided by the Cauchy–Schwarz inequality, because under the conditions of validity of this inequality, the integral of $f'\varphi$ is bounded in the following way:

$$\left| \int_K f'(x)\varphi(x)\,\mathrm{d}x \right| \leq \left[\int_K f'^2(x)\,\mathrm{d}x \right]^{1/2} \left[\int_K \varphi^2(x)\,\mathrm{d}x \right]^{1/2}. \qquad (1.41)$$

Hence, if f' and φ are two square-integrable functions on K, the integral of $f'\varphi$ over K will converge. By symmetry, if f and φ' are square-integrable functions on K, then the integral of $f\varphi'$ over K will also converge.

These considerations motivate the introduction of a weak derivative, or derivative in the sense of distributions, defined naively here by analogy with (1.40):

Definition 1.10 If f is a real-valued square-integrable function on K, we shall say that f is weakly differentiable if there is a real-valued square-integrable function g on K such that

$$\int_K f(x)\varphi'(x)\,dx = -\int_K g(x)\varphi(x)\,dx, \quad \forall \text{ "sufficiently" well behaved } \varphi.$$

$$(1.42)$$

Then, we will symbolically set: $g \equiv f'$.

Note that Definition 1.10 agrees with the usual notion of derivative when f is differentiable in the classical sense, since we retrieve the standard formula (1.40) for integration by parts. In this case, the function g is just the usual derivative of f.

This intuitive introduction remains incomplete with regard to several mathematical aspects that will be further developed in the rest of this chapter. The aim here is to extend the notion of differentiability, and hence also the notion of function.

Other aspects of functions will also be treated by introducing this definition of weak derivative, and more generally by introducing distributions, which generalize the notion of function.

Let us say at the outset regarding the naive Definition 1.10 that the weak derivative no longer describes a local feature of the function f, as would the classical derivative defined by a limiting process in the neighborhood of a given point, but rather an integral, hence global, property of the given function.

We shall thus see in the remainder of this chapter that a function whose classical derivative has a discontinuity of the first kind at a point, and which is not therefore differentiable, e.g., $|x|$ at $x = 0$, will be treated in a global manner with the help of this new definition of the weak derivative.

Laurent Schwartz was the inventor and prodigious developer of the theory of distributions (1944). The interested reader with a good background in the techniques of integration and who wishes to delve further into this topic is invited to refer to the excellent book by Schwartz entitled *Mathematics for the Physical Sciences* [2].

As far as the present book is concerned, the aim will be to present only the basic properties of distributions, motivating the introduction of these new ideas by an analogy that may at first seem surprising.

1.4.2 Characterizing a Vector in a Finite-Dimensional Vector Space

Consider a vector space E of finite dimension n equipped with a scalar product $(\cdot, \cdot)_E$, and let $(e_i)_{i=1,\dots,n}$ be a basis for E. Then every vector x in E can be decomposed as follows relative to this basis:

$$x = \sum_{i=1}^{n} x^i e_i. \tag{1.43}$$

This can be written more concisely using the Einstein summation, or repeated index convention:

$$x = x^i e_i. \tag{1.44}$$

The sequence of numbers $(x^i)_{i=1,\,...,\,n}$ constitutes the *contravariant* components of the vector x in the basis $(e_i)_{i=1,\,...,\,n}$.

One can associate a dual space E^* with the vector space E, consisting of the linear forms defined on E. It can be shown that the dual space is also n-dimensional.

Note that all linear forms belonging to E^* are necessarily continuous, because E is finite-dimensional.

We now consider the canonical dual basis $(e^{*i})_{i=1,\,...,\,n}$ of E^* defined by

$$e^{*i}(e_j) = \delta^i_j,$$

where δ^i_j is the Kronecker symbol, defined by $\delta^i_i = 1$ (no summation) and $\delta^i_j = 0$ if $i \neq j$.

Then every linear form ω in the dual space E^* can be decomposed in the following way relative to the dual basis $(e^{*i})_{i=1,\,...,\,n}$:

$$\omega = \omega_i e^{*i}, \tag{1.45}$$

where the quantities $(\omega_i)_{i=1,\,...,\,n}$ are the *covariant* components of the linear form ω in the dual basis $(e^{*i})_{i=1,\,...,\,n}$, and where we have used the Einstein convention once again.

For each *fixed* vector x in E, we now consider the linear form L_x defined by

$$L_x : E \longrightarrow \mathbb{R}$$
$$v \longmapsto L_x(v) \equiv (x, v)_E. \tag{1.46}$$

It is then straightforward to show that the form L_x belongs to the dual space E^*. This means that L_x can be expressed in terms of the basis $(e^{*i})_{i=1,\,...,\,n}$, viz., $L_x = (L_x)_i e^{*i}$.

Furthermore, the action of the linear form L_x on any basis vector e_j of E is given by

$$L_x(e_j) = (L_x)_i e^{*i}(e_j) = (L_x)_i \delta^i_j = (L_x)_j \equiv (x, e_j)_E. \tag{1.47}$$

We can thus define the *covariant* component x_j of the originally chosen vector x by

$$x_j \equiv (L_x)_j = (x, e_j)_E. \tag{1.48}$$

We can now characterize every vector x in E by considering the map I from E to E^* defined by

$$\begin{aligned} I : E &\longrightarrow E^* \\ x &\longmapsto L_x. \end{aligned} \tag{1.49}$$

Lemma 1.11 *The map I is linear and injective.*

Proof: We first prove linearity, then injectivity.

- *Linearity:* (a) Let (x_1, x_2) be any pair of vectors in $E \times E$. We have

$$I(x_1 + x_2) \equiv L_{x_1+x_2},$$

where

$$L_{x_1+x_2}(v) = (x_1 + x_2, v)_E = (x_1, v)_E + (x_2, v)_E = L_{x_1}(v) + L_{x_2}(v), \quad \forall v \in E.$$

In other words,

$$I(x_1 + x_2) = I(x_1) + I(x_2). \tag{1.50}$$

(b) Now consider an arbitrary vector x in E and any real number λ. We have

$$I(\lambda x) \equiv L_{\lambda x},$$

where

$$\forall v \in E, \quad L_{\lambda x}(v) = (\lambda x, v)_E = \lambda(x, v)_E = \lambda L_x(v).$$

Hence,

$$I(\lambda x) = \lambda I(x). \tag{1.51}$$

- *Injectivity:* Since the map I is linear, in order to show that it is one-to-one, it suffices to establish that its kernel contains only the zero vector:

$$I \text{ linear and injective} \iff \ker I = \{0\}. \tag{1.52}$$

We thus consider any x in $\ker I$. By definition of the kernel of I, we have

$$I(x) = L_x = 0 \iff \forall v \in E, \quad L_x(v) \equiv (x, v)_E = 0. \tag{1.53}$$

We deduce that the only vector x orthogonal under the scalar product $(.,.)_E$ to all vectors v in E is the zero vector $x = 0$.

In other words, the kernel of the map I contains only the zero vector, which implies that I is injective.

Interpretation: The linear map I is injective, which means that

$$L_x = L_y \implies x = y. \tag{1.54}$$

So specifying the vector x in E is equivalent to specifying the linear form L_x through the definition (1.46), that is, the projections of the vector x onto vectors v in the vector space E.

Now, the linear form L_x is characterized *solely* by specifying its n components relative to the dual basis $(e^{*i})_{i=1,\ldots,n}$. Naturally, this is a consequence of the scenario we have been considering throughout, namely, a vector space E of *finite* dimension n.

Note finally that this characterization is nothing but the projection of the vector x onto the n vectors of the basis $(e_i)_{i=1,\ldots,n}$ of E.

In conclusion, for every vector x in E, the n projections of x onto the basis $(e_i)_{i=1,\ldots,n}$ completely determine this vector and correspond *ipso facto* to specifying the n components $(L_x)_i$ $(i = 1, \ldots, n)$ of the linear form associated with the vector x.

1.4.3 Extension to Functions and Introduction of Regular Distributions

The results obtained above in finite dimensions suggest that it may be worthwhile trying to extend the ideas of vector characterization to functions belonging to *infinite-dimensional* function spaces.

In other words, considering the notation adopted above, we shall replace the vector x in the space E by a function f belonging to a function space for which we can introduce linear forms T_f corresponding to the forms L_x characteristic of each vector x in E.

We begin by introducing the idea of the support of a function v from $\mathbb{R}^n$ to $\mathbb{R}$, denoted by $supp\, v$. This is the smallest closed set containing all the points at which the given function v is nonzero (the closure of that set, for those with a background in elementary topology):

$$supp\, v \equiv \overline{\left\{ x = (x_1, x_2, \ldots, x_n) \in \mathbb{R}^n : v(x) \neq 0 \right\}}^{\mathbb{R}^n}, \tag{1.55}$$

where the bar denotes the closure in $\mathbb{R}^n$.

To illustrate this idea, consider the function H of one real variable x defined by

$$H(x) = \begin{cases} 1 & \text{if } 0 < x < 1, \\ 0 & \text{otherwise}. \end{cases} \tag{1.56}$$

The function H is nonzero on the open set $(0, 1)$, but its support is the closed interval $[0, 1]$:

$$supp\, H \equiv \overline{\left\{ x \in \mathbb{R} \mid H(x) \neq 0 \right\}}^{\mathbb{R}} = [0, 1].$$

We now introduce the space $\mathscr{D}(\Omega)$ as follows:

Definition 1.11 If Ω is a regular open subset of $\mathbb{R}^n$, we define the space $\mathscr{D}(\Omega)$ by

$$\mathscr{D}(\Omega) = \left\{ v : \Omega \subset \mathbb{R}^n \to \mathbb{R} \mid v \in C^\infty(\Omega),\ supp\, v \subset \Omega \right\}. \tag{1.57}$$

The space $\mathscr{D}(\Omega)$ is the space of C^∞ functions on Ω with compact support *strictly* contained in Ω. It is also denoted by $C_0^\infty(\Omega)$.

For any function f in $L^2(\Omega)$, consider the linear form T_f defined by

$$T_f : \mathscr{D}(\Omega) \longrightarrow \mathbb{R}$$

$$\varphi \longmapsto T_f(\varphi) \equiv \int_\Omega f\varphi \, d\Omega, \tag{1.58}$$

where $d\Omega$ is the differential element on $\mathbb{R}^n$, i.e., $d\Omega \equiv dx_1 dx_2 \cdots dx_n$, and the symbol $\int_\Omega$ denotes the n-dimensional multiple integral in the space $\mathbb{R}^n$.

We note that T_f is well defined, since

$$\left| T_f(\varphi) \right| \leq \int_\Omega |f\varphi| d\Omega \leq \left(\int_\Omega f^2 d\Omega \right)^{1/2} \left(\int_{supp\, \varphi} \varphi^2 d\Omega \right)^{1/2}, \tag{1.59}$$

where we have used the Cauchy–Schwarz inequality.

Whenever φ belongs to $\mathscr{D}(\Omega)$, the integral of its square over its support converges, and the definition of T_f is thus appropriate for every f in $L^2(\Omega)$.

Note that the definition (1.58) of the map T_f could be extended to functions in $L_{\text{loc}}^1(\Omega)$, the space of absolutely integrable functions on any closed bounded subset of Ω, but this function space requires some technical niceties that go beyond the scope of this book. Once again, the interested reader is referred to the book by Schwartz [2].

Furthermore, the C^∞ regularity required of the functions φ is stronger than really necessary. Indeed, continuity would be sufficient at this stage of the presentation.

However, we shall see later that the definition of a distribution requires that one build in this level of regularity for the functions φ on which the distribution will be defined. But we have not quite reached that stage yet.

Note and Intuitive Definition. The action of the linear form T_f on a function φ belonging to $\mathscr{D}(\Omega)$ can be interpreted as the scalar product in $L^2(\Omega)$, denoted by $(.,.)_{L^2(\Omega)}$, of f and φ:

$$\int_\Omega f\varphi \, dx \equiv (f, \varphi)_{L^2(\Omega)}. \tag{1.60}$$

This is one reason why we adopt the following notation:

$$T_f(\varphi) = (f, \varphi)_{L^2(\Omega)} \equiv \langle T_f, \varphi \rangle. \tag{1.61}$$

We then have, at least formally, the equivalent of the linear form L_x introduced in the finite-dimensional case, characterizing every vector x belonging to a vector space E.

To complete the analogy, we should check to what extent a function f in $L^2(\Omega)$ can be completely characterized by specifying the linear form T_f.

The aim then is to reconstruct the equivalent of the injection I defined in (1.49).

We thus introduce the map J defined by

$$\begin{aligned} J : L^2(\Omega) &\longrightarrow \mathscr{D}'(\Omega) \\ f &\longmapsto T_f, \end{aligned} \tag{1.62}$$

where $\mathscr{D}'(\Omega)$ can be taken, at this stage in the construction, as the set of linear forms defined on $\mathscr{D}(\Omega)$. We then have the following property:

Lemma 1.12 *The map J defined in (1.62) is linear, injective, and continuous.*

Proof: We consider the three properties in turn:

- Linearity is an immediate consequence of the fact that integration is linear.

- Now consider the injectivity of J.

  As for the map I defined by (1.49), we exploit the fact that J is linear. This means that proving injectivity is equivalent to showing that the kernel of the map J contains only the zero vector.

  Therefore, consider a function f in $L^2(\Omega)$ that is in the kernel of J. Then by definition,

$$J(f) \equiv T_f = 0 \iff \int_\Omega f\varphi \, d\Omega = 0, \quad \forall \varphi \in \mathscr{D}(\Omega). \tag{1.63}$$

In the present case, the main obstacle lies in the fact that the function f we are looking for is in the space $L^2(\Omega)$, which strictly contains $\mathscr{D}(\Omega)$.

In other words, it is not immediately obvious that we will find the function f we seek among all those functions φ satisfying the integral equation (1.63).

We cannot therefore directly choose the special case $\varphi = f$ in the formulation (1.63), which would allow us to conclude that f must be identically zero.

To get around this difficulty, we apply a theorem to be discussed below, Theorem 1.1, which concludes as follows: the space $\mathscr{D}(\Omega)$ is dense in $L^2(\Omega)$.

The implication is that for every function ψ in $L^2(\Omega)$, there is a sequence of functions ψ_n in $\mathscr{D}(\Omega)$ such that

$$\lim_{n \to \infty} \int_\Omega |\psi_n - \psi|^2 \, d\Omega = 0. \tag{1.64}$$

The point about this asymptotic proximity of the sequence of functions ψ_n and the function ψ lies in the fact that the function ψ is in a certain sense contaminated by the properties of the functions ψ_n.

Indeed, we may legitimately claim that for all elements of the sequence ψ_n that belong to $\mathscr{D}(\Omega)$, the integral equation (1.63) is satisfied:

$$\int_\Omega f \psi_n \, d\Omega = 0, \quad \forall n \in \mathbb{N}. \tag{1.65}$$

As a consequence, for the arbitrary function ψ in $L^2(\Omega)$, we have

$$\left| \int_\Omega f \psi \, d\Omega \right| = \left| \int_\Omega f(\psi - \psi_n) \, d\Omega \right| \le \|f\|_{L^2(\Omega)} \|\psi_n - \psi\|_{L^2(\Omega)}, \quad \forall n \in \mathbb{N}, \tag{1.66}$$

where we have used the Cauchy–Schwarz inequality once again. We then let n tend to $+\infty$ in the inequality (1.66) and conclude that

$$\int_\Omega f \psi \, d\Omega = 0, \quad \forall \psi \in L^2(\Omega). \tag{1.67}$$

Hence, with f and ψ in the same space $L^2(\Omega)$ in the integral equation (1.67), there is no longer any doubt that we may choose the special case $\psi = f$ in order to conclude that $f = 0$.

The linear map J thus has a kernel containing only the zero vector, and it is therefore injective:

$$T_{f_1} = T_{f_2} \implies f_1 = f_2. \tag{1.68}$$

- Finally, we show that the injection J is continuous.

In other words, if a sequence f_n tends to f in $L^2(\Omega)$, then $J(f_n)$ must tend to $J(f)$ in $\mathscr{D}'(\Omega)$.

This is formulated as follows:

$$J(f_n) \to J(f) \text{ in } \mathscr{D}'(\Omega) \text{ if } \int_\Omega (f_n - f)\varphi \, d\Omega \to 0, \quad \forall \varphi \in \mathscr{D}(\Omega). \quad (1.69)$$

However, we have the following bound:

$$\left| \int_\Omega (f_n - f)\varphi \, d\Omega \right| \leq \|f_n - f\|_{L^2(\Omega)} \|\varphi\|_{L^2(\Omega)}, \quad \forall \varphi \in \mathscr{D}(\Omega). \quad (1.70)$$

Whenever φ is in $\mathscr{D}(\Omega)$, we know that it is also in $L^2(\Omega)$.

Then, according to (1.70), it follows that $T(f_n)$ tends to $T(f)$ in $\mathscr{D}'(\Omega)$, and the injection J is continuous.

The functions f in $L^2(\Omega)$ are thus completely characterized by the injection J defined in (1.62), in the same way as every vector x in a finite-dimensional vector space E was completely characterized by the injection I defined in (1.49).

The main consequence of the injectivity result (1.68) is that we may henceforth identify a function f in $L^2(\Omega)$ with its equivalent T_f in $\mathscr{D}'(\Omega)$.

We can now define a first category of distributions defined on Ω.

Definition 1.12 For f belonging to $L^2(\Omega)$, the linear form T_f defined by (1.58) is called the regular distribution associated with f.

Remark: In order to arrive at a sufficiently general definition of a distribution, it should be borne in mind that the aim is to be able to differentiate in some sense functions whose derivatives have a discontinuity of the first kind, such as the function $|x|$ at $x = 0$.

The usual properties of differentiation must be preserved in this process, whence this generalized differentiation should imply "continuity" of distributions in a suitable sense.

For this reason, we now define what is meant by continuity of distributions. For the purposes of presentation, we consider the regular distributions, as defined in (1.58). The continuity at 0 of the linear form T_f is defined as follows:

Definition 1.13 The regular distribution T_f is said to be continuous at 0 if for every sequence of functions $(\varphi_n)_{n \in \mathbb{N}}$ in $\mathscr{D}(\Omega)$, we have

$$\varphi_n \to 0 \text{ in } \mathscr{D}(\Omega) \quad \Longrightarrow \quad T_f(\varphi_n) \to 0 \text{ in } \mathbb{R}. \tag{1.71}$$

Given that the form T_f is linear, continuity at 0 is equivalent to continuity at all points φ_0 of $\mathscr{D}(\Omega)$.

To see this, express the difference $T_f(\varphi) - T_f(\varphi_0)$ in the form $T_f(\varphi - \varphi_0)$, which focuses attention on the reference point $\psi \equiv \varphi - \varphi_0$.

In other words, continuity of the form T_f at the point $\varphi = \varphi_0$ is equivalent to continuity of T_f at the point $\psi \equiv \varphi - \varphi_0 = 0$.

To complete Definition 1.13, we have to say what we mean by convergence of a sequence of functions φ_n in $\mathscr{D}(\Omega)$.

Definition 1.14 A sequence of functions φ_n converges to 0 in $\mathscr{D}(\Omega)$ if:

1. $\exists$ a fixed compact set $K_0 \subset \Omega$ (independent of n) such that $\operatorname{supp} \varphi_n \subset K_0$,

2. $D^k \varphi_n$ converges uniformly to 0, $\quad \forall k \in \mathbb{N}$,

   where $D^k \varphi_n$ is the kth derivative of the function φ_n.

We can now examine the continuity of the regular distribution T_f defined by (1.58).

Consider, therefore, a sequence of functions φ_n belonging to $\mathscr{D}(\Omega)$ and converging to 0 in the sense of Definition 1.14.

It is clear that the uniform convergence of the sequence φ_n alone suffices to establish convergence of the sequence $T_f(\varphi_n)$ in $\mathbb{R}$, since for the regular distribution T_f defined by (1.58), we can write

$$
\begin{aligned}
|T_f(\varphi_n)| &\leq \sup_{x \in \Omega} |\varphi_n(x)| \int_{K_0} 1 \times |f| \, d\Omega \\
&\leq \sup_{x \in \Omega} |\varphi_n(x)| \sqrt{\mu(K_0)} \|f\|_{L^2(K_0)} \\
&\leq \sup_{x \in \Omega} |\varphi_n(x)| \sqrt{\mu(K_0)} \|f\|_{L^2(\Omega)}, \tag{1.72}
\end{aligned}
$$

where $\mu(K_0)$ is the measure of K_0, i.e., the area, volume, etc., depending on the dimension of the space $\mathbb{R}^n$, containing the open set Ω, and hence also the set K_0.

Therefore, if φ_n tends to 0 in $\mathscr{D}(\Omega)$, it follows that $\sup_{x \in \Omega} |\varphi_n(x)|$ tends to 0.

Since f is in $L^2(\Omega)$, the inequality (1.72) indeed implies that $T_f(\varphi_n)$ tends to 0 in $\mathbb{R}$.

Consequently, for the regular distribution T_f, the definition of convergence in $\mathscr{D}(\Omega)$ is sufficient to guarantee continuity in the sense of linear forms.

The reader may thus ask why it is necessary to extend the uniform convergence of the sequence φ_n to the successive derivatives $D^k \varphi_n$ as in the second point of Definition 1.14.

To understand this, we must first obtain a more global view of distributions.

1.4.4 Definition of Distributions

Definition 1.15 A distribution T is a linear form defined on $\mathscr{D}(\Omega)$ and continuous in the sense of Definition 1.13. The action of the distribution T on any function φ in $\mathscr{D}(\Omega)$ is then written

$$T(\varphi) \equiv \langle T, \varphi \rangle, \quad \forall \varphi \in \mathscr{D}(\Omega). \qquad (1.73)$$

Furthermore, the set of distributions on Ω is denoted by $\mathscr{D}'(\Omega)$.

Note also that in the general case of a distribution T in $\mathscr{D}'(\Omega)$, the duality bracket $\langle T, \varphi \rangle$ cannot be interpreted as the scalar product in $L^2(\Omega)$ of the distribution T and the function φ.

This is merely the notation adopted by analogy with the case of regular distributions T_f associated with functions f in $L^2(\Omega)$.

Remark: We still need to specify what is meant by the continuity of a distribution T in $\mathscr{D}'(\Omega)$, by analogy with Definition 1.13 characterizing the continuity of a regular distribution T_f.

Definition 1.16 If T is a distribution in $\mathscr{D}'(\Omega)$, we say that T is continuous at 0 if

$$\forall \varphi_n \in \mathscr{D}(\Omega), \ \varphi_n \to 0 \text{ in } \mathscr{D}(\Omega) \implies T(\varphi_n) \to 0 \text{ in } \mathbb{R}. \qquad (1.74)$$

Note that continuity of an arbitrary distribution T at 0 is equivalent to continuity of that distribution at all points φ_0 of $\mathscr{D}(\Omega)$, due to the linearity of distributions.

In addition, despite the strong formal resemblance between Definitions 1.13 and 1.16, the reader should be aware that for an arbitrary distribution T, the image $T(\varphi_n)$ in (1.74) can no longer be written as in the special case of regular distributions T_f as the integral of a function f multiplied by φ_n.

Indeed, Definition 1.15 of a distribution T involves new mathematical objects that can no longer be systematically associated with functions f *via* the regular distributions T_f.

The best-known example is undoubtedly the Dirac distribution δ, defined by

$$\delta : \mathscr{D}(\mathbb{R}) \longrightarrow \mathbb{R},$$
$$\varphi \longmapsto \delta(\varphi) \equiv \langle \delta, \varphi \rangle \equiv \varphi(0). \tag{1.75}$$

The definition of δ shows by simple inspection that it is a distribution in $\mathscr{D}'(\mathbb{R})$, i.e., a linear form on $\mathscr{D}(\mathbb{R})$ that is continuous in the sense of Definition 1.16.

It can then be shown that there is no function f in $L^2(\mathbb{R})$ with the property that

$$\delta(\varphi) \equiv \varphi(0) = \int_{\mathbb{R}} f\varphi \, dx, \quad \forall \varphi \in \mathscr{D}(\mathbb{R}). \tag{1.76}$$

Indeed, we apply *reductio ad absurdum*, assuming that there is such a function f in $L^2(\mathbb{R})$ that satisfies (1.76).

Consider the special case of functions φ belonging to $\mathscr{D}(\mathbb{R})$ such that $\varphi(0) = 0$. For each of these functions φ, there is a function ϕ in $\mathscr{D}(\mathbb{R})$ such that

$$\phi(x) = \frac{\varphi(x)}{x}. \tag{1.77}$$

Indeed, the only difficulty for the function ϕ concerns its regularity in the vicinity of $x = 0$.

But for every φ in $\mathscr{D}(\mathbb{R})$ that is zero at $x = 0$, the expression for ϕ can be rewritten in the form

$$\phi(x) = \frac{\varphi(0) + \int_0^x \varphi'(t) \, dt}{x} = \frac{\int_0^x \varphi'(t) \, dt}{x}. \tag{1.78}$$

Then as x tends to 0, we have

$$\phi(x) = \frac{\int_0^x \varphi'(t) \, dt}{x} \longrightarrow \varphi'(0), \tag{1.79}$$

by l'Hôpital's rule.

However, $\varphi'(0)$ is bounded, since φ is in $\mathscr{D}(\mathbb{R})$. Hence the function ϕ defined by (1.77) is bounded near $x = 0$.

Given that for $x \neq 0$, the function ϕ is C^∞ on $\mathbb{R}$, this is enough to ensure that it belongs to $\mathscr{D}(\mathbb{R})$.

We thus rewrite (1.76) using the function ϕ defined by (1.77):

$$\varphi(0) = 0 \times \phi(0) = 0 = \int_{\mathbb{R}} x f \phi \, \mathrm{d}t, \quad \forall \phi \in \mathscr{D}(\mathbb{R}). \tag{1.80}$$

Using density arguments, we deduce that xf is zero, which implies that the function f must itself be zero.

Hence, the original assumption (1.76) degenerates to

$$\varphi(0) = 0, \quad \forall \varphi \in \mathscr{D}(\mathbb{R}), \tag{1.81}$$

which is clearly absurd, since there are certainly functions in $\mathscr{D}(\mathbb{R})$ that are not zero at $x = 0$.

That original assumption was thus false, and we conclude that δ is not a regular distribution.

To end this section, we extend the definition of convergence of a sequence of distributions $(T_n)_{n \in \mathbb{N}}$ belonging to $\mathscr{D}'(\Omega)$ to an arbitrary distribution T in $\mathscr{D}'(\Omega)$.

Definition 1.17 Let $(T_n)_{n \in \mathbb{N}}$ be a sequence of distributions in $\mathscr{D}'(\Omega)$. We say that the sequence T_n converges to the distribution T in $\mathscr{D}'(\Omega)$ if

$$\langle T_n, \varphi \rangle \longrightarrow \langle T, \varphi \rangle, \quad \forall \varphi \in \mathscr{D}(\Omega). \tag{1.82}$$

We can now define differentiation in the sense of distributions.

1.4.5 Differentiation of Distributions

Definition 1.18 For a distribution T in $\mathscr{D}'(\Omega)$, we define another distribution $\partial T / \partial x_i$, the partial derivative in the sense of distributions in the direction x_i, $(i = 1, \ldots, n)$, of the distribution T by:

$$\left\langle \frac{\partial T}{\partial x_i}, \varphi \right\rangle \equiv -\left\langle T, \frac{\partial \varphi}{\partial x_i} \right\rangle, \quad \forall \varphi \in \mathscr{D}(\Omega). \tag{1.83}$$

Note immediately that according to (1.83), $\partial T / \partial x_i$ is indeed a distribution in $\mathscr{D}'(\Omega)$. The object $\partial T / \partial x_i$ is linear and continuous as defined by (1.74), thanks to these same properties of T.

Lemma 1.13 *The operator $\partial/\partial x_i$ mapping $\mathscr{D}'(\Omega)$ into $\mathscr{D}'(\Omega)$ that associates the partial derivative $\partial T/\partial x_i$ in the sense of distributions with a distribution T is linear and continuous.*

Proof: Let $(T_n)_{n\in\mathbb{N}}$ be a sequence of distributions in $\mathscr{D}'(\Omega)$ that converges to the distribution T. Let us show that the sequence $\partial T_n/\partial x_i$ converges to the distribution $\partial T/\partial x_i$ in $\mathscr{D}'(\Omega)$.

By definition of the partial derivative in the sense of distributions, for every function φ in $\mathscr{D}(\Omega)$, we have

$$\left\langle \frac{\partial T_n}{\partial x_i}, \varphi \right\rangle = -\left\langle T_n, \frac{\partial \varphi}{\partial x_i} \right\rangle \longrightarrow -\left\langle T, \frac{\partial \varphi}{\partial x_i} \right\rangle = \left\langle \frac{\partial T}{\partial x_i}, \varphi \right\rangle, \qquad (1.84)$$

where we have used Definition 1.17 characterizing the convergence of a sequence of distributions in $\mathscr{D}'(\Omega)$.

Remark: Definition 1.18 generalizes differentiation of functions in the usual sense. To see this, consider a function f in $L^2(\Omega)$ and the associated regular distribution T_f, and assume, moreover, that f is C^1 on Ω *in the classical sense* of differentiation of functions on Ω.

We now calculate its partial derivative $\partial T_f/\partial x_i$ in the sense of distributions.

For all $\varphi \in \mathscr{D}(\Omega)$, we have

$$\left\langle \frac{\partial T_f}{\partial x_i}, \varphi \right\rangle \equiv -\left\langle T_f, \frac{\partial \varphi}{\partial x_i} \right\rangle \equiv -\int_\Omega f \frac{\partial \varphi}{\partial x_i} \, \mathrm{d}\Omega. \qquad (1.85)$$

Since we have assumed that the function f is also C^1 on Ω in the usual sense, we can associate a regular distribution $T_{\partial f/\partial x_i}$ with each classical partial derivative $\partial f/\partial x_i$, since the latter also belongs to $L^2_{\mathrm{loc}}(\Omega)$.

To this end, note that the definition (1.58) of the regular distribution T_f is also valid for every function f belonging to $L^2_{\mathrm{loc}}(\Omega)$.

Now using Green's formula, which will be presented in the Sect. 1.5 on the Sobolev spaces $H^1(\Omega)$ (see Sect. 1.5), we obtain

$$\left\langle T_{\frac{\partial f}{\partial x_i}}, \varphi \right\rangle \equiv \int_\Omega \frac{\partial f}{\partial x_i} \varphi \, \mathrm{d}\Omega = -\int_\Omega f \frac{\partial \varphi}{\partial x_i} \, \mathrm{d}\Omega, \qquad (1.86)$$

where we have used the fact that the functions φ have compact support strictly contained in Ω.

In other words, such functions are zero on the boundary $\partial\Omega$ of Ω. This explains why there is no surface integral in the integration by parts (1.86). Finally, comparing (1.85) and (1.86), we obtain

$$\left\langle \frac{\partial T_f}{\partial x_i}, \varphi \right\rangle = \left\langle T_{\frac{\partial f}{\partial x_i}}, \varphi \right\rangle, \quad \forall \varphi \in \mathscr{D}(\Omega), \tag{1.87}$$

or again,

$$\frac{\partial T_f}{\partial x_i} = T_{\frac{\partial f}{\partial x_i}}, \quad \text{in } \mathscr{D}'(\Omega). \tag{1.88}$$

Equation (1.88) can then be interpreted as follows.

It is common practice to refer to $\partial T_f / \partial x_i$ as the "derivative of f" in the sense of distributions, but of course it would be meaningless to speak of differentiating the function f in this sense, since only the distribution T_f can be differentiated in the sense of distributions.

Furthermore, the distribution $T_{\partial f / \partial x_i}$ is characteristic of the usual partial derivative $\partial f / \partial x_i$, whenever the injection J defined by (1.62) allows us to associate this partial derivative with its regular distribution.

This is perfectly justified because we assumed that the function f was C^1 in the sense of functions. In other words, its first partial derivatives are continuous in Ω and thus belong to $L^2_{\text{loc}}(\Omega)$.

Therefore, (1.88) shows that the derivative in the sense of distributions of a function, i.e., its associated regular distribution T_f, coincides with the usual derivative of functions when the distribution is a "function," still identifying the function and the distribution associated with it by the operator J.

The distribution associated with the derivative, viz., $T_{\partial f / \partial x_i}$, and the derivative of the distribution, viz., $\partial T_f / \partial x_i$, are equal.

This proves the consistency of the new derivative and its generalization with regard to the classical derivative in the sense of functions.

Let us return now to the example of the function H defined in (1.56), which obviously belongs to $L^2(\mathbb{R})$, in order to calculate its derivative in the sense of distributions.

Since H trivially belongs to $L^2_{\text{loc}}(\mathbb{R})$, we may consider its regular distribution T_H defined by

$$\forall \varphi \in \mathscr{D}(\mathbb{R}), \quad \langle T_H, \varphi \rangle \equiv \int_{\mathbb{R}} H(x)\varphi(x)\,dx = \int_0^1 \varphi(x)\,dx. \tag{1.89}$$

The derivative T_H' of the regular distribution T_H is then found in the following way:

$$\forall \varphi \in \mathscr{D}(\mathbb{R}), \quad \langle T_H', \varphi \rangle \equiv -\langle T_H, \varphi' \rangle = -\int_0^1 \varphi'(x)\,dx = \varphi(0) - \varphi(1), \tag{1.90}$$

whence

$$\forall \varphi \in \mathscr{D}(\mathbb{R}), \quad \langle T'_H, \varphi \rangle \equiv \langle \delta_0 - \delta_1, \varphi \rangle, \tag{1.91}$$

where we have used an analogous notation to that adopted in (1.75) for the distribution δ, with δ_0 the Dirac distribution at 0 and δ_1 the Dirac distribution at 1.

It then follows that

$$\frac{dT_H}{dx} \equiv T'_H = \delta_0 - \delta_1, \quad \text{in } \mathscr{D}'(\mathbb{R}). \tag{1.92}$$

Generalization to the kth Derivative. The first partial derivative (1.83) in the sense of distributions can be extended to any order k. This yields the kth partial derivative, denoted by $\dfrac{\partial^k T}{\partial x_1^{k_1} \partial x_2^{k_2} \cdots \partial x_n^{k_n}}$, defined as follows:

$$\left\langle \frac{\partial^k T}{\partial x_1^{k_1} \partial x_2^{k_2} \cdots \partial x_n^{k_n}}, \varphi \right\rangle \equiv (-1)^{|k|} \left\langle T, \frac{\partial^k \varphi}{\partial x_1^{k_1} \partial x_2^{k_2} \cdots \partial x_n^{k_n}} \right\rangle, \quad \forall \varphi \in \mathscr{D}(\Omega), \tag{1.93}$$

with $|k| = k_1 + k_2 + \cdots + k_n$.

The definition (1.93) brings out the fact that the main burden of differentiation in the sense of distributions is borne by the functions φ belonging to $\mathscr{D}(\Omega)$.

This means that it is always possible to differentiate in the sense of distributions, even when the distribution in question is particularly irregular!

This, in turn, is the main reason for working in the functional framework of $\mathscr{D}(\Omega)$, which ensures the C^∞ regularity of the functions φ on which the action of any distribution T in $\mathscr{D}'(\Omega)$ is defined.

1.5 Sobolev Spaces

This section is devoted to the well-known Sobolev spaces introduced by Sergei Sobolev in 1935. Given the many successful applications of these function spaces in the mathematical analysis of partial differential equations, they have since become indispensable for anyone requiring robust and adaptable mathematical tools for the many problems that arise in the world of physics, mechanics, and more generally, engineering sciences, not to mention economics, marketing, communications, and the world of finance.

In the introductory spirit of the present book, we limit the discussion here to a modest account of the way these spaces are used, omitting most of the proofs, except those that have some immediate relevance to an application.

Unless otherwise stated, in the rest of this chapter, Ω will always denote an open subset of $\mathbb{R}^n$. In addition, given the intrinsic properties of the Sobolev spaces, any integration considered here will be taken in the sense of Lebesgue.

This amounts basically to having the following property at our disposal:

$$\int_{\Omega} |f(x)| \, d\Omega = 0 \iff f = 0 \text{ (a.e.) in } \Omega, \tag{1.94}$$

where a.e. is the usual abbreviation for the measure-theoretic "almost everywhere," i.e., with the exception of a set of measure zero.

In other words, the integral of a nonzero function over a set of measure zero is zero in the Lebesgue sense.

A classic example of a set of measure zero is the countably infinite set of rational numbers $\mathbb{Q}$, which has zero measure in $\mathbb{R}$.

1.5.1 The Space $L^2(\Omega)$

Many readers will be surprised to find that the Sobolev spaces are already familiar to them, at least in the most basic version, which is just the space $L^2(\Omega)$, sometimes denoted by $H^0(\Omega)$. This has the following definition:

$$L^2(\Omega) = \left\{ v : \Omega \subset \mathbb{R}^n \to \mathbb{R}, \int_{\Omega} |f(x)|^2 d\Omega < +\infty \right\}. \tag{1.95}$$

One consequence of the property (1.94) of the Lebesgue integral is that two functions that are equal in $L^2(\Omega)$ are equal almost everywhere.

This is why we will treat all elements of $L^2(\Omega)$ that are equal almost everywhere as belonging to one and the same class, and we will implicitly identify a given class with one of its representatives for notational purposes. The same will be done for the other Sobolev spaces.

The lemmas 1.14 and 1.15 sum up the key properties of the space $L^2(\Omega)$:

Lemma 1.14 $L^2(\Omega)$ *is a Hilbert space when equipped with the scalar product* $\langle \cdot, \cdot \rangle_{L^2(\Omega)}$ *and the associated norm* $\| \cdot \|_{L^2(\Omega)}$ *defined by:*

- $\forall (f, g) \in \left[L^2(\Omega) \right]^2, \quad \langle f, g \rangle_{L^2(\Omega)} \equiv \int_{\Omega} fg \, d\Omega.$
- $\forall f \in L^2(\Omega), \quad \| f \|^2_{L^2(\Omega)} = \langle f, f \rangle_{L^2(\Omega)}.$

Lemma 1.15 *Cauchy–Schwarz inequality and Sobolev injection theorem:*

- $\forall (f, g) \in \left[L^2(\Omega)\right]^2$, $\langle f, g \rangle_{L^2(\Omega)} \leq \|f\|_{L^2(\Omega)} \|g\|_{L^2(\Omega)}$.

- *If Ω is a bounded open subset of $\mathbb{R}^n$, we have*

$$\int_\Omega |f(x)| d\Omega \leq [\mu(\Omega)]^{1/2} \left[\int_\Omega |f(x)|^2 d\Omega\right]^{1/2}, \qquad (1.96)$$

where $\mu(\Omega)$ is the measure of Ω, i.e., its area, volume, etc., depending on the dimension of the space containing Ω.

The last property can be interpreted from the functional standpoint by the fact that every function whose square is integrable over a bounded open subset Ω is absolutely integrable over the same open set, so $L^2(\Omega) \subset L^1(\Omega)$.

The inequality (1.96) can also be interpreted by considering the natural injection defined by

$$i : L^2(\Omega) \longrightarrow L^1(\Omega)$$
$$f \longmapsto i(f) \equiv f.$$

The injection i reflects the fact that $L^2(\Omega)$ is a subset of $L^1(\Omega)$. Then, the inequality (1.96) expresses the continuity of the injection i, which is trivially continuous by construction.

Indeed, (1.96) can be rewritten in the form

$$\|i(f)\|_{L^1(\Omega)} \leq [\mu(\Omega)]^{1/2} \|f\|_{L^2(\Omega)}. \qquad (1.97)$$

The injection i is then said to be a continuous Sobolev injection.

The next result relates the space of functions $L^2(\Omega)$ to another function space that plays a key role in the mathematical analysis of partial differential equations, particularly in the theory of distributions (see Sect. 1.15).

Theorem 1.1 *The space $\mathscr{D}(\Omega)$ is dense in $L^2(\Omega)$.*

In other words, this theorem expresses the fact that for every function f in $L^2(\Omega)$, there is a sequence of functions $(f_n)_{n \in \mathbb{N}}$ in $\mathscr{D}(\Omega)$ such that

$$\lim_{n \to +\infty} \|f_n - f\| = 0. \qquad (1.98)$$

From a qualitative point of view, the density property says that every neighborhood of an element of $L^2(\Omega)$ contains an infinite number of functions of $\mathscr{D}(\Omega)$, which can thus "contaminate" the functions of $L^2(\Omega)$ with some of their properties.

This situation will be systematically exploited to implement the mathematical analysis of variational problems presented in the following chapters.

1.5.2 The Space $H^1(\Omega)$

The true power of the Sobolev spaces becomes apparent when we consider the space $H^1(\Omega)$. By analogy with the continuously differentiable functions $C^1(\Omega)$, this space characterizes the functions whose partial derivatives are not continuous on Ω but belong only to $L^2(\Omega)$.

Definition 1.19 If Ω is an open subset of $\mathbb{R}^n$, we define the Sobolev space $H^1(\Omega)$ by

$$H^1(\Omega) = \left\{ v : \Omega \subset \mathbb{R}^n \to \mathbb{R}, \ v \in L^2(\Omega), \ \frac{\partial v}{\partial x_i} \in L^2(\Omega), \ i = 1, \ldots, n \right\}. \tag{1.99}$$

Note that in Definition 1.19, the partial derivatives $\partial v/\partial x_i$ should be understood in the sense of distributions (see Definition 1.18).

Some comments are in order here. We should ask how the partial derivatives $\partial v/\partial x_i$ can simultaneously describe two quite different types of mathematical object, viz., a distribution belonging to $\mathscr{D}'(\Omega)$ and at the same time, a function belonging to $L^2(\Omega)$.

One answer to this is that when we speak of a distribution f belonging to $L^2(\Omega)$, this is an abuse of language, since we actually mean to refer to the regular distribution T_f associated with f by the injection J defined in (1.62), as discussed earlier.

In this context, and in order to make things more precise, we now introduce a second definition of the space $H^1(\Omega)$:

Definition 1.20 The space $H^1(\Omega)$ is defined by

$$H^1(\Omega) = \left\{ v \in L^2(\Omega) \text{ s.t. } \exists g_i \in L^2(\Omega) : \int_\Omega v \frac{\partial \varphi}{\partial x_i} \, d\Omega = - \int_\Omega g_i \varphi \, d\Omega \right.$$
$$\left. \forall \varphi \in \mathscr{D}(\Omega), \ i \in \{1, \ldots, n\} \right\}. \tag{1.100}$$

In this definition of the space $H^1(\Omega)$, the function g_i is called the weak derivative of v. It is denoted by

$$g_i \equiv \frac{\partial v}{\partial x_i}.$$

This is nothing other than the derivative in the sense of the regular distributions $T_{\partial v / \partial x_i}$, which we identify with the function $\partial v / \partial x_i$ in $L^2(\Omega)$ that it represents through the injection J defined in (1.62).

We then have the following result:

Lemma 1.16 *Definitions 1.19 and 1.20 of the Sobolev space* $H^1(\Omega)$ *are equivalent.*

Proof: Consider v belonging to $H^1(\Omega)$ according to Definition 1.19. Then v and $\partial v / \partial x_i$, $(i = 1, \ldots, n)$, are distributions in $L^2(\Omega)$ in the sense specified above.

Furthermore, we have

$$\left\langle \frac{\partial v}{\partial x_i}, \varphi \right\rangle = \int_\Omega \frac{\partial v}{\partial x_i} \varphi \, d\Omega, \quad \forall \varphi \in \mathscr{D}(\Omega), \tag{1.101}$$

where we have used the definition of regular distributions (see Definition 1.12), which is valid for every function in $L^2(\Omega)$.

Now, according to the definition of the derivative in the sense of distributions and using the fact that v is also associated with a regular distribution because it belongs to $L^2(\Omega)$, we have

$$\left\langle \frac{\partial v}{\partial x_i}, \varphi \right\rangle = -\left\langle v, \frac{\partial \varphi}{\partial x_i} \right\rangle = -\int_\Omega v \frac{\partial \varphi}{\partial x_i} \, d\Omega, \quad \forall \varphi \in \mathscr{D}(\Omega). \tag{1.102}$$

Comparing (1.101) and (1.102), we obtain

$$\int_\Omega \frac{\partial v}{\partial x_i} \varphi \, d\Omega = -\int_\Omega v \frac{\partial \varphi}{\partial x_i} \, d\Omega, \quad \forall \varphi \in \mathscr{D}(\Omega). \tag{1.103}$$

In other words, v indeed belongs to $H^1(\Omega)$ according to Definition 1.20.

Conversely, let v be an element of $H^1(\Omega)$ according to Definition 1.20, whence

$$\forall i = 1, \ldots, n, \ \exists g_i \in L^2(\Omega) : \int_\Omega v \frac{\partial \varphi}{\partial x_i} d\Omega = -\int_\Omega g_i \varphi d\Omega, \ \forall \varphi \in \mathscr{D}(\Omega), \tag{1.104}$$

where g_i is the weak derivative of v in the direction x_i, written symbolically

$$g_i \equiv \frac{\partial v}{\partial x_i}.$$

Hence, (1.104) can be written

$$\forall i = 1, \ldots, n, \ \exists g_i \in L^2(\Omega): \int_\Omega v \frac{\partial \varphi}{\partial x_i} d\Omega = - \int_\Omega \frac{\partial v}{\partial x_i} \varphi d\Omega, \ \forall \varphi \in \mathscr{D}(\Omega). \quad (1.105)$$

Given v and $\partial v / \partial x_i$ in $L^2(\Omega)$, we may rewrite (1.105) using distributions in the form

$$\forall i = 1, \ldots, n, \ \left\langle v, \frac{\partial \varphi}{\partial x_i} \right\rangle = - \left\langle \frac{\partial v}{\partial x_i}, \varphi \right\rangle, \ \forall \varphi \in \mathscr{D}(\Omega), \quad (1.106)$$

where we have used the injection J to identify the function v in $L^2(\Omega)$ with its associated regular distribution T_v, and also to identify $\partial v / \partial x_i$, another function in $L^2(\Omega)$, with its associated regular distribution $T_{\partial v / \partial x_i}$.

Therefore, (1.106) reflects the fact that v is in $H^1(\Omega)$ in the sense of Definition 1.19. This completes the proof.

The lemma 1.17 summarizes the structural properties of the space $H^1(\Omega)$.

Lemma 1.17

1. *The real-valued map on $H^1(\Omega) \times H^1(\Omega)$ denoted by $\langle .,. \rangle_{H^1}$ and defined by*

$$\langle u, v \rangle_{H^1} \equiv \int_\Omega uv \, d\Omega + \int_\Omega \nabla u \cdot \nabla u \, d\Omega \quad (1.107)$$

   provides a scalar product on $H^1(\Omega)$.

2. *$H^1(\Omega)$ is a Hilbert space equipped with the norm $\|.\|_{H^1}$ defined by*

$$\forall v \in H^1(\Omega), \quad \|v\|_{H^1}^2 \equiv \langle v, v \rangle_{H^1}. \quad (1.108)$$

   In other words,
$$\forall v \in H^1(\Omega), \quad \|v\|_{H^1}^2 = \|v\|_{L^2}^2 + \|\nabla v\|_{L^2}^2, \quad (1.109)$$

   where
$$\|\nabla v\|_{L^2}^2 = \left\| \frac{\partial v}{\partial x_1} \right\|_{L^2}^2 + \cdots + \left\| \frac{\partial v}{\partial x_n} \right\|_{L^2}^2. \quad (1.110)$$

Proof: There are several parts to the proof:

- We show that $\langle .,. \rangle_{H^1}$ is a scalar product on $H^1(\Omega)$ in three steps:

1. Bilinearity and symmetry are obvious by inspection straight from the definition of $\langle .,. \rangle_{H^1}$.

2. Positivity of the form $\langle .,. \rangle_{H^1}$ is also obvious, since

$$\forall v \in H^1(\Omega), \quad \langle v, v \rangle_{H^1} = \|v\|_{L^2}^2 + \|\nabla v\|_{L^2}^2 \geq 0.$$

3. Definiteness of the form $\langle .,. \rangle_{H^1}$ is obtained by considering

$$\langle v, v \rangle_{H^1} = 0 \quad \Longrightarrow \quad \|v\|_{L^2} = 0 \quad \Longrightarrow \quad v = 0 \text{ (a.e.) in } \Omega.$$

In other words, $v = 0$ in $L^2(\Omega)$, and hence also in $H^1(\Omega)$.

- Since the form $\langle .,. \rangle_{H^1}$ satisfies the scalar product axioms, the induced norm $\| \cdot \|_{H^1}$ satisfies all the norm axioms by construction, as was shown in Lemma 1.7.

- We now show that $H^1(\Omega)$ is a Hilbert space for the norm $\| \cdot \|_{H^1}$ associated with the scalar product $\langle .,. \rangle_{H^1}$.

Let $(u_n)_{n \in \mathbb{N}}$ be a Cauchy sequence in $H^1(\Omega)$. We must show that this sequence converges to some element u of $H^1(\Omega)$.

From the definition of the norm, $(u_n)_{n \in \mathbb{N}}$ is also a Cauchy sequence in $L^2(\Omega)$, and likewise for the sequence $(\nabla u_n)_{n \in \mathbb{N}}$.

Now, $L^2(\Omega)$ is a complete space, so

$$\exists u \in L^2(\Omega) \text{ such that } u_n \to u \text{ in } L^2(\Omega), \tag{1.111}$$

$$\exists u_i \in L^2(\Omega) \text{ such that } \frac{\partial u_n}{\partial x_i} \to u_i \text{ in } L^2(\Omega). \tag{1.112}$$

Put another way,

$$\lim_{n \to +\infty} \|u_n - u\|_{L^2} = 0 \text{ and } \lim_{n \to +\infty} \left\| \frac{\partial u_n}{\partial x_i} - u_i \right\|_{L^2} = 0, \quad \forall i = 1, \dots, n. \tag{1.113}$$

In order to show that $H^1(\Omega)$ is complete, we have to show that

$$\forall i = 1, \dots, n : \quad u_i = \frac{\partial u}{\partial x_i},$$

an equality to be understood in the sense of distributions.

Indeed, in this case, we will know that $u_n \to u$ in $H^1(\Omega)$, i.e., in the sense of the norm $\| \cdot \|_{H^1}$ defined by (1.109).

To do this, recall that the injection of $L^2(\Omega)$ into $\mathscr{D}'(\Omega)$ is continuous, by Lemma 1.12. This means that

$$\text{if } u_n \to u \text{ in } L^2(\Omega), \text{ then } u_n \to u \text{ in } \mathscr{D}'(\Omega), \qquad (1.114)$$

and

$$\text{if } \frac{\partial u_n}{\partial x_i} \to u_i \text{ in } L^2(\Omega), \text{ then } \frac{\partial u_n}{\partial x_i} \to u_i \text{ in } \mathscr{D}'(\Omega), \quad \forall i = 1, \ldots, n. \qquad (1.115)$$

Furthermore, the differential operator $\partial/\partial x_i$ is continuous from $\mathscr{D}'(\Omega)$ to itself (see Lemma 1.13). Hence, (1.114) implies that

$$\frac{\partial u_n}{\partial x_i} \longrightarrow \frac{\partial u}{\partial x_i} \text{ in } \mathscr{D}'(\Omega). \qquad (1.116)$$

By the uniqueness of the limit in $\mathscr{D}'(\Omega)$ and comparing (1.116) and (1.115), we thus have

$$u_i = \frac{\partial u}{\partial x_i}, \quad 1 \le i \le n.$$

Hence, $\partial u/\partial x_i$ is an element of $L^2(\Omega)$, and u belongs to $H^1(\Omega)$.

Finally, in terms of convergence, we have shown that

$$u_n \longrightarrow u \text{ in } L^2(\Omega) \text{ and } \frac{\partial u_n}{\partial x_i} \longrightarrow \frac{\partial u}{\partial x_i} \text{ in } L^2(\Omega). \qquad (1.117)$$

This means that u_n converges to u in $H^1(\Omega)$, with u belonging to $H^1(\Omega)$. This, in turn, proves that $H^1(\Omega)$ is a Hilbert space, and this completes the proof.

The Lemma 1.18 gathers the main results concerning the relationship between the space $H^1(\Omega)$ and the spaces $C^0(\overline{\Omega})$, $C^1(\overline{\Omega})$, and $C_0^\infty(\overline{\Omega})$.

When we refer to a *regular* bounded open set Ω of $\mathbb{R}^n$ in this and the following, we shall mean that the tangent to its boundary is at least piecewise C^1, i.e., the boundary has at most some angular points.

Lemma 1.18 *The lemma has three parts:*

1. *$C^1(\overline{\Omega}) \subset H^1(\Omega)$. This follows because the partial derivatives in the classical sense coincide with the derivatives in the sense of distributions (see the note just after Lemma 1.13).*

2. *In one dimension, $\Omega \subset \mathbb{R}$, and we have*

$$H^1([a, b]) \subset C^0([a, b]).$$

This is not true in the general case in which Ω is an open subset of $\mathbb{R}^n$, $n > 1$. In this case, we have

$$H^m(\Omega) \subset C^0(\Omega), \quad m > n/2.$$

3. *If Ω is a regular open subset of $\mathbb{R}^n$, then $C_0^\infty(\overline{\Omega})$, the space of C^∞ functions with compact support in $\overline{\Omega}$ as defined in (1.57), is dense in $H^1(\Omega)$.*

   Moreover, if Ω is bounded, then $C_0^\infty(\overline{\Omega}) = C^\infty(\overline{\Omega})$, because all the functions in $C^\infty(\overline{\Omega})$ necessarily have compact support.

   Note, finally, that the functions in $C_0^\infty(\overline{\Omega})$ do not necessarily vanish at the boundary of Ω.

In the next result, known in the literature as Rellich's theorem, compactness is essential, in particular to establish the Poincaré inequality, which will be proven in the Sect. 1.5.3, devoted to the space $H_0^1(\Omega)$.

Theorem 1.2 *If Ω is a regular bounded open subset of $\mathbb{R}^n$, then the injection of $H^1(\Omega)$ into $L^2(\Omega)$ is compact.*

Put another way, given any bounded sequence in $H^1(\Omega)$, we can always extract a convergent subsequence that converges in $L^2(\Omega)$.

Having introduced the Sobolev space $H^1(\Omega)$, we must now discuss another important question, namely the possibility of defining the values, also called the trace, of a function in $H^1(\Omega)$ on the boundary $\partial\Omega$ of Ω.

For a function in $L^2(\Omega)$, this operation is not always possible, as can be seen in the following one-dimensional example.

If the open set Ω is taken to be the open interval $[0, 1]$, and if $v(x) = \sin(1/x)$, then v belongs to $L^2((0, 1))$. Indeed, if we set $t = 1/x$ then $v^2(1/t)dt/t^2$ is bounded by $1/t^2$ which corresponds to a convergent Riemann integral at the neighborhood of infinity.

However, there is no natural way to define the trace of v at $x = 0$, one of the two points of the boundary $\partial\Omega$.

So what about the elements of $H^1(\Omega)$? The partial derivatives are required to belong to $L^2(\Omega)$, but will that suffice for the trace of a function in $H^1(\Omega)$ to be defined on the boundary $\partial\Omega$ of Ω? This is dealt with in the Theorem 1.2.

Theorem 1.3 *If Ω is a regular open bounded subset of $\mathbb{R}^n$ with boundary $\partial\Omega$, the trace map γ_0 defined by*

$$\gamma_0 : H^1(\Omega) \longrightarrow L^2(\partial\Omega)$$
$$v \longmapsto \gamma_0 v \equiv v|_{\partial\Omega},$$

is linear and continuous:

$$\exists\, C > 0 \text{ such that } \forall v \in H^1(\Omega), \quad \|\gamma_0 v\|_{L^2(\partial\Omega)} \leq C \|v\|_{H^1(\Omega)}. \qquad (1.118)$$

Notes.

- A function in $H^1(\Omega)$ is not necessarily continuous. However, Theorem 1.3 means that one can always define its trace on the boundary $\partial\Omega$ of Ω.

- The trace map γ_0 is surjective, and one can thus define the inverse of the trace of a function v on the boundary $\partial\Omega$ of the region Ω, this being referred to as an extension. It is an operation that extends a function defined on the boundary $\partial\Omega$ to the whole of Ω. Naturally, such an exercise cannot lead to a unique extension defined on Ω.

Theorem 1.4 *Let Ω be a regular open bounded subset of $\mathbb{R}^n$ with boundary $\partial\Omega$. For every function $w \in L^2(\partial\Omega)$, there is an extension v belonging to $H^1(\Omega)$ such that $\gamma_0 v \equiv v|_{\partial\Omega} = w$.*

We end this section by extending Green's formula to the functions in $H^1(\Omega)$.

Theorem 1.5 *Let Ω be a bounded open subset of $\mathbb{R}^n$ with continuous boundary $\partial\Omega = \Gamma$, and such that the tangent vector to the boundary has at worst discontinuities of the first kind, i.e., typically, angular points.*

Let u and v be two real-valued functions defined on Ω and belonging to $H^1(\Omega)$. Then we have

$$\int_\Omega \frac{\partial u}{\partial x_i} v \, d\Omega = -\int_\Omega u \frac{\partial v}{\partial x_i} \, d\Omega + \int_\Gamma uv \, n_i \, d\Gamma, \qquad (1.119)$$

where $\mathbf{n}$ is the outward-pointing unit normal vector on the boundary of the open set Ω (see Fig. 1.1) and n_i is its component in the direction of the coordinate x_i.

Fig. 1.1 Region of integration Ω and outward-pointing normal $\boldsymbol{n}$

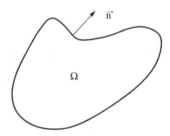

Proof: This is a classical result when u and v are two functions belonging to $C^1(\overline{\Omega})$. In the present case, we exploit the fact that $C_0^\infty(\overline{\Omega})$ is dense in $H^1(\Omega)$.

So, for all $(u, v) \in H^1(\Omega) \times H^1(\Omega)$, there exists $(u_n, v_n) \in C_0^\infty(\overline{\Omega}) \times C_0^\infty(\overline{\Omega})$ such that

$$u_n \to u \text{ in } H^1(\Omega), \qquad v_n \to v \text{ in } H^1(\Omega). \tag{1.120}$$

In $L^2(\Omega)$ and for the $L^2(\Omega)$ norm, this becomes

$$u_n \to u \text{ in } L^2(\Omega), \quad \frac{\partial u_n}{\partial x_i} \to \frac{\partial u}{\partial x_i} \text{ in } L^2(\Omega),$$

$$v_n \to v \text{ in } L^2(\Omega), \quad \frac{\partial v_n}{\partial x_i} \to \frac{\partial v}{\partial x_i} \text{ in } L^2(\Omega). \tag{1.121}$$

Furthermore, for the sequences of functions u_n and v_n in $C_0^\infty(\overline{\Omega})$, we have Green's formula (1.119), viz.,

$$\int_\Omega \frac{\partial u_n}{\partial x_i} v_n \, d\Omega = -\int_\Omega u_n \frac{\partial v_n}{\partial x_i} \, d\Omega + \int_\Gamma u_n v_n \, n_i \, d\Gamma. \tag{1.122}$$

It remains only to take the limit as n tends to $+\infty$ in (1.122) to obtain the result. To this end, we observe the asymptotic behavior of the first integral in (1.122), noting that the other terms have a similar structure and can be treated in the same way. We thus have

$$\left| \int_\Omega \frac{\partial u_n}{\partial x_i} v_n \, d\Omega - \int_\Omega \frac{\partial u}{\partial x_i} v \, d\Omega \right| = \left| \int_\Omega \left(\frac{\partial u_n}{\partial x_i} - \frac{\partial u}{\partial x_i} \right) v_n \, d\Omega + \int_\Omega \frac{\partial u}{\partial x_i} (v_n - v) \, d\Omega \right|,$$

$$\leq \left\| \frac{\partial u_n}{\partial x_i} - \frac{\partial u}{\partial x_i} \right\|_{L^2(\Omega)} \|v_n\|_{L^2(\Omega)} + \left\| \frac{\partial u}{\partial x_i} \right\|_{L^2(\Omega)} \|v_n - v\|_{L^2(\Omega)}, \tag{1.123}$$

where we have used the Cauchy–Schwarz inequality.

In Sect. 1.2 (see the properties listed after Definition 1.3), we saw that in a Banach space, every sequence converging for a given norm is necessarily bounded for that norm.

In other words, there exists $M > 0$ *independent* of n such that for all $n \in \mathbb{N}$, $\|v_n\|_{L^2(\Omega)} \leq M$.

Given the convergence properties of the sequences u_n and v_n in $L^2(\Omega)$, along with those of their partial derivatives, it follows, taking the limit as n tends to $+\infty$ in (1.123), that

$$\lim_{n \to +\infty} \int_\Omega \frac{\partial u_n}{\partial x_i} v_n \, d\Omega = \int_\Omega \frac{\partial u}{\partial x_i} v \, d\Omega. \qquad (1.124)$$

Notes.

- As already mentioned, the asymptotic behavior of the two other integrals in (1.122) can be treated in the same way.

However, for the surface integral on the boundary Γ, we can use the trace theorem (Theorem 1.3), noting that

$$\exists C > 0 \text{ such that } \forall n \in \mathbb{N}, \quad \|\gamma_0(v_n - v)\|_{L^2(\Gamma)} \leq C \|v_n - v\|_{H^1(\Omega)}.$$

Then v_n tends to v in $L^2(\Gamma)$, and it is a straightforward matter to show, as in (1.124), that

$$\lim_{n \to +\infty} \int_\Gamma u_n v_n \, n_i \, d\Gamma = \int_\Gamma u v \, n_i \, d\Gamma.$$

- Green's formula (1.119) is nothing other than a generalization of the formula for integration by parts in one dimension.

1.5.3 The Space $H_0^1(\Omega)$

Definition 1.21 Let Ω be a regular open subset of $\mathbb{R}^n$. Then we define the Sobolev space $H_0^1(\Omega)$ by

$$H_0^1(\Omega) = \left\{ v \in H^1(\Omega) : \gamma_0 v \equiv v|_{\partial\Omega} = 0 \right\}. \qquad (1.125)$$

These are therefore the functions in $H^1(\Omega)$ that vanish on the boundary of Ω.

Lemma 1.19 *The space $C_0^\infty(\Omega)$ is dense in $H_0^1(\Omega)$. Put another way, $H_0^1(\Omega)$ is the closure of $C_0^\infty(\Omega)$ in $H^1(\Omega)$, and we have*

$$H_0^1(\Omega) \equiv \overline{C_0^\infty(\Omega)}^{H^1(\Omega)}. \tag{1.126}$$

This lemma is the counterpart of the one noted above that shows that $C_0^\infty(\overline{\Omega})$ is dense in $H^1(\Omega)$.

We now consider the structural properties of the new space $H_0^1(\Omega)$.

Lemma 1.20 *$H_0^1(\Omega)$ is a Hilbert space for the norm and scalar product induced from $H^1(\Omega)$.*

Proof: Let $(v_n)_{n \in \mathbb{N}}$ be a Cauchy sequence in $H_0^1(\Omega)$, hence in $H^1(\Omega)$, which we already know to be a Hilbert space. There is then v in $H^1(\Omega)$ such that $v_n \to v$ in $H^1(\Omega)$.

Let us show that v must also belong to $H_0^1(\Omega)$, i.e., that $\gamma_0 v = 0$.

Now,

$$\gamma_0 v = \gamma_0(v - v_n + v_n) = \gamma_0(v - v_n), \tag{1.127}$$

since $v_n \in H_0^1(\Omega)$, whence $\gamma_0 v_n = 0$.

Moreover, since the trace map γ_0 is continuous from $H^1(\Omega)$ into $L^2(\Gamma)$ (see Theorem 1.3), we have

$$\exists C > 0 \text{ such that } \forall w \in H^1(\Omega), \quad \|\gamma_0 w\|_{L^2(\partial\Omega)} \leq C\|w\|_{H^1(\Omega)}.$$

Then using this property in (1.127), it follows that

$$\|\gamma_0 v\|_{L^2(\partial\Omega)} = \|\gamma_0(v - v_n)\|_{L^2(\partial\Omega)} \leq C\|v - v_n\|_{H^1(\Omega)}. \tag{1.128}$$

Taking the limit as n tends to $+\infty$ in this last inequality, we have finally

$$\gamma_0 v = 0 \implies v \in H_0^1(\Omega). \tag{1.129}$$

Put another way, $H_0^1(\Omega)$ is closed in $H^1(\Omega)$. This completes the proof.

A particularly useful characterization of the functions of $H_0^1(\Omega)$ is given by the following lemma, a proof of which can be found in [3]:

Lemma 1.21 *Consider a function u in $L^2(\Omega)$ and let $\bar{u}$ be the extension of u by 0 on $\mathbb{R}^n \setminus \Omega$. Then u belongs to $H_0^1(\Omega)$ if and only if the function $\bar{u}$ is an element of $H^1(\mathbb{R}^n)$. In this case, we have $\nabla \bar{u} = 1_\Omega \nabla u$, where 1_Ω is the indicator function of Ω equal to 1 on Ω and 0 on $\mathbb{R}^n \setminus \Omega$.*

The following result deals with the well-known Poincaré inequality, which will prove essential for the first few applications considered in Chap. 2, on the finite-element method.

Lemma 1.22 *If Ω is a regular bounded open subset of $\mathbb{R}^n$, then*

$$\exists C > 0 \text{ such that } \forall v \in H_0^1(\Omega), \quad \int_\Omega |v(x)|^2 \, d\Omega \leq C \int_\Omega |\nabla v(x)|^2 \, d\Omega. \tag{1.130}$$

Proof: We apply *reductio ad absurdum* and assume that there is no constant $C > 0$ such that

$$\forall v \in H_0^1(\Omega), \quad \int_\Omega |v(x)|^2 \, d\Omega \leq C \int_\Omega |\nabla v(x)|^2 \, d\Omega.$$

This implies that there is a sequence of functions $(v_n)_{n \in \mathbb{N}}$ in $H_0^1(\Omega)$ such that

$$\int_\Omega |v_n(x)|^2 \, d\Omega \geq n \int_\Omega |\nabla v_n(x)|^2 \, d\Omega, \quad \forall n \geq 1. \tag{1.131}$$

We now consider the sequence of functions $(w_n)_{n \in \mathbb{N}}$ in $H_0^1(\Omega)$ that normalizes the sequence $(v_n)_{n \in \mathbb{N}}$ for the $L^2(\Omega)$ norm, viz.,

$$w_n \equiv \frac{v_n}{\|v_n\|_{L^2(\Omega)}}. \tag{1.132}$$

Substituting w_n for v_n in the inequality (1.131), we find that

$$1 = \int_\Omega |w_n(x)|^2 \, d\Omega \geq n \int_\Omega |\nabla w_n(x)|^2 \, d\Omega. \tag{1.133}$$

The sequence $(w_n)_{n \in \mathbb{N}}$ is bounded for the $L^2(\Omega)$ norm, since by construction, $\|w_n\|_{L^2(\Omega)} = 1$, and the sequence $(\nabla w_n)_{n \in \mathbb{N}}$ is also bounded for the $L^2(\Omega)$ norm, according to (1.133).

It thus follows that the sequence $(w_n)_{n \in \mathbb{N}}$ is bounded in $H_0^1(\Omega)$ for the $H^1(\Omega)$ norm, since

$$\|w_n\|_{H^1(\Omega)}^2 \equiv \|w_n\|_{L^2(\Omega)}^2 + \|\nabla w_n\|_{L^2(\Omega)}^2 \leq 1 + \frac{1}{n^2} \leq 2.$$

By Rellich's compactness theorem (Theorem 1.2), we can extract a subsequence $(w_{n'})_{n' \in \mathbb{N}}$ of $(w_n)_{n \in \mathbb{N}}$ that converges to w in $L^2(\Omega)$ and that also satisfies the inequality (1.133):

$$\|\nabla w_{n'}\|_{L^2(\Omega)} \leq \frac{1}{n'}. \tag{1.134}$$

Put another way, the sequence $(\nabla w_{n'})_{n' \in \mathbb{N}}$ converges to 0 in $L^2(\Omega)$.

Now, since the sequence $(w_{n'})_{n' \in \mathbb{N}}$ converges to w in $L^2(\Omega)$, it must also converge to w in $\mathscr{D}'(\Omega)$, because the injection of $L^2(\Omega)$ into $\mathscr{D}'(\Omega)$ is continuous (see Lemma 1.12).

It follows that the sequence $(\nabla w_{n'})_{n' \in \mathbb{N}}$ also converges to ∇w in $\mathscr{D}'(\Omega)$, since the differential operator is itself continuous from $\mathscr{D}'(\Omega)$ into $\mathscr{D}'(\Omega)$ (see Lemma 1.13).

We thus have the following two limits:

$$\nabla w_{n'} \to 0 \text{ in } \mathscr{D}'(\Omega) \text{ and } \nabla w_{n'} \to \nabla w \text{ in } \mathscr{D}'(\Omega). \tag{1.135}$$

By the uniqueness of the limit of a sequence of distributions in $\mathscr{D}'(\Omega)$, we deduce that

$$\nabla w = 0 \text{ in } \mathscr{D}'(\Omega). \tag{1.136}$$

As a consequence, ∇w belongs to $L^2(\Omega)$, since the zero distribution can be considered to be associated with the identically zero function on Ω, and finally, w belongs to $H^1(\Omega)$.

Let us summarize the properties established for the subsequence $(w_{n'})_{n' \in \mathbb{N}}$:

1. $w_{n'}$ belongs to $H_0^1(\Omega)$.

2. $w_{n'}$ converges to w in $H^1(\Omega)$.

Since $H_0^1(\Omega)$ is closed in $H^1(\Omega)$ (see the conclusion of Lemma 1.20), we may conclude that w is also an element of $H_0^1(\Omega)$.

In conclusion, the limit w of the sequence $(w_{n'})_{n' \in \mathbb{N}}$ has zero gradient a.e. in Ω, and as a consequence, w is therefore constant a.e. in Ω.

We now consider the extension $\bar{w}$ of w to the whole of $\mathbb{R}^n$, equal to zero on $\mathbb{R}^n \setminus \Omega$. Since w belongs to $H_0^1(\Omega)$, Lemma 1.21 implies that $\bar{w}$ belongs to $H^1(\mathbb{R}^n)$ and $\nabla \bar{w}$ is zero on the whole of $\mathbb{R}^n$.

Hence, $\bar{w}$ is piecewise constant on $\mathbb{R}^n$, in principle an arbitrary constant on Ω and zero on its complement in $\mathbb{R}^n$. Thus $\bar{w}$ can be identically zero only on the whole of $\mathbb{R}^n$, since it belongs to $H^1(\mathbb{R}^n)$.

Indeed, one discontinuity in the piecewise constant $\bar{w}$ would introduce a Dirac distribution for the gradient on the boundary $\partial\Omega$ of Ω (see the example of differentiation of the Heaviside function in the sense of distributions in Sect. 1.4.5), and $\bar{w}$ could no longer belong to $H^1(\mathbb{R}^n)$, since its gradient would no longer be a function in $L^2(\Omega)$.

We deduce finally that $w = 0$ (a.e.) in Ω. But this contradicts the fact that w also satisfies

$$\|w\|_{L^2(\Omega)} = \lim_{n'\to+\infty} \|w_{n'}\|_{L^2(\Omega)} = 1. \tag{1.137}$$

This ends the proof by *reductio ad absurdum*, and the Poincaré inequality (1.130) is necessarily satisfied by every function v in $H_0^1(\Omega)$.

Notes.

- What we should retain from the proof of Poincaré's inequality is that if a function v is constant on Ω and if it belongs to $H_0^1(\Omega)$, then it must be identically zero on the whole of Ω.

- The constant C arising in the Poincaré inequality (1.130) depends exclusively on Ω.

- The Poincaré inequality is also valid whenever the functions being considered in $H^1(\Omega)$ vanish on part of the boundary of Ω (see, for example, [1] or [3]).

The next property exploits Poincaré's inequality and concerns the equivalence of norms on $H_0^1(\Omega)$.

Lemma 1.23 *Let Ω be a regular bounded open subset of $\mathbb{R}^n$.*

Then $\exists\,(\alpha, \beta) \in (\mathbb{R}_+^*)^2$ such that

$$\forall v \in H_0^1(\Omega), \quad \alpha\|v\|_{H^1(\Omega)} \leq |v|_1 \leq \beta\|v\|_{H^1(\Omega)}, \tag{1.138}$$

where $|\,.\,|_1$ is defined by

$$|v|_1 \equiv \left(\int_\Omega |\nabla v|^2 d\Omega\right)^{1/2}. \tag{1.139}$$

Proof: Once again, we proceed in stages:

- To begin with, note that $|\,.\,|_1$ as defined by (1.139) is indeed a norm on $H_0^1(\Omega)$. In particular, we have

$$|v|_1 = 0 \implies \nabla v = 0 \text{ (a.e.) in } \Omega,$$
$$\implies v = K \text{ (a.e.) in } \Omega,$$

where K is an arbitrary real constant. But for any constant function K belonging to $H_0^1(\Omega)$, the Poincaré inequality implies that

$$K^2 \mu(\Omega) = \int_\Omega K^2 d\Omega \leq C \int_\Omega \nabla K d\Omega = 0, \qquad (1.140)$$

where $\mu(\Omega)$ is the measure of Ω. We deduce that $K = 0$, and we thus obtain the first property required of a norm, viz.,

$$|v|_1 = 0 \implies v = 0. \qquad (1.141)$$

- The right-hand inequality of (1.138) is obtained as follows. By definition of the natural norm on $H^1(\Omega)$, we have

$$\|v\|^2_{H^1(\Omega)} \equiv \|v\|^2_{L^2(\Omega)} + \|\nabla v\|^2_{L^2(\Omega)} \geq \|\nabla v\|^2_{L^2(\Omega)}, \qquad (1.142)$$

whence $\|v\|_{H^1(\Omega)} \geq |v|_1$ and $\beta \equiv 1$.

- Concerning the left-hand inequality of (1.138), since we are working in $H_0^1(\Omega)$, the Poincaré inequality implies that

$$\exists\, C > 0 \text{ such that } \forall v \in H_0^1(\Omega), \quad \|v\|^2_{L^2(\Omega)} \leq C \|\nabla v\|^2_{L^2(\Omega)}, \qquad (1.143)$$

whence

$$\|v\|^2_{H^1(\Omega)} \equiv \|v\|^2_{L^2(\Omega)} + \|\nabla v\|^2_{L^2(\Omega)} \leq (1 + C) \|\nabla v\|^2_{L^2(\Omega)}. \qquad (1.144)$$

This therefore implies that

$$\frac{1}{\sqrt{1 + C}} \|v\|_{H^1(\Omega)} \leq \|\nabla v\|_{L^2(\Omega)} \equiv |v|_1. \qquad (1.145)$$

Therefore, α in (1.138) can be taken as

$$\alpha \equiv \frac{1}{\sqrt{1 + C}}.$$

Notes.

- This result concerning equivalence of norms on $H_0^1(\Omega)$ generalizes to the Sobolev space $H_{\Gamma_1}^1(\Omega)$ defined by

$$H_{\Gamma_1}^1(\Omega) \equiv \left\{ v \in H^1(\Omega) : v|_{\Gamma_1} = 0 \right\}, \qquad (1.146)$$

where Γ_1 is part of the boundary of $\partial\Omega$ with nonzero measure. As in $H_0^1(\Omega)$, we then have

$$\exists\, C > 0 \text{ such that } \forall\, v \in H^1_{\Gamma_1}(\Omega), \quad \|v\|^2_{L^2(\Omega)} \leq C\,\|\nabla v\|^2_{L^2(\Omega)}. \tag{1.147}$$

- When the function v does not vanish on part of the boundary $\partial\Omega$, we have the generalized Poincaré inequality, a direct consequence of the Petree–Tartar lemma (see, for example, [4]):

$$\exists\, C > 0 \text{ such that } \forall v \in H^1(\Omega), \quad \|v\|^2_{H^1(\Omega)} \leq C\left(\|v\|^2_{L^2(\partial\Omega)} + \|\nabla v\|^2_{L^2(\Omega)}\right). \tag{1.148}$$

Consequently, on $H^1(\Omega)$, the norm $\|.\|_{H^1(\Omega)}$ is equivalent to the norm $|\,.\,|_*$ defined by

$$|v|_* \equiv \left[\|v\|^2_{L^2(\partial\Omega)} + \|\nabla v\|^2_{L^2(\Omega)}\right]^{1/2}. \tag{1.149}$$

1.5.4 The Space $H^2(\Omega)$

We now extend the Definition 1.19 of $H^1(\Omega)$ to the space $H^2(\Omega)$:

Definition 1.22 If Ω is an open subset of $\mathbb{R}^n$, we define the Sobolev space $H^2(\Omega)$ by

$$H^2(\Omega) = \left\{v \in L^2(\Omega) : \frac{\partial v}{\partial x_i} \in L^2(\Omega), \ \frac{\partial^2 v}{\partial x_i \partial x_j} \in L^2(\Omega), \ 1 \leq i, j \leq n\right\}, \tag{1.150}$$

where the partial derivatives are taken in the sense of distributions.

Lemma 1.24 The space $H^2(\Omega)$ is a Hilbert space for the scalar product $\langle \cdot, \cdot \rangle_{H^2(\Omega)}$ and the associated norm $\|.\|_{H^2(\Omega)}$ defined by

$$\langle u, v \rangle_{H^2(\Omega)} = \int_\Omega uv \, d\Omega + \sum_{i=1}^n \int_\Omega \frac{\partial u}{\partial x_i}\frac{\partial v}{\partial x_i}\,d\Omega + \sum_{i,j=1}^n \int_\Omega \frac{\partial^2 u}{\partial x_i \partial x_j}\frac{\partial^2 v}{\partial x_i \partial x_j}\,d\Omega, \tag{1.151}$$

$$\|u\|_{H^2(\Omega)} = \sqrt{\|u\|^2_{L^2(\Omega)} + \sum_{i=1}^n \left\|\frac{\partial u}{\partial x_i}\right\|^2_{L^2(\Omega)} + \sum_{i,j=1}^n \left\|\frac{\partial^2 u}{\partial x_i \partial x_j}\right\|^2_{L^2(\Omega)}}. \tag{1.152}$$

Lemma 1.24 is proved in a similar way to Lemma 1.17, where we established that the space $H^1(\Omega)$ is a Hilbert space for the appropriate norm $\|.\|_{H^1(\Omega)}$ and scalar product $\langle .,.\rangle_{H^1(\Omega)}$.

Here we present the counterpart of the first trace theorem (Theorem 1.3) for functions in $H^2(\Omega)$.

Theorem 1.6 *Let Ω be a regular bounded open subset of $\mathbb{R}^n$ with boundary $\partial\Omega$. Then the trace map γ_1 is defined by*

$$\gamma_1 : H^2(\Omega) \longrightarrow L^2(\partial\Omega)$$

$$v \longmapsto \gamma_1 v \equiv \left.\frac{\partial v}{\partial n}\right|_{\partial\Omega} \equiv \nabla v|_{\partial\Omega}\cdot n,$$

where $\boldsymbol{n}$ is the outward-pointing unit normal vector on the boundary Ω. Then γ_1 is linear and continuous:

$$\exists\, C > 0 \text{ such that } \forall\, v \in H^2(\Omega), \quad \|\gamma_1 v\|_{L^2(\partial\Omega)} \leq C\|v\|_{H^2(\Omega)}. \quad (1.153)$$

Proof: This is a consequence of the trace theorem, Theorem 1.3, in $H^1(\Omega)$. Indeed, if v is an element of $H^2(\Omega)$, then each component $\partial v/\partial x_i$ of ∇v belongs to $H^1(\Omega)$, and we can define the trace $\gamma_0(\partial v/\partial x_i)$ on the boundary $\partial\Omega$ of Ω as a function of $L^2(\partial\Omega)$.

In addition, since we consider only regular open subsets Ω of $\mathbb{R}^n$, the normal $\mathbf{n}$ is a bounded function on $\partial\Omega$, which we may assume to be continuous in the context of this proof, given that the present trace theorem remains valid for open sets that may, for example, exhibit angular points where the normal would have a discontinuity of the first kind.

As a consequence, $\partial v/\partial n$ belongs to $L^2(\partial\Omega)$, and we have

$$\|\gamma_1 v\|^2_{L^2(\partial\Omega)} = \left\|\sum_{i=1}^n \gamma_0\left(\frac{\partial v}{\partial x_i}\right) n_i\right\|^2_{L^2(\partial\Omega)} = \int_{\partial\Omega}\left[\sum_{i=1}^n \gamma_0\left(\frac{\partial v}{\partial x_i}\right) n_i\right]^2 d\Gamma$$

$$\leq \max_{1\leq i\leq n}\left(\|n_i\|^2_{C^0(\Omega)}\right)\int_{\partial\Omega}\left[\sum_{i=1}^n \gamma_0\left(\frac{\partial v}{\partial x_i}\right)\right]^2 d\Gamma$$

$$\leq \max_{1\leq i\leq n}\left(\|n_i\|^2_{C^0(\Omega)}\right) 2\sum_{i=1}^n \int_{\partial\Omega}\left[\gamma_0\left(\frac{\partial v}{\partial x_i}\right)\right]^2 d\Gamma$$

$$\leq 2\max_{1\leq i\leq n}\left(\|n_i\|^2_{C^0(\Omega)}\right)\sum_{i=1}^n \left\|\gamma_0\left(\frac{\partial v}{\partial x_i}\right)\right\|^2_{L^2(\partial\Omega)}, \quad (1.154)$$

where $\|.\|_{C^0(\Omega)}$ is the standard notation for the *supremum* norm defined by

$$\|n_i\|_{C^0(\Omega)} \equiv \sup_{x \in \partial\Omega} |n_i(x)|.$$

Note that we have used the following property:

$$\forall (\alpha_1, \ldots, \alpha_n) \in \mathbb{R}^n, \quad \left(\sum_{i=1}^n \alpha_i \right)^2 \leq 2 \sum_{i=1}^n \alpha_i^2.$$

Finally, for v in $H^2(\Omega)$, the partial derivatives $\partial v / \partial x_i$ are in $H^1(\Omega)$, so according to Theorem 1.3, the trace theorem in $H^1(\Omega)$, we have

$$\exists D > 0 \text{ such that } \left\| \gamma_0 \left(\frac{\partial v}{\partial x_i} \right) \right\|_{L^2(\partial\Omega)}^2 \leq D^2 \left\| \frac{\partial v}{\partial x_i} \right\|_{H^1(\Omega)}^2. \tag{1.155}$$

Hence, using (1.155) in (1.154), we obtain

$$\|\gamma_1 v\|_{L^2(\partial\Omega)}^2 \leq 2D^2 \max_{1 \leq i \leq n} \left(\|n_i\|_{C^0(\Omega)}^2 \right) \sum_{i=1}^n \left(\left\| \frac{\partial v}{\partial x_i} \right\|_{H^1(\Omega)}^2 \right)$$

$$\leq 2D^2 \max_{1 \leq i \leq n} \left(\|n_i\|_{C^0(\Omega)}^2 \right) \|v\|_{H^2(\Omega)}^2. \tag{1.156}$$

To obtain this, we used the following algebraic property:

$$\forall i = 1, \ldots, n, \quad \left\| \frac{\partial v}{\partial x_i} \right\|_{H^1(\Omega)}^2 = \left\| \frac{\partial v}{\partial x_i} \right\|_{L^2(\Omega)}^2 + \sum_{j=1}^n \left\| \frac{\partial^2 v}{\partial x_i \partial x_j} \right\|_{L^2(\Omega)}^2. \tag{1.157}$$

Summing both sides of (1.157) over i, we obtain

$$\sum_{i=1}^n \left\| \frac{\partial v}{\partial x_i} \right\|_{H^1(\Omega)}^2 = \sum_{i=1}^n \left\| \frac{\partial v}{\partial x_i} \right\|_{L^2(\Omega)}^2 + \sum_{i,j=1}^n \left\| \frac{\partial^2 v}{\partial x_i \partial x_j} \right\|_{L^2(\Omega)}^2 \leq \|v\|_{H^2(\Omega)}^2. \tag{1.158}$$

The constant C in (1.153) can then be taken as

$$C \equiv \sqrt{2} D \max_{1 \leq i \leq n} \left(\|n_i\|_{C^0(\Omega)} \right),$$

and this completes the proof of Theorem 1.6, the trace theorem in $H^2(\Omega)$.

We end this section by stating the second Green's formula relating to the Laplacian:

Lemma 1.25 *Let Ω be a regular bounded open subset of $\mathbb{R}^n$. If u and v are two elements of $H^2(\Omega) \times H^1(\Omega)$, then*

$$\int_\Omega (\Delta u)v \, d\Omega = -\int_\Omega \nabla u \cdot \nabla v \, d\Omega + \int_{\partial\Omega} \frac{\partial u}{\partial n} v \, d\Gamma, \qquad (1.159)$$

where n is the outward-pointing unit normal vector on the boundary of the open set Ω, and $\partial u/\partial n$ is the projection of the gradient vector in the direction of the normal n.

Proof: As in the Green's formula of Theorem 1.5 for $H^1(\Omega)$, we use the fact that $C_0^\infty(\overline{\Omega})$ is dense in $H^2(\Omega)$, together with the continuity of the trace map γ_1.

Note. There is a vast literature discussing and extending the ideas developed here concerning Sobolev spaces (see, for example, [1, 5], or [6]). As already mentioned, the aim of this book is to set out mathematical tools that will help us to develop certain applications later on.

That is why we have not attempted to present the generalization of $L^2(\Omega)$, $H^1(\Omega)$, and $H^2(\Omega)$ to the spaces $H^m(\Omega)$ and $W^{m,p}(\Omega)$. The interested reader is referred to the above literature for a detailed discussion of these developments.

1.6 Fundamental Theorems for the Functional Analysis of PDEs

In order to present the main theorems guaranteeing the existence and uniqueness of solutions to the variational or minimization problems, we now introduce a formalism that generalizes the example discussed in Sect. 1.1.3.

To do this, we introduce a Hilbert space V equipped with a scalar product $\langle .,. \rangle_V$ and the associated norm $\|.\|_V$. We also consider a linear form $L(.)$ defined on V and a bilinear form $a(.,.)$ defined on V^2.

The general formalism to be considered now is expressed in terms of the variational problem (**VP**) and the minimization problem (**MP**), defined as follows:

$$(\textbf{VP}) \begin{cases} \text{Find } u \in V \text{ solution of} \\ a(u,v) = L(v), \ \forall v \in V \end{cases}, \quad (\textbf{MP}) \begin{cases} \text{Find } u \in V \text{ solution of} \\ J(u) \equiv \min_{v \in V} J(v) \\ \text{where } J(v) = \frac{1}{2}a(v,v) - L(v) \end{cases}.$$

The first fundamental result in this chapter is Riesz's theorem, which shows us how to represent any linear form defined on a Hilbert space V.

Theorem 1.7 *Let V be a Hilbert space equipped with a scalar product $\langle .,.\rangle_V$ and the associated norm $\|.\|_V$. For every linear form $L(.)$ defined on V, i.e., $L \in \mathscr{L}(V, \mathbb{R})$,*

$$\exists! \, u_L \in V \text{ such that } L(v) = \langle u_L, v\rangle_V, \quad \forall v \in V.$$

There is an extension of this theorem to bilinear forms that will allow us later on in the chapter to obtain the Lax–Milgram theorem, among other things.

Theorem 1.8 *Let V be a Hilbert space equipped with a scalar product $\langle .,.\rangle_V$ and the associated norm $\|.\|_V$. Let $a(.,.)$ be a bilinear form, defined and continuous on V^2. Then there is a unique linear and continuous operator A from V to V such that*

$$\forall (u, v) \in V^2, \quad a(u, v) = \langle Au, v\rangle_V.$$

Before stating the main theorems of this chapter guaranteeing the existence and uniqueness of the solution to the variational problem (**VP**) or the minimization problem (**MP**), we introduce the notion of a V-elliptic bilinear form:

Definition 1.23 Let V be a Hilbert space equipped with a scalar product $\langle .,.\rangle_V$ and the associated norm $\|.\|_V$. Let $a(.,.)$ be a bilinear form defined on V^2. This form will be said to be V-elliptic or coercive if

$$\exists \alpha > 0 \text{ such that } \forall u \in V, \quad a(u, u) \geq \alpha \|u\|_V^2.$$

The first version of Stampacchia's theorem can then be stated as follows:

Theorem 1.9 *We make the following hypotheses:*

1. *Let V be a Hilbert space equipped with a scalar product $\langle .,.\rangle_V$ and the associated norm $\|.\|_V$.*

2. *Let K be a closed convex subset of V, i.e., $K \subset V$.*

3. *Let $a(.,.)$ be a symmetric, continuous, V-elliptic bilinear form defined on V^2.*

4. *Finally, let $L(.)$ be a continuous linear form on V.*

Then there is a unique element $u \in K$ solving the minimization problem (**MP**):

$$J(u) \leq J(v), \quad \forall \, v \in K, \quad \text{where } J(v) \equiv \frac{1}{2}a(v, v) - L(v). \qquad (1.160)$$

When the closed convex set K is a vector subspace of the Hilbert space V, Theorem 1.9 takes the following form:

Theorem 1.10 *We make the following hypotheses:*

1. *Let V be a Hilbert space equipped with a scalar product $\langle .,. \rangle_V$ and the associated norm $\| . \|_V$.*

2. *Let K be a closed vector subspace of V.*

3. *Let $a(.,.)$ be a symmetric, continuous, V-elliptic bilinear form defined on V^2.*

4. *Finally, let $L(.)$ be a continuous linear form on V.*

Then, the following three results are equivalent:

* $\exists ! \, u \in K$ *such that* $J(u) \leq J(v), \quad \forall v \in K.$

* $\exists ! \, u \in K$ *such that* $a(u, v) = L(v), \quad \forall v \in K.$

* $\exists ! \, u \in K$ *such that* $a(u, v - u) = L(v - u), \quad \forall v \in K.$

The next result concerns sufficient conditions for the existence and uniqueness of a solution to the variational formulation (**VP**). This is the celebrated Lax–Milgram theorem:

Theorem 1.11 *We make the following hypotheses:*

1. *Let V be a Hilbert space equipped with a scalar product $\langle .,. \rangle_V$ and the associated norm $\| . \|_V$.*

2. *Let $a(.,.)$ be a continuous, V-elliptic bilinear form defined on V^2.*

3. *Finally, let $L(.)$ be a continuous linear form on V.*

*Then, there is a unique u in V solving the variational problem (**VP**):*

$$\text{Find } u \in V \text{ such that } a(u, v) = L(v), \quad \forall v \in V.$$

The next result characterizes the continuous dependence of the solution to the variational problem (**VP**) on the problem data. In the literature, one speaks of *a priori* estimation of the solution u.

Theorem 1.12 *With the hypotheses of the Lax–Milgram theorem, the unique solution u of the problem, viz.,*

$$\text{find } u \in V \text{ solution of } a(u, v) = L(v), \quad \forall v \in V, \tag{1.161}$$

depends continuously on the map L:

$$\exists \, M > 0 \text{ such that } \|u\|_V \leq M \, |||L|||, \tag{1.162}$$

where $|||.|||$ is the norm defined by

$$|||L||| \equiv \sup_{x \in V \setminus \{0\}} \frac{|L(x)|}{\|x\|_V}. \tag{1.163}$$

This is proven as follows. From the V-ellipticity assumption,

$$\exists \, \alpha > 0 \text{ such that } a(v, v) \geq \alpha \|v\|_V^2, \quad \forall v \in V. \tag{1.164}$$

If u solves the variational problem (1.161), then choosing $v = u$, we have

$$a(u, u) = L(u). \tag{1.165}$$

Hence, by the inequality (1.164),

$$L(u) \geq \alpha \|u\|_V^2. \tag{1.166}$$

- It follows that for $u \neq 0$, we have

$$\|u\|_V \leq \frac{1}{\alpha} \frac{L(u)}{\|u\|_V} \leq \frac{1}{\alpha} |||L|||. \tag{1.167}$$

Put another way, we can take the constant M in (1.162) to be $1/\alpha$.

- If $u = 0$, then trivially $0 \equiv \|0\|_V \leq M|||L|||$.

Notes.

1. If V has finite dimension N, e.g., $V = \mathbb{R}^N$, or V is isomorphic to $\mathbb{R}^N$, and if we introduce a basis $(\varphi)_{i=1,\ldots,N}$ of $\mathbb{R}^N$, we can decompose the solution u of the variational formulation (1.161) relative to this basis:

$$u = \sum_{i=1,\ldots,N} u_i \varphi_i. \tag{1.168}$$

Furthermore, in the formulation (1.161), we can also choose $v = \varphi_j$, for any j between 1 and N. This implies that

$$\sum_{j=1,\,...,\,N} a(\varphi_i, \varphi_j)u_i = L(\varphi_i), \quad \forall i = 1, \ldots, N. \qquad (1.169)$$

If we introduce the matrix A with elements $A_{ij} \equiv a(\varphi_i, \varphi_j)$ and the vector v_L with components $(v_L)_i \equiv L(\varphi_i)$, the formulation (1.161) becomes

$$\text{find } u \in \mathbb{R}^N \text{ solution of } Au = v_L. \qquad (1.170)$$

This last formulation is a linear system of equations in which the matrix A describes the operator appearing in Theorem 1.8, the extension of Riesz's theorem.

Indeed, using the definition of the matrix A, it is straightforward to show that we indeed have

$$\forall (u, v) \in \mathbb{R}^{2N}, \quad a(u, v) = \langle Au, v \rangle.$$

The V-ellipticity of the bilinear form $a(.,.)$ then implies that the matrix A is invertible. To see this, let us rewrite the ellipticity inequality in Definition 1.23 using the extension of Riesz's theorem and choosing $v = u$:

$$\exists \alpha > 0 \text{ such that } \alpha \|u\|_V^2 \leq a(u, u) = \langle Au, u \rangle \leq \|Au\|_V \|u\|_V, \qquad (1.171)$$

where we have used the Cauchy–Schwarz inequality stated in Property 1.6 to establish the last upper bound.

Dividing through by the norm of u, we obtain finally

$$\alpha \|u\|_V \leq \|Au\|_V. \qquad (1.172)$$

Showing that the operator A is invertible amounts in the present case, i.e., in finite dimensions, since we are effectively working in $\mathbb{R}^N$, to establishing that the operator A is injective, according to the rank theorem, viz., if $Au = 0$, then $u = 0$.

Now this can be shown using the inequality (1.172). Indeed, if $Au = 0$, then $\alpha \|u\|_V \leq 0$, and finally, u is also zero. The operator A is injective, so the linear system (1.170) involves an invertible matrix, and this in turn implies that there is one and only one solution u to this linear system.

This property underpins the numerical approximation of variational formulations, which will be more fully developed in Chap. 2.

2. In both Stampacchia's theorem and the Lax–Milgram theorem, the properties of continuity, ellipticity, and Hilbert structure must be established either for a given norm, or, if necessary, for equivalent norms, i.e., satisfying the following property:

if $\|.\|_1$ and $\|.\|_2$ are two suitable norms for the space V, these norms will be said to be equivalent if there are two strictly positive constants α and β, independent of v, such that

$$\forall v \in V, \quad \alpha \|v\|_2 \leq \|v\|_1 \leq \beta \|v\|_2.$$

3. In the context of the Lax–Milgram theorem, the bilinear form $a(\cdot, \cdot)$ need not be symmetric. However, when it is symmetric, we can then show equivalence of the variational formulation (**VP**) and the minimization problem (**MP**) introduced at the beginning of this chapter. This is dealt with in the theorem 1.13.

> **Theorem 1.13** *Let V be a Hilbert space equipped with a scalar product $\langle .,. \rangle_V$ and the associated norm $\|.\|_V$. Let $a(.,.)$ be a bilinear form defined on V^2 and $L(.)$ a linear form on V. We assume that $a(.,.)$ and $L(.)$ satisfy the conditions of the Lax–Milgram theorem (Theorem 1.11).*
>
> *If the bilinear form $a(.,.)$ is also symmetric, then the formulations (VP) and (MP) are equivalent. In other words, the two formulations (VP) and (MP) have one and the same unique solution u in V.*

Proof: We begin by showing that if u is a solution of the variational problem (**VP**), then u will also be a solution of the minimization problem (**MP**):

- (**VP**) $\implies$ (**MP**). Let u be a solution to the variational problem (**VP**). Since the space V is a vector space, for every v in V, $u + v$ also belongs to V. We may thus evaluate the functional J at the point $u + v$:

$$J(u + v) = \frac{1}{2}a(u + v, u + v) - L(u + v)$$

$$= \frac{1}{2}a(u, u) - L(u) + \frac{1}{2}a(v, v) + \left[a(u, v) - L(v)\right] \quad (1.173)$$

$$= J(u) + \frac{1}{2}a(v, v). \quad (1.174)$$

Note that the symmetry of the bilinear form $a(.,.)$ is used to identify $a(u, v)$ and $a(v, u)$ and hence give $a(u, v)$ on the right-hand side of (1.173).

Furthermore, we use the fact that u is a solution of the variational problem (**VP**) to eliminate the quantity $a(u, v) - L(v)$ appearing in (1.173).

Now, since the bilinear form $a(.,.)$ satisfies the V-ellipticity property, the quantity $a(v, v)$ is clearly positive or zero.

We thus have

$$J(u + v) \geq J(u), \quad \forall v \in V. \quad (1.175)$$

Put another way, if u is a solution of the variational problem (**VP**), then u realizes the minimum of the functional J. It is therefore a solution of the minimization problem (**MP**).

- (**MP**) $\implies$ (**VP**). Suppose that u solves the minimization problem (**MP**). In the present case, the idea is to show that the family of inequalities

$$J(u) \leq J(v), \quad \forall v \in V, \tag{1.176}$$

implies that u solves the family of equations

$$a(u, v) = L(v), \quad \forall v \in V.$$

The fact that a family of inequalities should lead to a family of equalities looks significantly less intuitive than the opposite process described in the first part of this proof, viz., the demonstration that (**VP**) $\Rightarrow$ (**MP**).

This is why we introduce an arbitrary real number λ and an arbitrary function v in V. If u is a solution of the minimization problem (**MP**), then the vector space structure of V once again guarantees that the quantity $u + \lambda v$ also belongs to V.

We can thus evaluate the functional J at the point $u + \lambda v$:

$$J(u + \lambda v) = \frac{1}{2} a(u + \lambda v, u + \lambda v) - L(u + \lambda v) \tag{1.177}$$

$$= J(u) + \frac{\lambda^2}{2} a(v, v) + \lambda \big[a(u, v) - L(v) \big]. \tag{1.178}$$

Put another way,

$$J(u + \lambda v) - J(u) = \frac{\lambda^2}{2} a(v, v) + \lambda \big[a(u, v) - L(v) \big]. \tag{1.179}$$

Hence, according to (1.176), the polynomial $P(\lambda)$ defined by

$$P(\lambda) \equiv \frac{\lambda^2}{2} a(v, v) + \lambda \big[a(u, v) - L(v) \big] \tag{1.180}$$

does not change sign and must remain positive for all real values of λ. It follows that the discriminant of the second-degree equation associated with $P(\lambda)$ is negative or zero, whence

$$\big[a(u, v) - L(v) \big]^2 \leq 0. \tag{1.181}$$

This, in turn, implies that only equality can hold, and we conclude that

$$a(u, v) - L(v) = 0,$$

and u is therefore a solution of the variational problem **(VP)**.

We end this section by stating the existence and uniqueness of the solution of the variational inequality when the bilinear form $a(.,.)$ is not symmetric.

Theorem 1.14 *Let V be a Hilbert space and K a closed convex subset of V. We make the hypotheses of the Lax–Milgram theorem:*

1. *$a(.,.)$ is a given continuous, V-elliptic, and nonsymmetric bilinear form, defined on V^2.*

2. *$L(.)$ is a given continuous linear form defined on V.*

Then it follows that

$$\exists! \; u \in K \; such \; that \; a(u, v - u) \geq L(v - u), \; \forall v \in K.$$

References

1. H. Brézis, Analyse fonctionnelle : théorie et applications (Masson, Paris, 1983)
2. L. Schwartz, Méthodes mathématiques pour les sciences physiques (Hermann, Paris, 1983)
3. P.A. Raviart, J.M. Thomas, Introduction à l'analyse numérique des équations aux dérivées partielles (Masson, Paris, 1988)
4. A. Ern, J.L. Guermond, Eléments finis: théorie, applications, mise en œuvre, (Springer/SMAI, Heidelberg, 2002)
5. R.A. Adams, J.J.F. Fournier, in *Sobolev Spaces*. Pure and Applied Mathematics Series (Elsevier, Amsterdam, 2003)
6. R. Dautray, J.-L. Lions, Analyse mathématique et calcul numérique pour les sciences et les techniques (Masson, Paris, 1987)

Chapter 2
Finite-Element Method

2.1 Basic Idea

2.1.1 Mathematical Modeling

Developments in numerical analysis in the twentieth century generated many methods for obtaining approximate solutions to partial differential equations. Compared with the finite-difference method, spectral methods, the finite-volume method, or the singularity method, the successes of the finite-element method are unquestionable, and its supremacy is quite justifiably acclaimed.

It goes without saying that the other methods have their specific fields of application, but the finite-element method literally revolutionized our ability to handle the often complex features of partial differential equations.

It is undoubtedly this extraordinary adaptability with regard to the solution of equations - whose intrinsic complexity is partly due to the shape of the regions of integration whenever one has to deal with real problems arising in industry - that led to the spectacular development of the finite-element method in the second half of the twentieth century.

Indeed, the problem here is no longer to find an analytic solution. These are cases in which the engineer will know immediately that standard techniques are sure to fail.

Let us just summarize what is involved in numerical approximation of partial differential equations, whichever method one may choose. In practice, when it comes to studying complex systems, which may be extremely varied in nature, the common approach is to resort to what is typically referred to as a model.

In the engineering sciences, many such models lead inevitably to differential or partial differential equations.

The problem then is to find a sufficiently reliable tool for simulation, able to predict the behavior of a system under a range of different conditions, rather than to stand by and measure the degree of damage in a real trial run.

J. Chaskalovic, *Mathematical and Numerical Methods for Partial Differential Equations*, Mathematical Engineering, DOI: 10.1007/978-3-319-03563-5_2, © Springer International Publishing Switzerland 2014

However, in this first stage of modeling, when we select the key mechanisms governing the life and evolution of the given system and translate them into mathematical language, some information must necessarily be left by the wayside.

By its very essence, a model cannot reproduce every aspect of reality, rather as the mirror on the bathroom wall can produce only a two-dimensional picture of what is inevitably a three-dimensional body, even though it may do so with accuracy and elegance!

But this first level of approximation may prove to be fatal in the context of a mathematical model. Indeed, analysis of the model may well lead to the conclusion that there is no solution, in which case the model must be revised, and probably enriched with one or more mechanisms neglected on the first attempt.

The question of uniqueness must also be examined with great care. If the model generates several possible solutions, that too raises a general question about the legitimacy of such a situation with regard to the behavior of a real system.

In addition, the numerical methods implemented downstream must also integrate this question of multiple approximate solutions.

It is thus important to set up a global methodology for approximate solution that incorporates all the necessary precautions if we hope to obtain final results that make sense in the real-world system under investigation.

2.1.2 Formalism and Functional Framework for Partial Differential Equations

Once a model has been set up, the next task is to select a method for solution that best takes advantage of the mathematical aspects of the problem. In particular, the mathematical model must be manipulated into a form as well suited as possible for numerical approximation.

To exemplify this aspect of things, let us consider the two-dimensional Laplace–Dirichlet problem, which will provide a good illustration for later discussion.

So, let Ω be a bounded open subset of $\mathbb{R}^2$. The problem is to find a real-valued function u defined on Ω that solves

$$\textbf{(CP)} \qquad \begin{cases} -\Delta u = f & \text{in } \Omega, \\ u = 0 & \text{on } \partial\Omega, \end{cases} \qquad (2.1)$$

where f is a given function.

At this stage, it is important to note that this formulation is not complete, since the regularity of the boundary $\partial\Omega$ of the integration domain Ω is not specified, and neither is the regularity of f, while it is clear that these features will have a significant bearing on the regularity of the solution u of the continuous problem (CP), and hence on the space V within which one expects to find u.

For reasons to be explained shortly, we thus assume that the domain of integration Ω has a boundary $\partial\Omega$ with C^2 regularity. That is to say, the curvature is a continuous function of the curvilinear coordinate describing the boundary $\partial\Omega$.

Furthermore, if we assume that f belongs to $C^0(\Omega)$, it makes sense to look for solutions to the continuous problem (CP) in $C^2(\Omega)$, since this would guarantee that the Laplacian itself was continuous. One then speaks of classical solutions.

In this case, Poisson's equation can be reinterpreted, not as a functional equation, but at each point M of Ω in the following form:

Find u in $C^2(\Omega)$ solving

$$(\mathbf{CP}) \quad \begin{cases} -\Delta u(M) = f(M), \ \forall\, M \in \Omega, \\ u(M) = 0, \qquad \forall\, M \in \partial\Omega. \end{cases} \tag{2.2}$$

Naturally, the function f on the right-hand side does not always satisfy the regularity requirements of the space C^0.

Consider, for instance, a case in which f belongs to $L^2(\Omega)$. The Laplacian of the solution u (equal to $-f$) must then also be an element of $L^2(\Omega)$.

For this reason, we look for the solution u of the continuous problem (CP) in the Sobolev space $H^2(\Omega)$, because if this is the case, the Laplacian of u is then an element of $L^2(\Omega)$.

However, we should stress once again that this is merely a reasonable choice of space in which to seek solutions u, because the only obligation here, imposed by the assumed regularity of f, is to find a solution whose Laplacian belongs to $L^2(\Omega)$.

In principle, Poisson's equation can now no longer be considered pointwise, as it could when f was C^0, but must be treated as a functional equation. In fact, it must be considered as an equality in $L^2(\Omega)$, i.e., a mean squared equality, or an "energy" balance:

$$\Delta u + f = 0 \ \text{in} \ L^2(\Omega) \quad \Longleftrightarrow \quad \int_\Omega (\Delta u + f)^2 \, d\Omega = 0. \tag{2.3}$$

To end this discussion, note that treating Poisson's equation as a functional equation in $L^2(\Omega)$ nevertheless implies that this equation can be considered at each point M of Ω, apart from a set of measure zero (see Sect. 1.5.1).

For the reader unfamiliar with the idea of a set of measure zero, a first approach would be to say that Poisson's equation is satisfied at all points M of Ω with the exception of a countably infinite number of points of Ω. But we shall nevertheless consider Poisson's equation as a global equation like (2.3) expressed in $L^2(\Omega)$, rather than a local equation like (2.2).

2.1.3 Constructing a Variational Formulation

We now discuss the basic principles underlying the finite-element method. The main idea is to consider the unknown u, not as a scalar field that associates a real number $u(M)$ (to be determined) with each point M of Ω, but rather as an element of a space of functions V in which various search paths will be explored in order to identify the solution.

Concerning the notion of approximation, the problem is no longer to determine a numerical sequence $(\tilde{u}_1, \ldots, \tilde{u}_N)$ providing an approximation, as in the finite-difference method, to the values $(u_1, \ldots, u_N)$ of the solution u to the continuous problem (**CP**) at points M_j, $(j = 1, \ldots, N)$, chosen on a suitable mesh covering the domain of integration Ω.

The idea now is rather to devise a procedure that allows us to obtain an approximate function $\tilde{u}$. Naturally, in the end, knowing the solution u, or rather its approximation $\tilde{u}$, we will be able to evaluate $\tilde{u}$ at any point of the domain Ω, and not only at a limited set of points sitting on a predetermined mesh, as happens with the finite-difference method.

A second key feature of the finite-element method is the integral formulation of the continuous problem (**CP**), known as a variational problem (**VP**), as discussed in Chap. 1.

To obtain this, we consider a real-valued function v on Ω called the *test function*, which is not specified *a priori*, lying in a function space V to be constructed later.

We then multiply (2.1) by the *test function* v and integrate both sides of the equation over Ω to obtain

$$-\int_\Omega \Delta u \cdot v \, d\Omega = \int_\Omega f v \, d\Omega, \quad \forall v \in V. \tag{2.4}$$

This transformation is motivated by the historical traditions of the finite-element method, which was originally introduced as a generalization of the principle of virtual work in continuum mechanics (see, for example, [1]).

Indeed, the partial differential Eq. (2.1) of the continuous formulation (**CP**) can be construed as the fundamental principle of statics expressing the equilibrium of an elastic membrane subjected to a density of transverse forces f generating a displacement field u perpendicular to the membrane.

On the other hand, (2.4) expresses an "energy" formulation, where even the non-specialist will recognize that the right-hand side of (2.4) can be interpreted as the work done by external forces f causing a displacement field v, which remains arbitrary at this stage.

For its part, the left-hand side of (2.4) corresponds to the work done by internal forces, intrinsic to the deformation of the elastic medium Ω.

Furthermore, the transformation of the local expression of the problem (CP) into a global or integral formulation (VP) is motivated by the need for a formalism amenable to the idea of search paths in a function space V.

But this is precisely the situation in an integral formulation, in the sense that the values of the function at specific points M of Ω no longer appear directly, since only its average is apparent through the integral.

Formally, we may then apply Green's formula (1.159) from Lemma 1.25 to rewrite (2.4) in the form

$$\int_\Omega \nabla u \cdot \nabla v \, d\Omega - \int_{\partial\Omega} \frac{\partial u}{\partial n} v \, d\Gamma = \int_\Omega f v \, d\Omega, \quad \forall v \in V. \tag{2.5}$$

We are now in a position to determine the characteristics of the space V.

The first point to make concerns the total conservation of information between the formulation of the continuous problem (CP) and the formulation of the variational problem (VP).

For example, the Dirichlet condition $u = 0$ on the boundary $\partial\Omega$ of Ω cannot be taken into account directly in the integral formulation (2.5). Given that the future solution u of the variational problem (VP) must be one of the functions v of V, we simply require all functions v in V to satisfy the Dirichlet condition:

$$v = 0 \quad \text{on} \quad \partial\Omega. \tag{2.6}$$

Then (2.5) can be written

$$\int_\Omega \nabla u \cdot \nabla v \, d\Omega = \int_\Omega f v \, d\Omega, \quad \forall v \in V. \tag{2.7}$$

The second point concerns the existence of the integrals in the formulation (2.7). Indeed, it is essential to impose adequate convergence conditions on the integrals in (2.7).

Since we are thinking of sufficient conditions for convergence, several functional frameworks may meet our needs. For reasons that will become clear later on, we consider the framework provided by the Sobolev spaces, which ensure all the desired properties, even beyond the issues we are dealing with here.

Convergence of the right-hand side of (2.7) is ensured simply by the upper bound guaranteed by the Cauchy–Schwarz inequality:

$$\left| \int_\Omega f v \, d\Omega \right| \le \int_\Omega |fv| \, d\Omega \le \left(\int_\Omega |f|^2 \, d\Omega \right)^{1/2} \left(\int_\Omega |v|^2 \, d\Omega \right)^{1/2}. \tag{2.8}$$

Therefore, since f is a given function in $L^2(\Omega)$, it suffices also to take v in $L^2(\Omega)$, in order to ensure convergence of the right-hand side of (2.7).

Regarding the convergence of the first integral on the left-hand side of (2.7), we once again consider absolute convergence of the integral and apply the Cauchy–Schwarz inequality as before:

$$\left| \int_\Omega \nabla u \cdot \nabla v \, d\Omega \right| \le \int_\Omega |\nabla u \cdot \nabla v| \, d\Omega \le \left(\int_\Omega |\nabla u|^2 \, d\Omega \right)^{1/2} \left(\int_\Omega |\nabla v|^2 \, d\Omega \right)^{1/2}.$$
(2.9)

Convergence of the left-hand side of (2.7) is thus ensured if we require the *test functions* v in V to have gradients belonging to $L^2(\Omega)$.

To sum up, we have established that the following are sufficient conditions for convergence of the integrals in (2.7):

$$v \in L^2(\Omega) \ \text{ and } \ \nabla v \in [L^2(\Omega)]^2.$$

This explains why we choose the variational space V to be the Sobolev space $H^1(\Omega)$ introduced in Chap. 1.

We must then also add the homogeneous Dirichlet condition (2.6). Put another way, we set the functional space V as follows:

$$V \equiv H_0^1(\Omega) \equiv \left\{ v : \Omega \to \mathbb{R}, \ v \in L^2(\Omega), \ \nabla v \in [L^2(\Omega)]^2, \ v = 0 \text{ on } \partial\Omega \right\}.$$
(2.10)

We can now bring together all our results and state in full the variational problem (**VP**) that we will be discussing in the rest of the book:

$$(\textbf{VP}) \quad \begin{cases} \text{Find } u \in H_0^1(\Omega) \text{ solution of} \\ \displaystyle\int_\Omega \nabla u \cdot \nabla v \, d\Omega = \int_\Omega f v \, d\Omega, \quad \forall v \in H_0^1(\Omega). \end{cases} \quad (2.11)$$

2.2 Existence, Uniqueness, and Regularity of a Weak Solution

2.2.1 Application to the Homogeneous Laplace–Dirichlet Problem

General results regarding the existence and uniqueness of solutions of differential equations or partial differential equations, or indeed, variational equations, remain a completely open question. The complexity of such results depends on the nature and structure of the equation or system of equations.

As far as variational formulations are concerned, there is a rather general formalism, which we discussed in Sect. 1.6, that provides theorems guaranteeing the existence and uniqueness of the solution of the differential problem under certain conditions.

This is precisely what happens for the variational problem (**VP**) defined by (2.11), for which we shall now propose a first application of the Lax–Milgram theorem to establish the existence and uniqueness of the solution whenever the data f belongs to $L^2(\Omega)$.

In order to apply the Lax–Milgram theorem, one must identify the space V, the bilinear form $a(.,.)$, and the linear form $L(.)$. The variational problem (**VP**) defined by (2.11) suggests introducing the following quantities.

Let V be the space in which we seek the solution u of the variational problem:

$$V \equiv H_0^1(\Omega). \tag{2.12}$$

The space $H_0^1(\Omega)$ is equipped with the natural norm $\| . \|_{H^1(\Omega)}$ on functions belonging to $H^1(\Omega)$.

Then, $\forall v \in H^1(\Omega)$, we set

$$\|v\|_{H^1(\Omega)}^2 \equiv \int_{\Omega} v^2 \, d\Omega + \int_{\Omega} \left(\frac{\partial v}{\partial x} \right)^2 d\Omega + \int_{\Omega} \left(\frac{\partial v}{\partial y} \right)^2 d\Omega. \tag{2.13}$$

This is a Hilbert norm for the space $H^1(\Omega)$ (see Sect. 1.20), and also for $H_0^1(\Omega)$, since it is a closed vector subspace of $H^1(\Omega)$.

Let $a(.,.)$ be the bilinear form defined by

$$a : V \times V \longrightarrow \mathbb{R}$$
$$(u, v) \longmapsto a(u, v) \equiv \int_{\Omega} \nabla u \cdot \nabla v \, d\Omega. \tag{2.14}$$

Likewise, let $L(.)$ be the linear form defined by

$$L : V \longrightarrow \mathbb{R}$$
$$v \longmapsto L(v) \equiv \int_{\Omega} f v \, d\Omega. \tag{2.15}$$

Then, the variational problem (**VP**) specified by (2.11) can be written in the following form:
$$\text{Find } u \in V \text{ solution of } a(u, v) = L(v), \quad \forall v \in V. \tag{2.16}$$

We now check the premises of the Lax–Milgram theorem (Theorem 1.11):

1. $a(.,.)$ is a continuous bilinear form. Bilinearity is obvious. Concerning continuity, consider any two elements u and v of $H_0^1(\Omega)$. Then,

$$|a(u, v)| \leq \int_{\Omega} |\nabla u \cdot \nabla v| \, d\Omega \leq \left(\int_{\Omega} |\nabla u|^2 \, d\Omega \right)^{1/2} \left(\int_{\Omega} |\nabla v|^2 \, d\Omega \right)^{1/2}, \tag{2.17}$$

where we have used the Cauchy–Schwarz inequality. Now,

$$\int_{\Omega} |\nabla u|^2 \, d\Omega = \int_{\Omega} \left[\left(\frac{\partial u}{\partial x} \right)^2 + \left(\frac{\partial u}{\partial y} \right)^2 \right] d\Omega = \left\| \frac{\partial u}{\partial x} \right\|_{L^2(\Omega)}^2 + \left\| \frac{\partial u}{\partial y} \right\|_{L^2(\Omega)}^2, \tag{2.18}$$

where $\|.\|_{L^2(\Omega)}$ is the natural norm on $L^2(\Omega)$, defined by

$$\forall u \in L^2(\Omega), \quad \|u\|_{L^2(\Omega)} \equiv \left(\int_{\Omega} |u|^2 \, d\Omega \right)^{1/2}. \tag{2.19}$$

This implies that

$$\int_{\Omega} |\nabla u|^2 \, d\Omega \le \|u\|_{H^1(\Omega)}^2. \tag{2.20}$$

The inequality (2.17) then gives

$$|a(u, v)| \le \|u\|_{H^1(\Omega)} \|v\|_{H^1(\Omega)}, \tag{2.21}$$

and the continuity constant of the map $a(., .)$ is clearly equal to 1.

2. $a(., .)$ is a V-elliptic form. To establish this property, we must find a lower bound for the quantity $a(v, v)$ defined in (2.14). Now, for any function $v \in H_0^1(\Omega)$, we have

$$a(v, v) = \int_{\Omega} |\nabla v|^2 \, d\Omega = \left\| \frac{\partial v}{\partial x} \right\|_{L^2(\Omega)}^2 + \left\| \frac{\partial v}{\partial y} \right\|_{L^2(\Omega)}^2. \tag{2.22}$$

To find a lower bound for $a(v, v)$ relative to the $H^1(\Omega)$ norm, recall that for every function v in $H_0^1(\Omega)$, we have the Poincaré inequality (1.130).

This tells us that there is a constant $C(\Omega) > 0$ such that

$$\int_{\Omega} |v|^2 \, d\Omega \le C(\Omega) \int_{\Omega} |\nabla v|^2 \, d\Omega. \tag{2.23}$$

We now add the square of the $L^2(\Omega)$ norm of the modulus of ∇v on each side of the Poincaré inequality (2.23), which brings in the square of the $H^1(\Omega)$ norm of the function v:

$$\|v\|_{H^1(\Omega)}^2 \equiv \|v\|_{L^2(\Omega)}^2 + \left\| \frac{\partial v}{\partial x} \right\|_{L^2(\Omega)}^2 + \left\| \frac{\partial v}{\partial y} \right\|_{L^2(\Omega)}^2 \tag{2.24}$$

$$\le [1 + C(\Omega)] \left[\left\| \frac{\partial v}{\partial x} \right\|_{L^2(\Omega)}^2 + \left\| \frac{\partial v}{\partial y} \right\|_{L^2(\Omega)}^2 \right] \tag{2.25}$$

$$\le [1 + C(\Omega)] a(v, v). \tag{2.26}$$

It thus follows that

$$a(v, v) \geq C'\|v\|_{H^1(\Omega)}^2, \tag{2.27}$$

where the coercivity constant C' is given by

$$C' = \frac{1}{1 + C(\Omega)}.$$

3. $L(.)$ is a continuous linear form. Once again, the linearity of $L(.)$ is obvious. It is particularly easy to find a bound for L because the data f is a function in $L^2(\Omega)$:

$$|L(v)| \leq \int_\Omega |fv| \, d\Omega \leq \|f\|_{L^2(\Omega)}\|v\|_{L^2(\Omega)} \leq \|f\|_{L^2(\Omega)}\|v\|_{H^1(\Omega)}. \tag{2.28}$$

The continuity constant of the linear form L is thus $\|f\|_{L^2(\Omega)}$.

Summary: Since $H_0^1(\Omega)$ is closed in $H^1(\Omega)$ for the norm $\|.\|_{H^1(\Omega)}$, it is a Hilbert space for this norm. According to the Lax–Milgram theorem, there thus exists one and only one function in $H_0^1(\Omega)$ that solves the variational problem **(VP)** specified by (2.16).

Notes:

- When the space $H_0^1(\Omega)$ was given the norm induced from $H^1(\Omega)$, all the conditions of the Lax–Milgram theory were satisfied. However, if we consider the norm $|.|_1$ defined by

$$|v|_1 \equiv \left(\int_\Omega |\nabla v|^2 \, d\Omega \right)^{1/2}, \tag{2.29}$$

it can also be shown that the conditions of the Lax–Milgram theorem are satisfied. To do this, we exploit the fact that the natural norm on $H^1(\Omega)$ and the one defined by (2.29) are equivalent, according to Lemma 1.23.

- When the data f is less regular than we have supposed, i.e., $L^2(\Omega)$, the tools required for functional analysis of the variational problem go beyond the scope of this book. The interested reader is referred to more-specialized literature, such as the book by Brézis [2] or the series by Dautray and Lions [3].

We now state a lemma that gives an *a priori* estimate for the solution u of the variational problem **(VP)** specified by (2.11), by establishing a result concerning the continuous dependence of the solution u on the data f.

Lemma 2.1 *If Ω is a bounded open set with sufficiently regular boundary $\partial\Omega$, and if f is a given function in $L^2(\Omega)$ and u_f solves the variational problem **(VP)** specified by (2.11), the map Δ defined by*

$$\Delta : L^2(\Omega) \longrightarrow H_0^1(\Omega),$$
$$f \longmapsto \Delta(f) \equiv u_f,$$

is linear and continuous. Put another way,

$$\exists\, C > 0 \ \text{such that} \ \forall f \in L^2(\Omega), \ \ \|u_f\|_{H^1(\Omega)} \leq C\|f\|_{L^2(\Omega)}. \tag{2.30}$$

Proof:

1. Linearity of Δ is clear by inspection.

   - Let (f_1, f_2) be a pair of functions in $L^2(\Omega) \times L^2(\Omega)$. Let us show that $\Delta(f_1 + f_2) = \Delta(f_1) + \Delta(f_2)$, or again, that $u_{f_1+f_2} = u_{f_1} + u_{f_2}$, if u_{f_1} and u_{f_2} are solutions of

$$a(u_{f_1}, v) = L_{f_1}(v), \quad \forall v \in H_0^1(\Omega), \tag{2.31}$$

$$a(u_{f_2}, v) = L_{f_2}(v), \quad \forall v \in H_0^1(\Omega), \tag{2.32}$$

   where $L_f(.)$ is the linear form $L(.)$ of the variational problem (**VP**) defined by (2.11), but specifying its dependence on f:

$$L_f(v) \equiv \int_\Omega f v \, d\Omega. \tag{2.33}$$

   Then using the bilinearity of the form $a(.,.)$ and the linearity of $L_f(.)$, we have

$$a(u_{f_1} + u_{f_2}, v) = L_{f_1}(v) + L_{f_1}(v) = L_{f_1+f_2}(v)$$
$$= a(u_{f_1+f_2}, v), \quad \forall v \in H_0^1(\Omega).$$

   This, in turn, implies that $u_{f_1} + u_{f_2}$ corresponds to the solution $u_{f_1+f_2}$.

   - Likewise, for every real number λ and every function f in $L^2(\Omega)$, we have

$$a(u_{\lambda f}, v) = L_{\lambda f}(v) = \lambda L_f(v)$$
$$= \lambda a(u_f, v) = a(\lambda u_f, v), \quad \forall v \in H_0^1(\Omega).$$

   Therefore, $u_{\lambda f} = \lambda u_f$, and hence $\Delta(\lambda f) = \lambda \Delta(f)$.

2. Continuity of the map Δ is shown as follows. It results from the ellipticity of the bilinear form $a(.,.)$, whence

$$\forall v \in H_0^1(\Omega), \ \exists\, \alpha > 0 \ \text{such that} \ a(v, v) \geq \alpha \|v\|_{H^1(\Omega)}^2. \tag{2.34}$$

So, if u solves the variational problem **(VP)** specified by (2.11), then by choosing the generic element $v \in H_0^1(\Omega)$ as solution u_f, we obtain

$$\alpha \|u_f\|_{H^1(\Omega)}^2 \le a(u_f, u_f) = L(u_f) = \int_\Omega f u_f \, d\Omega$$
$$\le \|f\|_{L^2(\Omega)} \|u_f\|_{L^2(\Omega)} \le \|f\|_{L^2(\Omega)} \|u_f\|_{H^1(\Omega)}, \qquad (2.35)$$

where we have used the Cauchy–Schwarz inequality. After dividing both sides of (2.35) by $\|u_f\|_{H^1(\Omega)}$, we obtain finally

$$\|\Delta(f)\|_{H^1(\Omega)} \equiv \|u_f\|_{H^1(\Omega)} \le \frac{\|f\|_{L^2(\Omega)}}{\alpha}, \qquad (2.36)$$

which expresses the continuity of the linear map Δ.

The inequality (2.36) is called an *a priori* estimate of the solution u for the data f.

It puts an upper bound on the "energy" of the solution u as defined by the H^1 norm relative to the "energy" of the data f as defined by the L^2 norm.

We end this section with an equivalence result for the variational problem **(VP)** of (2.11) and a suitable minimization problem **(MP)**.

Lemma 2.2 *Let Ω be a bounded open set with sufficiently regular boundary $\partial\Omega$, and f a given function in $L^2(\Omega)$. If u is the unique solution of the variational problem **(VP)** of (2.11) belonging to $H_0^1(\Omega)$, then u is also the unique solution of the minimization problem **(MP)** specified by the following:*

$$\textbf{(MP)} \quad \begin{cases} \text{Find } u \in H_0^1(\Omega) \text{ solution of} \\[1mm] J(u) \equiv \min_{v \in H_0^1(\Omega)} J(v), \\[2mm] \text{with } J(v) = \dfrac{1}{2} a(v, v) - L(v), \end{cases}$$

*where the bilinear form $a(.,.)$ and the linear form $L(.)$ are those specifying the variational problem **(VP)** in (2.11).*

Proof: This is a direct application of Theorem 1.13, where we note that the bilinear form $a(.,.)$ of the variational problem **(VP)** specified by (2.11) is symmetric.

Note. The existence and uniqueness of the solution u of the minimization problem **(MP)** can be obtained directly as a consequence of Stampacchia's theorem (Theorem 1.9) by observing that the space $H_0^1(\Omega)$ is a closed convex subset of $H^1(\Omega)$.

2.2.2 *Application to the Inomogeneous Laplace–Dirichlet Problem*

In this section, we consider the problem of the Laplacian with an inhomogeneous Dirichlet condition. More precisely, let Ω be a bounded open subset of $\mathbb{R}^2$ with a sufficiently regular boundary $\partial\Omega$. The aim will be to find a real-valued function u on Ω that solves

$$\textbf{(CP)} \quad \begin{cases} -\Delta u = f \ \text{ in } \ \Omega, \\ \quad u = g \ \text{ on } \ \partial\Omega, \end{cases} \tag{2.37}$$

where f and g are given functions in $L^2(\Omega)$ and $L^2(\partial\Omega)$, respectively.

If we hope to implement the Lax–Milgram theorem (Theorem 1.11), we cannot adapt our analysis to the Laplacian problem with an inhomogeneous Dirichlet condition if we take as functional context the space V defined by

$$V \equiv \left\{ v : \Omega \longrightarrow \mathbb{R}, \quad v = g \ \text{ on } \ \partial\Omega \right\}. \tag{2.38}$$

Indeed, if this were the case, the definition of V would not be consistent with the first premise of the Lax–Milgram theorem, which assumes that the set V has a vector space structure.

But it is clear that the space defined by (2.38) does not have a vector space structure, since if v_1 and v_2 are two elements of the space V, the linear combination $v_1 - v_2$ is zero on the boundary $\partial\Omega$. But this means that $v_1 - v_2$ does not belong to V, which contradicts the requirement of closure under linear combination that must be satisfied by any vector space.

To get around this difficulty, we use the technique of extending a function defined on the boundary $\partial\Omega$ to the whole of Ω. Naturally, the extension will not be unique, and the whole problem now is to obtain a sufficiently regular extension to be able to carry out the ensuing differential operations.

We shall thus assume the following result:

Lemma 2.3 *If g is in $L^2(\partial\Omega)$, there is a real-valued function G on Ω that is in $H^2(\Omega)$ and satisfies $G(x) = g(x)$ (a.e.) on $\partial\Omega$. The function G is an extension of g to the whole of Ω and is not unique.*

If we take this as given, we can introduce the following real-valued function U on Ω :

$$U(x) \equiv u(x) - G(x), \tag{2.39}$$

where u solves the Laplacian problem with an inhomogeneous Dirichlet condition specified by g, and G is the extension of g given by Lemma 2.3.

Substituting the change of variable (2.39) into the Laplace–Dirichlet problem (2.37), it is easy to see that the function U solves

$$\textbf{(CP)} \qquad \begin{cases} -\Delta U = f + \Delta G \equiv F \text{ in } \Omega, \\ \qquad U = 0 \text{ on } \partial\Omega, \end{cases} \tag{2.40}$$

where F is the new right-hand side of the Laplace equation.

Since we have assumed that f belongs to $L^2(\Omega)$, and also that the extension G is an element of $H^2(\Omega)$, it follows that ΔU belongs to $L^2(\Omega)$, and as a consequence, the function F on the right-hand side is also in $L^2(\Omega)$.

Finally, U solves the Laplacian problem with a homogeneous Dirichlet condition. We may thus conclude from the discussion in Sect. 2.2.1 that $H_0^1(\Omega)$ contains one and only one solution U of the variational problem **(VP)** specified by (2.11), provided that we replace the function f on the right-hand side by the function F defined in (2.40).

Let us end this investigation with an *a priori* estimate of the solution u of the continuous problem **(CP)** specified by (2.37):

Lemma 2.4 *If u is a solution of the continuous problem **(CP)** specified by (2.37), then*

$$\exists\, C > 0, \quad such\ that\ \|u\|_{H^1(\Omega)} \le C\left[\|f\|_{L^2(\Omega)} + \|g\|_{L^2(\partial\Omega)}\right]. \tag{2.41}$$

Proof: Let V be a test function in $H_0^1(\Omega)$. Multiplying both sides of the partial differential Eq. (2.40) of the continuous problem **(CP)** and integrating the resulting equation over the whole of Ω, we obtain

$$-\int_\Omega (\Delta U)V \, \mathrm{d}\Omega - \int_\Omega (\Delta G)V \, \mathrm{d}\Omega = \int_\Omega fV \, \mathrm{d}\Omega. \tag{2.42}$$

We then use Green's formula (1.159) for each integral on the left-hand side of (2.42):

$$\int_\Omega \nabla U \cdot \nabla V \, \mathrm{d}\Omega - \int_{\partial\Omega} \frac{\partial U}{\partial n} V \, \mathrm{d}\Gamma + \int_\Omega \nabla G \cdot \nabla V \, \mathrm{d}\Omega - \int_{\partial\Omega} \frac{\partial G}{\partial n} V \, \mathrm{d}\Gamma = \int_\Omega fV \, \mathrm{d}\Omega. \tag{2.43}$$

Now V is in $H_0^1(\Omega)$, which means that the integrals over $\partial\Omega$ in (2.43) both vanish. The formulation **(VP)** associated with the continuous problem **(CP)** specified by (2.40) can thus be written as follows:

$$\textbf{(VP)} \quad \begin{cases} \text{Find } U \in H_0^1(\Omega) \text{ solution of} \\[4pt] \displaystyle\int_\Omega \nabla U \cdot \nabla V \, \mathrm{d}\Omega = \int_\Omega fV \, \mathrm{d}\Omega - \int_\Omega \nabla G \cdot \nabla V \, \mathrm{d}\Omega, \quad \forall v \in H_0^1(\Omega). \end{cases}$$

$$(2.44)$$

Choosing $V = U$ in the variational formulation (2.44), it follows that

$$\int_\Omega |\nabla U|^2 \, \mathrm{d}\Omega = \int_\Omega fU \, \mathrm{d}\Omega - \int_\Omega \nabla G \cdot \nabla U \, \mathrm{d}\Omega. \qquad (2.45)$$

However, U belongs to $H_0^1(\Omega)$, so we can use the Poincaré inequality, viz.,

$$\exists \, \alpha > 0 \ \text{ such that } \ \alpha \|\nabla U\|_{L^2(\Omega)}^2 \geq \|U\|_{L^2(\Omega)}^2. \qquad (2.46)$$

The bound on U for the H^1 norm is obtained from

$$(1 + \alpha) \|\nabla U\|_{L^2(\Omega)}^2 \geq \|U\|_{H^1(\Omega)}^2. \qquad (2.47)$$

Equation (2.45) then yields

$$\frac{1}{1+\alpha} \|U\|_{H^1(\Omega)}^2 \leq \int_\Omega fU \, \mathrm{d}\Omega - \int_\Omega \nabla G \cdot \nabla U \, \mathrm{d}\Omega. \qquad (2.48)$$

We now apply the Cauchy–Schwarz inequality to obtain an upper bound for the right-hand side of the inequality (2.48):

$$\frac{1}{1+\alpha} \|U\|_{H^1(\Omega)}^2 \leq \|f\|_{L^2(\Omega)} \|U\|_{L^2(\Omega)} + \|\nabla G\|_{L^2(\Omega)} \|\nabla U\|_{L^2(\Omega)}$$

$$\leq \left[\|f\|_{L^2(\Omega)} + \|\nabla G\|_{L^2(\Omega)} \right] \|U\|_{H^1(\Omega)}.$$

Dividing both sides by $\|U\|_{H^1(\Omega)}$, we then have

$$\|U\|_{H^1(\Omega)} \leq (1 + \alpha) \left[\|f\|_{L^2(\Omega)} + \|\nabla G\|_{L^2(\Omega)} \right]$$

$$\leq (1 + \alpha) \left[\|f\|_{L^2(\Omega)} + \|G\|_{H^1(\Omega)} \right].$$

We substitute the expression (2.39) for U to obtain

$$\|u - G\|_{H^1(\Omega)} \leq (1 + \alpha) \left[\|f\|_{L^2(\Omega)} + \|G\|_{H^1(\Omega)} \right]. \qquad (2.49)$$

However, we also have

$$\|u\|_{H^1(\Omega)} - \|G\|_{H^1(\Omega)} \leq \left| \|u\|_{H^1(\Omega)} - \|G\|_{H^1(\Omega)} \right| \leq \|u - G\|_{H^1(\Omega)}, \qquad (2.50)$$

whence (2.49) implies that

$$\|u\|_{H^1(\Omega)} - \|G\|_{H^1(\Omega)} \le (1+\alpha)\left[\|f\|_{L^2(\Omega)} + \|G\|_{H^1(\Omega)}\right]. \tag{2.51}$$

This, in turn, means that we can write

$$\|u\|_{H^1(\Omega)} \le (2+\alpha)\left[\|f\|_{L^2(\Omega)} + \|G\|_{H^1(\Omega)}\right]. \tag{2.52}$$

To conclude here, we need to replace the H^1 norm of G by the L^2 norm of g for the measure on the boundary $\partial\Omega$. Now, since g is the trace γ_0 of G on the boundary $\partial\Omega$, the trace theorem (Theorem 1.3) tells us that

$$\exists\, \beta > 0 \ \text{ such that } \ \|G\|_{H^1(\Omega)} \le \beta\, \|g\|_{L^2(\partial\Omega)}. \tag{2.53}$$

Using the inequality (2.53) in (2.52), we finally have

$$\|u\|_{H^1(\Omega)} \le (2+\alpha)\left[\|f\|_{L^2(\Omega)} + \beta\,\|g\|_{L^2(\partial\Omega)}\right] \le C\left[\|f\|_{L^2(\Omega)} + \|g\|_{L^2(\partial\Omega)}\right], \tag{2.54}$$

where C can be defined by $C \equiv (2+\alpha)(1+\beta)$.

2.2.3 Application to the Laplace–Neumann–Dirichlet Problem

When the continuous problem **(CP)** specified by (2.1) is replaced by the Laplace–Neumann–Dirichlet problem, the boundary Γ of Ω comprises two complementary parts, Γ_1 and Γ_2, on which the Dirichlet condition and the Neumann condition are defined respectively.

In this case, the continuous problem **(CP)** is formulated as follows:

$$\textbf{(CP)} \quad \begin{cases} -\Delta u = f \ \text{ in } \ \Omega, \\[4pt] u = 0 \ \text{ on } \ \Gamma_1, \\[4pt] \dfrac{\partial u}{\partial n} = g \ \text{ on } \ \Gamma_2, \end{cases} \tag{2.55}$$

where it is assumed that f and g belong to $L^2(\Omega)$ and $L^2(\Gamma_2)$, respectively.

As a consequence, it is not difficult to show that the new associated variational problem **(VP)** can be written thus:

$$\textbf{(VP)} \quad \begin{cases} \text{Find } u \in H^1_{\Gamma_1}(\Omega) \text{ solution of} \\[6pt] \displaystyle\int_\Omega \nabla u \cdot \nabla v \, d\Omega = \int_\Omega f v \, d\Omega + \int_{\Gamma_2} g v \, d\Gamma, \quad \forall v \in H^1_{\Gamma_1}(\Omega), \end{cases} \tag{2.56}$$

where the Sobolev space $H^1_{\Gamma_1}(\Omega)$ is defined by

$$H^1_{\Gamma_1}(\Omega) \equiv \left\{ v : \Omega \to \mathbb{R}, \ v \in L^2(\Omega), \ \nabla v \in [L^2(\Omega)]^2, \ v = 0 \ \text{on} \ \Gamma_1 \right\}. \quad (2.57)$$

Notes:

1. When the Dirichlet data on Γ_1 is not homogeneous, the extension technique [3], although sometimes delicate, can be used to transform the inhomogeneous problem to a homogeneous one, the same as the one presented in (2.55).

2. The Lax–Milgram theorem is applied to the variational formulation (2.56) in an analogous way to what was discussed for the variational formulation (2.11) associated with the Laplace–Dirichlet problem.

   However, quite substantial changes are required to establish continuity of the linear form $L(.)$. Indeed, in this case, the action of the form $L(.)$ on any function v belonging to $H^1_{\Gamma_1}(\Omega)$ is

$$L(v) \equiv \int_\Omega f v \, d\Omega + \int_{\Gamma_2} g v \, d\Gamma, \quad \forall v \in H^1_{\Gamma_1}(\Omega). \quad (2.58)$$

The bound on $L(v)$ is then obtained as follows:

$$|L(v)| \leq \int_\Omega |fv| \, d\Omega + \int_{\Gamma_2} |gv| \, d\Gamma$$

$$\leq \|f\|_{L^2(\Omega)} \|v\|_{L^2(\Omega)} + \|g\|_{L^2(\Gamma_2)} \|v\|_{L^2(\Gamma_2)}. \quad (2.59)$$

A new difficulty arises here because of the Neumann condition g defined on the boundary Γ_2. The problem is that the bound for $L(v)$ must be expressed solely in terms of the $H^1(\Omega)$ norm of the function v.

For this reason, the term resulting from the Neumann condition, which involves the $L^2(\Gamma_2)$ norm of v, must be modified accordingly.

In fact, the trace theorem (Theorem 1.3) stated in Sect. 1.5.2 can be used to obtain a bound on $L(v)$ solely in terms of the $H^1(\Omega)$ norm of v. Let C_{tr} be the continuity constant for the trace map γ_0 defined in Theorem 1.3 of Sect. 1.5.2.

We can then modify the inequality (2.59) as follows:

$$|L(v)| \leq \|f\|_{L^2(\Omega)} \|v\|_{L^2(\Omega)} + C_{\text{tr}} \|g\|_{L^2(\Gamma_2)} \|v\|_{H^1(\Omega)}$$

$$\leq C \|v\|_{H^1(\Omega)}, \quad \forall v \in H^1_{\Gamma_1}(\Omega),$$

where we have set $C = \|f\|_{L^2(\Omega)} + C_{\text{tr}} \|g\|_{L^2(\Gamma_2)}$.

These are the main remarks that need to be made about the extension of the homogeneous Laplace–Dirichlet problem (2.1) to the Laplace–Neumann–Dirichlet problem (2.55).

Other less important modifications concern adaptations of the result in going from the context of $H_0^1(\Omega)$ to that of $H_{\Gamma_1}^1(\Omega)$, but these present no major difficulties.

For this reason, once we have dealt with the bound on the linear form $L(.)$ defined by (2.58), application of the Lax–Milgram theorem guarantees existence and uniqueness of the solution $u \in H_{\Gamma_1}^1(\Omega)$ of the variational problem **(VP)** specified by (2.56).

2.3 Equivalence of Weak and Strong Formulations

There is one further important point regarding the transformation discussed in the last section, and in particular the question of the equivalence of the two formulations, i.e., the continuous problem **(CP)** and the associated variational problem **(VP)**.

For it should not be forgotten that the process of numerical approximation will produce an approximate solution to the variational formulation **(VP)**.

But then, what can be said about the solution to the continuous problem **(CP)** if no equivalence result is established between the two formulations?

The point is that it is no easy matter to show that a solution of a variational problem **(VP)** is a solution of the associated continuous problem **(CP)**. Worse, in many cases, it is not even true.

To grasp the subtlety of this notion of equivalence between the two formulations, let us return to the example of the Laplace–Dirichlet problem **(CP)** as specified by (2.1) and the associated variational formulation **(VP)** as specified by (2.11), and assume as before that f on the right-hand side is a function in $L^2(\Omega)$.

It suffices to point out that if the solution u of the continuous problem **(CP)** can be sought in the Sobolev space $H^2(\Omega)$, then the solution of the variational problem **(VP)** is sought in $H^1(\Omega)$ and $H^2(\Omega) \subset H^1(\Omega)$.

In other words, every solution of the continuous problem **(CP)** could be a solution of the variational problem as far as its regularity is concerned, whereas there is no *a priori* reason why a solution of the variational problem **(VP)** should be a solution of the continuous problem **(CP)**.

In fact, the notion of equivalence between the two formulations depends entirely on the function spaces in which one seeks solutions to the continuous problem **(CP)** on the one hand and the variational formulation **(VP)** on the other.

So let us examine in the context of the Laplace–Dirichlet problem how one might arrange for equivalence between the variational formulation **(VP)** and the continuous formulation **(CP)**.

Clearly, by construction, any solution of the continuous problem **(CP)** belonging to $H^2(\Omega)$ will be a solution of the variational problem **(VP)**. *A priori*, i.e., regardless of whether there are solutions, the regularity properties of a solution u of the variational problem (**VP**) depend on the regularity of the function f on the right-hand side and also on the geometric properties of the boundary $\partial\Omega$ of the integration domain Ω, and this leaves us two possibilities for establishing that a solution of the variational problem (**VP**) is a solution of the continuous problem (**CP**).

First Method

The first method is a partial converse, which can be stated as follows: a solution of the variational problem (**VP**) with the same regularity as the solutions of the continuous problem (**CP**), i.e., belonging to $H^2(\Omega)$ and not just having the regularity ensured by $H^1(\Omega)$, is a solution of the continuous problem (**CP**).

We thus consider u belonging to $H^2(\Omega)$, a solution of the variational problem (**VP**). By Green's formula, we have

$$-\int_{\Omega} \Delta u \cdot v \, \mathrm{d}\Omega + \int_{\partial\Omega} \frac{\partial u}{\partial n} v \, \mathrm{d}\Gamma = \int_{\Omega} f v \, \mathrm{d}\Omega, \quad \forall v \in H_0^1(\Omega). \tag{2.60}$$

Since we know that v belongs to $H_0^1(\Omega)$, the integral over the boundary $\partial\Omega$ vanishes in (2.60). It follows that

$$\int_{\Omega} (\Delta u + f) v \, \mathrm{d}\Omega = 0, \quad \forall v \in H_0^1(\Omega). \tag{2.61}$$

Note how the $H^2(\Omega)$ regularity of the solution u has already served us in the application of Green's formula by guaranteeing convergence of the integral involving the Laplacian of u.

Indeed, applying the Cauchy–Schwarz inequality, we obtain the following upper bound:

$$\left| \int_{\Omega} \Delta u \cdot v \, \mathrm{d}\Omega \right| \leq \int_{\Omega} |\Delta u \cdot v| \, \mathrm{d}\Omega \leq \left(\int_{\Omega} |\Delta u|^2 \, \mathrm{d}\Omega \right)^{1/2} \left(\int_{\Omega} |v|^2 \, \mathrm{d}\Omega \right)^{1/2}. \tag{2.62}$$

Let us now examine the problem raised by (2.61). Looking at this family of equations, and there are as many equations here as there are functions v in $H_0^1(\Omega)$, we would like to be able to conclude that

$$\Delta u + f = 0 \text{ in } \Omega. \tag{2.63}$$

Now, the obvious way to justify the step from the integral Eq. (2.61) to the partial differential Eq. (2.63) would be to choose the specific function $v^* = \Delta u + f$ among all the functions v of $H_0^1(\Omega)$, because in that case, the integral in (2.61) would become

$$\int_{\Omega} (\Delta u + f)^2 \, d\Omega = 0, \tag{2.64}$$

which would, in turn, imply Poisson's equation (2.63), since the integral of a non-negative function can vanish only if its integrand is identically zero.

But the problem is that we cannot necessarily choose the function v in $H_0^1(\Omega)$ to be the specific function $v^* = \Delta u + f$, because we do not know whether this choice lies in $H_0^1(\Omega)$, only that it belongs to $L^2(\Omega)$.

There are two ways to get around this problem. The first is to use a density technique, appealing to "contamination" by proximity. To do this, the key is to have a density theorem that allows us to transfer the desired property, namely the integral Eq. (2.61), to suitable functions v in $L^2(\Omega)$.

To be precise, we will implement a density theorem to show that (2.61) is valid, not only for every v in $H_0^1(\Omega)$, but also for every function v in $L^2(\Omega)$, whence we will be able to choose the specific function v^* equal to $\Delta u + f$.

Be warned, however! This is not a trivial result, because the inclusion of the function spaces is not in the sense that would allow us to apply the following kind of reasoning: he who can do more can do less.

Indeed, given that $H_0^1(\Omega)$ is contained in $L^2(\Omega)$ and not the opposite, we certainly cannot claim directly that (2.61) holds for every v in $L^2(\Omega)$.

We thus use the density theorem on page 35 (Theorem 1.1) to assert that for every function w in $L^2(\Omega)$, there is a sequence of functions w_n in $C_0^\infty(\Omega)$ that converges in the sense of the $L^2(\Omega)$ norm to the function w:

$$\lim_{n \to \infty} \int_{\Omega} |w_n - w|^2 \, d\Omega = 0. \tag{2.65}$$

Furthermore, we also have the Sobolev embedding $C_0^\infty(\Omega) \subset H_0^1(\Omega)$. We can thus write (2.61) for each function in the sequence w_n, since these do belong to $H_0^1(\Omega)$:

$$\int_{\Omega} (\Delta u + f) w_n \, d\Omega = 0, \quad \forall n \in \mathbb{N}. \tag{2.66}$$

We can now establish this same property for the functions w in $L^2(\Omega)$:

$$\left| \int_{\Omega} (\Delta u + f) w \, d\Omega \right| = \left| \int_{\Omega} (\Delta u + f)(w - w_n) \, d\Omega \right|$$

$$\leq \left(\int_{\Omega} |\Delta u + f|^2 \, d\Omega \right)^{1/2} \left(\int_{\Omega} |w_n - w|^2 \, d\Omega \right)^{1/2}. \tag{2.67}$$

We then let n tend to $+\infty$ in the inequality (2.67) and use the convergence property (2.65) to conclude that for every function w in $L^2(\Omega)$, we have

$$\int_{\Omega} (\Delta u + f) w \, d\Omega = 0, \quad \forall w \in L^2(\Omega). \tag{2.68}$$

The desired conclusion is now immediate, as pointed out earlier, since we can now choose the specific function w^* given by $w^* \equiv \Delta u + f$ in (2.68).

It follows that Poisson's equation is satisfied in $L^2(\Omega)$ for every solution u of the variational problem (**VP**) that has the further regularity implied by its belonging also to $H^2(\Omega)$.

Note once again the implications of this last property, for it is indeed the fact that Δu is a function of $L^2(\Omega)$ that allows us to apply the density theorem (Theorem 1.1).

The second way of concluding from (2.61) is to use the theory of distributions. We first note that if (2.61) is satisfied for every function v in $H_0^1(\Omega)$, then it must in particular remain true for every function v in $C_0^\infty(\Omega)$, since $C_0^\infty(\Omega)$ is obviously included in $H_0^1(\Omega)$.

In this case, (2.61) becomes

$$\int_{\Omega} (\Delta u + f) v \, d\Omega = 0, \quad \forall v \in C_0^\infty(\Omega). \tag{2.69}$$

Now (2.69) can be interpreted in the sense of distributions as follows:

$$\langle \Delta u + f, v \rangle = 0, \quad \forall v \in C_0^\infty(\Omega). \tag{2.70}$$

Put another way,
$$\Delta u + f = 0 \text{ in } \mathscr{D}'(\Omega). \tag{2.71}$$

However, f and Δu both belong to $L^2(\Omega)$, since we assume that the solution u of the variational formulation (**VP**) belongs to $H^2(\Omega)$.

Hence, (2.71) also holds in $L^2(\Omega)$, and we have

$$\Delta u + f = 0 \text{ in } L^2(\Omega). \tag{2.72}$$

It follows that
$$\Delta u + f = 0 \text{ (a.e.) in } \Omega. \tag{2.73}$$

This clinches the partial converse once again. Indeed, if we have a solution u of the variational problem (**VP**) in $H_0^1(\Omega) \cap H^2(\Omega)$, then u must vanish on the boundary $\partial \Omega$ of Ω.

Fig. 2.1 Domain of
integration Ω whose boundary
has a discontinuous tangent

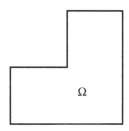

Second Method

The second converse regarding equivalence of the formulations **(VP)** and **(CP)** no
longer considers each solution u of the variational problem **(VP)** to be in $H^2(\Omega)$,
but starts from the assumption that the boundary $\partial\Omega$ of the open set Ω is C^2.

Indeed, the idea is to show that if the function f on the right-hand side is in $L^2(\Omega)$,
then every solution u of the variational problem **(VP)** belonging to $H_0^1(\Omega)$ is also
in $H^2(\Omega)$ for an integration domain Ω whose boundary $\partial\Omega$ has C^2 regularity (see,
for example, [2]).

This is precisely what was assumed in Sect. 2.1.2. In this situation, we can, in fact,
establish a full converse between the two formulations.

However, when the boundary contains corners of the kind depicted in Fig. 2.1,
for example, it can be shown that the solution u of the variational problem **(VP)** no
longer belongs to $H^2(\Omega)$.

Worse still, for certain boundary geometries, there are infinitely many solutions to
the continuous problem **(CP)** (see, for example, [4]), while the variational problem
(VP) has one and only one solution belonging to $H_0^1(\Omega)$.

Therefore, in situations in which the boundary $\partial\Omega$ exhibits geometric singularities, if we wish to maintain the equivalence of the two formulations, we must restrict
the search for solutions of the continuous problem **(CP)** to the space $H_0^1(\Omega)$.

It can then be shown that there is one and only one function in $H_0^1(\Omega)$ that
simultaneously solves the continuous problem **(CP)** and the variational problem
(VP).

In other words, since f belongs to $L^2(\Omega)$, Poisson's equation will be satisfied
in $L^2(\Omega)$, hence almost everywhere. As a consequence, the solution u will have a
Laplacian in $L^2(\Omega)$, while this does not mean that u will belong to $H^2(\Omega)$.

To end this converse, note that the homogeneous Dirichlet condition on $\partial\Omega$ is
automatically satisfied for each solution u of the variational problem **(VP)**: because
they lie in the space $H_0^1(\Omega)$, these functions have identically zero trace on $\partial\Omega$.

The aim of the above demonstration was to bring home to the reader, perhaps more
forcefully than is usual, the importance of this kind of issue, which is sometimes

neglected for various reasons, but which, in fact, constitutes the backbone of mathematical modeling in the engineering sciences and ensures its credibility.

And make no mistake: it is only by careful treatment of these questions of methodological consistency that the numerical scientist will be able to build up a genuinely scientific approach to modeling for the purposes of decision-making and predicting the behavior of real systems.

2.4 Methodology and a Series of Approximations

We now examine the structure of the variational problem (**VP**) expressed in (2.11) and compare it with that of the continuous problem (**CP**) specified by (2.1) in order to identify and understand the mechanisms that might prevent us from applying analytic methods of solution.

For this would be a situation in which numerical approximation might prove fruitful if it were possible to eliminate some or all of the structural constraints that prevent solution, regardless of the mathematical formulation that might be adopted.

We begin with a first remark concerning the continuous problem (**CP**). Two related mechanisms can prevent an analytic solution. The first is immediate and obvious. It lies in the complexity of the differential operations acting on the unknown function u.

Indeed, the combination of two second-order partial derivatives is undoubtedly a major problem for analytic solution, whatever the expression and complexity of the function f on the right-hand side.

Naturally, readers with experience in the standard techniques for solving partial differential equations (see, for example, [5]) might consider applying one or more miraculous transformations, such as those of Laplace, Fourier, and so on.

However, this would be to neglect the second factor entering into consideration for this type of problem, namely the shape of the integration domain, which may be fairly complex.

Indeed, when Ω has a regular shape, such as a square, circle, or ellipse, for instance, this tends to conceal the main difficulty.

For generally speaking, the problem here is to understand the formulation of the continuous problem (**CP**) from a different, perhaps unusual, angle, and in particular to realize that the continuous problem (**CP**) is a nonalgebraic system of equations, comprising an infinite number of equations in an infinite number of unknowns.

Indeed, Poisson's partial differential equation (2.1) has to hold at each point M of the domain of integration Ω, and as everyone knows, there are sure to be an infinite number of points within such an open set.

This means that we are actually considering, without explicitly acknowledging the fact, an infinite number of equations in an infinite number of unknowns that are nothing other than the values of the function u at each point M of Ω.

This is why the numerical scientist transforms the continuous problem into a finite-dimensional problem, for the human mind is not well equipped to comprehend the infinite.

The finite-element method, presented in many textbooks (see, for example, [5]), achieves this transformation by restriction, introducing a mesh, so that one need consider only the finite number of points M_i on a grid and approximations $\tilde{u}_i$ at these points that solve a system of algebraic equations obtained by approximating the partial derivatives, essentially with the help of Taylor's formula.

But in that case, does the transformation of the continuous problem (CP) to the variational formulation (VP) discussed above really generate a viable alternative? And if so, what do we gain by this new formulation, which, at first glance, seems to complicate the original continuous problem (CP).

At the present stage, let us just say that the difficulty inherent in the continuous problem (CP) is carried over wholesale to the variational formulation (VP).

Indeed, as we have just explained, it is the infinite number of unknowns and equations associated with the Poisson equation (2.1) that constitutes one of the main problems for solution.

And in the present case of the variational formulation (VP) as specified in (2.11), this problem of infinite dimensions is still extant, in the dimension of the search space V, here $H_0^1(\Omega)$, and as a consequence, in the infinite number of equations constituting the variational formulation (2.11).

For this reason, the approximation process adopted in the finite-element method, which is based on Galerkin's method, consists in considering a subspace $\tilde{V} \subset V$ that is in fact finite-dimensional. Let K_h be the dimension of this space $\tilde{V}$.

The transition from the variational problem (VP) to the approximate variational problem $\widetilde{(VP)}$ is made by replacing the pair of functions (u, v) in $V \times V$ by their approximations $(\tilde{u}, \tilde{v})$ in $\tilde{V} \times \tilde{V}$.

Hence, $\widetilde{(VP)}$ is expressed as follows:

$$\widetilde{(VP)} \quad \begin{cases} \text{Find } \tilde{u} \in \tilde{V} \text{ solution of} \\ \displaystyle\int_{\Omega} \nabla\tilde{u}\cdot\nabla\tilde{v}\,d\Omega = \int_{\Omega} f\tilde{v}\,d\Omega, \quad \forall \tilde{v} \in \tilde{V}. \end{cases} \quad (2.74)$$

The reader should not be beguiled by the apparent simplicity of the approximation procedure. The approximate variational formulation $\widetilde{(VP)}$ is not a simple rewriting of the exact formulation (VP). On the contrary, it represents real progress with regard to the possibility of finding an approximate solution to the variational problem (VP), while of course it also corresponds to a loss of information that will be important to estimate later on.

In order to get a feel for the advantages brought about by this decisive step, let us introduce a basis $(\varphi_i)_{i=1,\ldots,K_h}$ for the approximation space $\tilde{V}$, which we consider to have finite dimension K_h.

In this case, the unknown $\tilde{u}$ can be decomposed in the following form relative to the basis of functions φ_i:

$$\tilde{u} = \sum_{j=1}^{K_h} \tilde{u}_j \varphi_j. \tag{2.75}$$

Put another way, since (2.74) is valid for all $\tilde{v} \in \tilde{V}$, we are free to choose, among the approximate test functions $\tilde{v}$, each of the basis functions φ_i, $(i = 1, \ldots, K_h)$, whereupon we set $\tilde{v} = \varphi_i$.

The approximate variational Eq. (2.74) can then be written as follows:

$$\widetilde{\textbf{(VP)}} \quad \begin{cases} \text{Find } \tilde{u}_j, \quad j = 1, \ldots, K_h \text{ solution of} \\ \sum_{j=1}^{K_h} \left(\int_\Omega \nabla \varphi_i \cdot \nabla \varphi_j \, d\Omega \right) \tilde{u}_j = \int_\Omega f \varphi_i \, d\Omega, \quad \forall i = 1, \ldots, K_h. \end{cases} \tag{2.76}$$

We thus set

$$A_{ij} = \int_\Omega \nabla \varphi_i \cdot \nabla \varphi_j \, d\Omega, \qquad B_i = \int_\Omega f \varphi_i \, d\Omega. \tag{2.77}$$

The approximate variational problem $\widetilde{\textbf{(VP)}}$ can then be expressed in the following form:

$$\widetilde{\textbf{(VP)}} \quad \begin{cases} \text{Find } \tilde{u}_j, \quad (j = 1, \ldots, K_h) \text{ solution of} \\ \sum_{j=1}^{K_h} A_{ij} \tilde{u}_j = B_i, \quad \forall i = 1, \ldots, K_h. \end{cases} \tag{2.78}$$

This last form clearly shows how the variational formulation **(VP)** is reduced by the approximation process to a finite-dimensional problem consisting of a system of K_h linear equations in K_h unknowns.

In order to implement the finite-element method, we must therefore specify the basis functions φ_i in (2.78), decide how to calculate the integrals in the matrix elements A_{ij} and the coefficients B_i on the right-hand side, and set up the algorithm for inverting the linear system, provided, of course, that the matrix with entries A_{ij} is in fact invertible.

From the theoretical standpoint, it is clear that one must check for invertibility of the matrix before applying the inversion algorithm to the linear system (2.78) (see, for example, [5]).

Returning now to the general problem of the approximation process described at the beginning of this section, we may summarize the successive transformations leading up to these approximations in the following diagram:

Real system $\longrightarrow$ Mathematical model or continuous problem **(CP)**

               $\longrightarrow$ Variational formulation **(VP)**

               $\longrightarrow$ Approximate variational formulation $\widetilde{\textbf{(VP)}}$

               $\longrightarrow$ Method for inverting linear system $\widetilde{\widetilde{\textbf{(VP)}}}$.

Viewing the process as a whole in this way should persuade the numerical scientist of the need for humility and caution when publishing final results. It is true that there are various theorems concerning error estimation in the context of the finite-element method, but as is often the case, this type of result is never all-encompassing, but concerns only a part of the above process.

In general, these theorems deal only with errors introduced when the variational problem **(VP)** is replaced by its approximate formulation $\widetilde{\textbf{(VP)}}$ (see the Bramble–Hilbert lemma in the next section).

2.5 Variational Formulations and Approximations

Having laid down the fundamental principles underpinning the overall methodology, we now turn to the approximation of the variational formulations and the various choices available in the finite-element method.

It is this process as a whole that leads to the estimation of an approximate solution $\tilde{u}$ for both the variational problem **(VP)** and the continuous problem **(CP)** that inspired it.

As we have seen in (2.78), the variational problem **(VP)** associated with the Laplace–Dirichlet problem leads by Galerkin's method to an approximate formulation $\widetilde{\textbf{(VP)}}$ that is nothing but a linear system to be inverted.

It is the solution of this linear system that generates the approximation $\tilde{u}$ to the solution of the variational problem **(VP)**, and hence an approximation to the solution of the continuous problem **(CP)**, whenever the two formulations **(CP)** and **(VP)** are indeed equivalent.

As a matter of fact, many mathematical models in engineering science lead to a formalism that is analogous to the one we have described for the Laplace–Dirichlet problem. This formalism can be viewed as a generic family of variational problems **(VP)** with the following abstract description:

$$\text{Find } u \text{ in } V \text{ which solves } a(u, v) = L(v), \quad \forall v \in V, \qquad (2.79)$$

where

- V is a vector space of functions,
- $a(.,.)$ is a bilinear form on $V \times V$,
- $L(.)$ is a linear form on V.

Naturally, as already discussed in Sect. 2.3, further investigations involving suitable techniques of functional analysis will be needed if we are to obtain a variational formulation (**VP**) with one and only one solution that is, in addition, equivalent to the solution of the continuous problem (**CP**).

However, in ensuring a good match between the continuous and variational problems, the approximation to the variational formulation (2.79) is unavoidable. This observation is intimately related to the infinite dimension of the function spaces arising in most mathematical models in the engineering sciences.

In order to get around the fact that we cannot solve formulations with the structure of (2.79), the method put forward by Galerkin considers a subspace $\widetilde{V} \subset V$ with finite dimension K_h. The abstract variational formulation (2.79) is then transformed to the following approximation $\widetilde{(\mathbf{VP})}$:

$$\text{Find } \tilde{u} \text{ in } \widetilde{V} \text{ which solves } a(\tilde{u}, \tilde{v}) = L(\tilde{v}), \quad \forall \tilde{v} \in \widetilde{V}. \tag{2.80}$$

Now that we have introduced an approximation space $\widetilde{V}$ of finite dimension K_h, it becomes possible, and indeed natural, to consider a basis of functions φ_i, $(i = 1, \ldots, K_h)$, and seek the approximation $\tilde{u}$, which replaces the solution u belonging to V, in the form

$$\tilde{u} = \sum_{j=1,\ldots,K_h} \tilde{u}_j \varphi_j. \tag{2.81}$$

Note that the decomposition (2.81) would not be possible without first making our transition from the space V, of infinite dimension, to its inner approximation $\widetilde{V}$, of finite dimension K_h.

Indeed, in the case of an infinite-dimensional function space V, apart from specific vector spaces in which every element can be decomposed relative to a basis containing a *countable* infinity of elements (as happens for separable Hilbert spaces, for instance), this kind of decomposition (2.81) remains impossible and so could not contribute to solving the variational problem (**VP**).

But returning to the approximate variational formulation $\widetilde{(\mathbf{VP})}$ specified by (2.80), by choosing the functions $\tilde{v}$ equal to the basis functions φ_i, $(i = 1, \ldots, K_h)$, we may rewrite the formulation (2.80) in the following way:

Find $\tilde{u} = [\tilde{u}_1, \ldots, \tilde{u}_{K_h}]^t$ in $\widetilde{V}$ that solves

$$a\left(\sum_{j=1,K_h} \tilde{u}_j \varphi_j, \varphi_i \right) = L(\varphi_i), \quad \forall i = 1, \ldots, K_h. \tag{2.82}$$

We then exploit the bilinearity of the form $a(.,.)$ and the linearity of the form $L(.)$. The variational formulation $\widetilde{(\mathbf{VP})}$ can thus be expressed in the following form:

Find $\tilde{u} = [\tilde{u}_1, \ldots, \tilde{u}_{K_h}]^t$ in $\widetilde{V}$ that solves

$$\sum_{j=1,K_h} a(\varphi_j, \varphi_i)\tilde{u}_j = L(\varphi_i), \quad \forall i = 1, \ldots, K_h. \tag{2.83}$$

Finally, we introduce the quantities A_{ij} and b_i defined by

$$A_{ij} = a(\varphi_j, \varphi_i), \qquad b_i = L(\varphi_i). \tag{2.84}$$

The approximate variational formulation $\widetilde{(\mathbf{VP})}$ then assumes the following final form:

Find $\tilde{u} = [\tilde{u}_1, \ldots, \tilde{u}_{K_h}]^t$ in $\widetilde{V}$ that solves

$$\sum_{j=1,K_h} A_{ij}\tilde{u}_j = b_i, \quad \forall i = 1, \ldots, K_h. \tag{2.85}$$

At this point, we observe that the formulation (2.85) is none other than a linear system involving a matrix A with entries A_{ij} and a right-hand side b with components b_i.

Put another way, we have just shown that every variational formulation (**VP**) that can be written in the form (2.79), and in which the forms $a(.,.)$ and $L(.)$ are respectively bilinear and linear, can be solved by an approximation with solution $\tilde{u}$ given by the linear system (2.85).

The problem at present is to select suitable parameters to proceed to an effective solution of the linear system (2.85) and thereby obtain an approximation to the variational problem (**VP**) specified by (2.79).

In order to calculate the coefficients A_{ij} and b_i, one must know the basis functions φ_i, $(i = 1, \ldots, K_h)$, of the approximation space $\widetilde{V}$. Naturally, this will depend intimately on the definition of the space $\widetilde{V}$ with finite dimension K_h.

For this reason, a first approach that allows us to fix the dimension K_h of the space $\widetilde{V}$ consists in relating this dimension K_h to a finite number of values of the functions $\tilde{v}$ belonging to $\widetilde{V}$ at preselected points or nodes M_k, $(k = 1, \ldots, K)$, of the integration domain Ω.

We now introduce an elementary geometry G_m, $(m = 1, \ldots, M)$, that generates a mesh on the domain of integration Ω, and hence also a set of nodes allowing a discretization of the problem (see Fig. 2.2). We thus arrive at the Lagrange finite-element method, defined as follows:

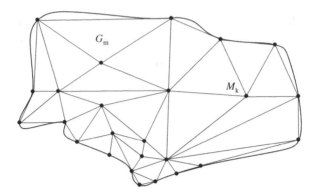

Fig. 2.2 Mesh consisting of triangular elements

Definition 2.1 The triplet $\big(G, \Sigma, P(G)\big)$ specifies a Lagrange finite-element method, where:

- G specifies the geometry of the primitive element of the mesh (segment, triangle, square, polyhedron, etc.).

- $\Sigma = (M_1, \ldots, M_{K'})$, $K' < K$, are the nodes at the vertices of, or otherwise delimiting, the primitive element G.

- $P(G)$ is the approximation space containing polynomials defined on G.

Finally, the triplet (G, Σ, P) must satisfy the property of *unisolvence*, defined as follows:

$$\forall \xi_1, \ldots, \xi_{K'} \in \mathbb{R}^{K'}, \ \exists! \ p \in P(G) \text{ such that } p(M_k) = \xi_k, \ \forall k = 1, \ldots, K'. \tag{2.86}$$

In other words, there is one and only one function p in $P(G)$ that passes through the K' data points $(\xi_1, \ldots, \xi_{K'})$ at the K' nodes on the primitive element G.

Concerning unisolvence, it follows that $P(G)$ is isomorphic to $\mathbb{R}^{K'}$, and we have $\dim P(G) = K'$.

Once we have specified the functions in $P(G)$, defined on a generating element G, the approximation space $\widetilde{V}$ is constructed in the framework of the Lagrange finite-element method by setting

$$\widetilde{V} \equiv \left\{ \tilde{v} : \Omega \to \mathbb{R}, \ \tilde{v} \in C^0(\Omega), \ \tilde{v}|_G \in P(G) \right\}, \tag{2.87}$$

where we have set aside the question of the boundary conditions that may be imposed on the functions $\tilde{v}$ in $\widetilde{V}$, depending on the problem under consideration.

Consequently, if we ignore for the moment the boundary conditions, which will vary from one problem to another, the dimension of the space $\widetilde{V}$ specified by (2.87) can be deduced from the dimension K' of $P(G)$, the number of elements, and the number of nodes arising in the geometric discretization of the integration domain Ω.

The Lagrange finite-element method can be generalized. A general finite-element method is defined by the triplet $\left(G, \Sigma, P(G)\right)$, where:

- G is a primitive element of the geometric discretization of $\mathbb{R}^n$, $n = 1, 2$, or 3.

- Σ is a set of degrees of freedom σ_k, $k = 1, \ldots, K'$, consisting of linear forms on the space of functions defined on G.

- $P(G)$ is a vector space of finite dimension K'.

- Unisolvence: for a K'-tuplet of real numbers in $\mathbb{R}^{K'}$, there is a unique element p of $P(G)$ with $\sigma_k(p) = \xi_k$, $(\forall k = 1, \ldots, K')$.

2.6 Convergence of the Finite-Element Method

As emphasized in Sect. 2.4, the different levels of approximation arising in the series of steps that lead from the model to its numerical approximation require the user to treat the estimation of error with great care and humility.

In the present case, the finite-element method can furnish a body of theoretical results for estimating the approximation error between the solution u of a variational problem (**VP**) and its approximation $\tilde{u}$, which solves the approximate variational problem $\widetilde{(\textbf{VP})}$.

Given the kind of mathematical objects involved here, i.e., the functions u and $\tilde{u}$, we shall describe in this section a family of results that allow us to estimate, relative to a suitable norm, the distance between the solution u and its approximation $\tilde{u}$, which we shall denote by $\|u - \tilde{u}\|$.

To illustrate the discussion, we shall refer to the family of variational problems (**VP**) with abstract form:

$$\text{Find } u \in V \text{ solution of } a(u, v) = L(v), \quad \forall v \in V. \tag{2.88}$$

As in the last section, $\widetilde{V} \subset V$ will denote the finite-dimensional approximation space contained in the space V, and $\tilde{v}$ the generic element of that space.

The approximation $\tilde{u}$ to the solution u will be a special case among these approximation functions $\tilde{v} \in \widetilde{V}$.

In other words, the approximate formulation $\widetilde{(\textbf{VP})}$ of the variational problem (**VP**) can be written in the following form:

$$\text{Find } \tilde{u} \in \widetilde{V} \text{ solution of } a(\tilde{u}, \tilde{v}) = L(\tilde{v}), \quad \forall \tilde{v} \in \widetilde{V}. \tag{2.89}$$

Under the hypotheses of the Lax–Milgram theorem (Theorem 1.11 in Sect. 1.6), with
Hilbert norm $\| \cdot \|$, we have the following lemma:

Lemma 2.5 *The variational problem* $\widetilde{(\mathbf{VP})}$ *specified by (2.89) admits a unique
solution* $\tilde{u}$*. Further, this solution satisfies the orthogonality relation*

$$a(u - \tilde{u}, \tilde{v}) = 0, \quad \forall \tilde{v} \in \widetilde{V}. \tag{2.90}$$

Proof: The existence and uniqueness of the solution $\tilde{u}$ in $\widetilde{V}$ are immediate, precisely
because $\widetilde{V} \subset V$. Indeed, we begin by noting that the finite-dimensional approximation space $\widetilde{V}$ contained in V is necessarily a closed vector subspace of V, and so
inherits a Hilbert structure from V.

The fact that $\widetilde{V}$ is a subspace of V thus means that all the conditions are fulfilled
for applying the Lax–Milgram theorem in $\widetilde{V}$.

Regarding the orthogonality relation (2.90), we simply rewrite the variational
equation of **(VP)** with $\tilde{v}$ in the place of v, whence

$$a(u, \tilde{v}) = L(\tilde{v}), \quad \forall \tilde{v} \in \widetilde{V}. \tag{2.91}$$

The difference between (2.91) and (2.89) leads immediately to the orthogonality
relation (2.90).

The first error estimate $\|u - \tilde{u}\|$ is provided by Céa's lemma:

Lemma 2.6 *Under the hypotheses of the Lax–Milgram theorem (Theorem
1.11), and if we assume in addition that the approximation* $\tilde{u}$ *of the exact
solution* u *lies in* $\widetilde{V} \subset V$*, we have the following error estimate:*

$$\|u - \tilde{u}\| \leq C \inf_{\tilde{v} \in \widetilde{V}} \|u - \tilde{v}\|. \tag{2.92}$$

Proof: The proof exploits the double bound on $a(u - \tilde{u}, u - \tilde{u})$, obtained from the
coercivity property, on the one hand, and the continuity of the bilinear form $a(., .)$,
on the other.

First of all, by the orthogonality relation (2.90) and choosing $\tilde{v} = \tilde{u}$, we have

$$a(u - \tilde{u}, \tilde{u}) = 0. \tag{2.93}$$

We now rewrite $a(u - \tilde{u}, u - \tilde{u})$ as follows:

$$\forall \tilde{v} \in \widetilde{V}, \quad a(u - \tilde{u}, u - \tilde{u}) = a(u - \tilde{u}, u) - a(u - \tilde{u}, \tilde{u}) = a(u - \tilde{u}, u)$$
$$= a(u - \tilde{u}, u) - a(u - \tilde{u}, \tilde{v}) = a(u - \tilde{u}, u - \tilde{v}).$$

But since $a(.,.)$ is continuous and V-elliptic, $\exists\, (\alpha, \beta) \in \mathbb{R}_+^* \times \mathbb{R}_+^*$ such that

$$\alpha\, \|u - \tilde{u}\|^2 \leq a(u - \tilde{u}, u - \tilde{u}) = a(u - \tilde{u}, u - \tilde{v}) \leq \beta\, \|u - \tilde{u}\|\, \|u - \tilde{v}\|. \quad (2.94)$$

Dividing through by $\|u - \tilde{u}\|$, it then follows that

$$\|u - \tilde{u}\| \leq \frac{\beta}{\alpha}\, \|u - \tilde{v}\|, \quad \forall \tilde{v} \in \widetilde{V}, \quad (2.95)$$

whereupon the constant C in the statement of the lemma can be taken as the ratio of β and α.

Naturally, (2.95) is all the more useful as the bound on the norm $\|u - \tilde{u}\|$ can be made smaller. This is why the conclusion of Céa's lemma refers to the lower bound of the quantities $\|u - \tilde{v}\|$ for all functions $\tilde{v}$ in $\widetilde{V}$, i.e.,

$$\|u - \tilde{u}\| \leq C \inf_{\tilde{v} \in \widetilde{V}} \|u - \tilde{v}\|. \quad (2.96)$$

The next step in specifying the error estimate produced by Céa's lemma is to characterize the approximation space $\widetilde{V}$.

As mentioned in Sect. 2.5, the Lagrange finite-element method provides a simple solution for systematically producing a finite-dimensional approximation space $\widetilde{V}$. This process depends on the *unique* determination of an approximation function from the set of values it takes at a finite number of points M_k, $(k = 1, \ldots, K)$, arranged on a given mesh on the integration domain Ω.

At this point, the reader should recall the discussion of the Lagrange finite-element method in Sect. 2.5, and in particular, the fact that the dimension of the approximation space $\widetilde{V}$ corresponds to the number of nodes in the mesh on the region Ω, neglecting for the moment the boundary conditions that may have to be imposed on the approximation functions $\tilde{v}$.

More generally, we thus introduce the interpolation operator π_h defined by

$$\pi_h : C^0(\bar{\Omega}) \longrightarrow \widetilde{V}$$
$$v \longmapsto \pi_h v \equiv \sum_{k=1,\ldots,K} v(M_k)\varphi_k, \quad (2.97)$$

where φ_k is the basis function in the approximation space $\widetilde{V}$ characterizing the node M_k, i.e., satisfying

$$\varphi_k(M_l) = \delta_{kl}, \quad (2.98)$$

with δ_{kl} the Kronecker symbol.

It is easy to check that the function $\pi_k v$, interpolated from v to the K nodes M_k of the mesh in the integration domain Ω, is the unique function of $\widetilde{V}$ satisfying

$$\pi_h v(M_k) = v(M_k), \quad \forall k = 1, \ldots, K. \tag{2.99}$$

We can then write the inequality of Céa's lemma in the particular case that $\tilde{v} = \pi_h u$, obtaining

$$\|u - \tilde{u}\| \leq C \|u - \tilde{v}\| = C \|u - \pi_h u\|. \tag{2.100}$$

According to the bound (2.100), the approximation error and the interpolation error will be of the same order of magnitude. For this reason, estimating the interpolation error provides an adequate method for measuring the approximation error, which depends on the nature and properties of each Lagrange finite-element method.

In order to make full use of Céa's lemma, we now discuss the Bramble–Hilbert lemma, which is based on these considerations. For present purposes, we limit the discussion to straight-edged but nonflat finite elements, and we take the variational space V to be the Sobolev space $H^1(\Omega)$, where Ω is a regular bounded open subset of $\mathbb{R}^2$.

Indeed, many problems arising in the engineering sciences correspond well to this functional framework, or are even more regular, but the reader should note that applications that cannot be formulated within this framework would require mathematical techniques of functional analysis that go well beyond the scope of the present book.

Lemma 2.7 *Let h measure the size of the primitive element of a Lagrange finite-element mesh. If the approximation space $\tilde{V}$ contains the space P_k of polynomials of order less than or equal to k in the pair of variables (x, y), then for any sufficiently regular solution u, let us say at least in $H^1(\Omega)$, of a variational problem (VP) of the form (2.88), we have*

$$\|u - \pi_h u\|_{H^1(\Omega)} = O(h^k), \qquad \|u - \tilde{u}\|_{H^1(\Omega)} = O(h^k), \tag{2.101}$$

where $O(h^k)$ is Landau's notation indicating that there is a positive constant C such that

$$|O(h^k)| \leq C h^k.$$

Naturally, the technical interest of this lemma lies in estimating the norm measuring the difference between the solution u and its interpolation $\pi_h u$.

The whole point of the preamble following Céa's lemma was to emphasize the need to estimate the latter norm if we are to draw conclusions about the approximation error in the finite-element analysis, at least in the context that we have described here.

2.7 Description of Commonly Used Finite-Element Methods

In this section we present the main finite-element geometries commonly encountered in the engineering sciences. We shall systematically present each finite-element method according to the following scheme:

1. Definition of the geometry G of the mesh element.

2. Definition of the approximation space $P(G)$, along with its dimension.

3. Definition of the set of linear forms σ_i on the function space defined on G.

4. Determination of the canonical basis functions for the space $P(G)$, i.e., the functions $p_1, \ldots, p_{\dim P(G)}$ satisfying $\sigma_i(p_j) = \delta_{ij}$, where δ_{ij} is the Kronecker symbol.

Note. The fact that there exists a collection of functions $p_1, \ldots, p_{\dim P(G)}$ in $P(G)$ satisfying the canonical property

$$\sigma_i(p_j) = \delta_{ij}, \quad \forall\, (i, j) \in \{1, \ldots, \dim P(G)\},$$

implies that this system of functions constitutes a basis for $P(G)$.

Indeed, let us show that the functions $p_1, \ldots, p_{\dim P(G)}$ are independent in $P(G)$.

Suppose then that there are $(\alpha_1, \ldots, \alpha_{\dim P(G)}) \in \mathbf{R}^{\dim P(G)}$ such that

$$\sum_{i=1,\ldots,\dim P(G)} \alpha_i\, pi = 0. \tag{2.102}$$

Let us show that the condition (2.102) implies that the coefficients α_i are all zero. To do this, for some preselected j, we apply the jth linear form σ_j to (2.102):

$$\sigma_j \left[\sum_{i=1,\ldots,\dim P(G)} \alpha_i\, pi \right] = \sigma_j(0). \tag{2.103}$$

We then use the linearity of the form σ_j and the fact that $\sigma_j(0) = 0$. Equation (2.103) can thus be written

$$\sum_{i=1,\ldots,\dim P(G)} \alpha_i \sigma_j(p_i) = \sum_{i=1,\ldots,\dim P(G)} \alpha_i \delta_{ij} = \alpha_j = 0, \ \forall\, j = 1, \ldots, \dim P(G).$$

$$\tag{2.104}$$

It follows that the coefficients α_j all vanish and that the family $\{p_1, \ldots, p_{\dim P(G)}\}$ is therefore independent in this space of finite dimension $\dim P(G)$.

This same family must therefore span the approximation space $P(G)$ and thus constitutes a basis for it.

Note that we use the word "canonical" to express the fact that each function p_i in this particular basis is characteristic of a favored linear form σ_j, in the sense that the other linear forms are zero on this function p_i of the canonical basis.

In particular, in the case of Lagrange finite-element methods, the linear forms reflect a number of specific values of the functions of $P(G)$ at certain points (the discretization nodes) of the integration domain.

In this case, each function in the canonical basis corresponds to the unique function equaling 1 at a given node of the discretization and 0 at all the other nodes.

2.8 Fundamental Classes of Finite-Element Methods

2.8.1 Finite-Element Analysis in One Spatial Dimension

For the finite element methods discussed in this section, the mesh element is just the interval $G \equiv [0, 1]$.

- **P_0 Finite-Element Analysis**

1. The space $P(G) \equiv P_0$ comprises the polynomials p defined and constant on the interval $[0, 1]$. The dimension of P_0 is clearly 1.

2. We consider the linear form σ defined by

$$\sigma : p \longrightarrow \int_0^1 p(x)\,\mathrm{d}x. \tag{2.105}$$

3. The only function in the canonical basis is the constant function equal to 1 on the interval $[0, 1]$. To see this, we turn to the definition of the functions in the canonical basis discussed above, viz.,

$$\sigma(p) = 1 \iff \int_0^1 p(x)\,\mathrm{d}x = 1, \quad \text{where } p(x) = \text{constant}, \quad \forall x \in [0, 1]. \tag{2.106}$$

We deduce immediately that $p(x) = 1$, for all $x \in [0, 1]$.

For this first finite element, the functions $\tilde{v}$ belonging to $\widetilde{V}$ are constant functions on each mesh element. Note also that the constant on each mesh element corresponds to the average value of the function $\tilde{v}$ on the corresponding element.

- **P_1 Finite-Element Analysis**

1. The approximation space $P(G) \equiv P_1$ comprises the affine functions defined on the primitive mesh element $[0, 1]$. The dimension of the space P_1 is 2.

2. The two linear forms are

$$\sigma_1 : p \longrightarrow p(0), \qquad \sigma_2 : p \longrightarrow p(1). \qquad (2.107)$$

3. To determine the functions in the canonical basis of the space P_1, we express the basic property of the two basis functions p_1, p_2:

$$\sigma_1(p_1) = 1 \iff p_1(0) = 1, \qquad \sigma_1(p_2) = 0 \iff p_2(0) = 0,$$
$$\sigma_2(p_1) = 0 \iff p_1(1) = 0, \qquad \sigma_2(p_2) = 1 \iff p_2(1) = 1. \qquad (2.108)$$

It is then straightforward to deduce that the basis functions p_1, p_2 solving (2.108) in the space P_1 of affine functions on the interval $[0, 1]$ are

$$p_1(x) = 1 - x, \qquad p_2(x) = x. \qquad (2.109)$$

- **P_2 Finite-Element Analysis**

1. The approximation space $P(G) \equiv P_2$ comprises the polynomials of degree less than or equal to two defined on the mesh element $[0, 1]$. The dimension of the space P_2 is 3.

2. Consider the three linear forms defined by

$$\sigma_1 : p \longrightarrow p(0), \qquad \sigma_2 : p \longrightarrow p(1/2), \qquad \sigma_3 : p \longrightarrow p(1). \qquad (2.110)$$

3. We now express the defining property of the functions p_1, p_2, p_3 in the canonical basis for P_2:

$$\sigma_1(p_1) = 1 \iff p_1(0) = 1, \quad \sigma_1(p_2) = 0 \iff p_2(0) = 0,$$
$$\sigma_1(p_3) = 0 \iff p_3(0) = 0, \quad \sigma_2(p_1) = 0 \iff p_1(1/2) = 0,$$
$$\sigma_2(p_2) = 1 \iff p_2(1/2) = 1, \sigma_2(p_3) = 0 \iff p_3(1/2) = 0, \quad (2.111)$$
$$\sigma_3(p_1) = 0 \iff p_1(1) = 0, \quad \sigma_3(p_2) = 0 \iff p_2(1) = 0,$$
$$\sigma_3(p_3) = 1 \iff p_3(1) = 1.$$

We then exploit the fact that each of the polynomials p_i, of degree less than or equal to two must have the form $ax^2 + bx + c$.

The nine relations (2.111) can be used to determine the nine coefficients of the three polynomials p_1, p_2, p_3. The result is

$$p_1(x) = (2x - 1)(x - 1), \quad p_2(x) = 4x(1 - x), \quad p_3(x) = x(2x - 1). \qquad (2.112)$$

- **Hermite Finite-Element Analysis**

1. The approximation space $P(G) \equiv P_3$ comprises the polynomials of degree less than or equal to three, defined on the primitive mesh element $[0, 1]$. The dimension of the space P_3 is 4.

2. Consider the four linear forms defined by

$$\sigma_1 : p \to p(0), \quad \sigma_2 : p \to \frac{\mathrm{d}p}{\mathrm{d}x}(0), \quad \sigma_3 : p \to p(1), \quad \sigma_4 : p \to \frac{\mathrm{d}p}{\mathrm{d}x}(1). \tag{2.113}$$

3. We now determine the four functions p_1, p_2, p_3, p_4 in the canonical basis for P_3. To do this, we write down the 16 defining relations of the form $\sigma_i(p_j) = \delta_{ij}$:

$$\begin{aligned}
\sigma_1(p_1) = 1 &\iff p_1(0) = 1, \quad \sigma_1(p_2) = 0 \iff p_2(0) = 0, \\
\sigma_1(p_3) = 0 &\iff p_3(0) = 0, \quad \sigma_1(p_4) = 0 \iff p_4(0) = 0, \\
\sigma_2(p_1) = 0 &\iff p_1'(0) = 0, \quad \sigma_2(p_2) = 1 \iff p_2'(0) = 1, \\
\sigma_2(p_3) = 0 &\iff p_3'(0) = 0, \quad \sigma_2(p_4) = 0 \iff p_4'(0) = 0, \\
\sigma_3(p_1) = 0 &\iff p_1(1) = 0, \quad \sigma_3(p_2) = 0 \iff p_2(1) = 0, \\
\sigma_3(p_3) = 1 &\iff p_3(1) = 1, \quad \sigma_3(p_4) = 0 \iff p_4(1) = 0, \\
\sigma_4(p_1) = 0 &\iff p_1'(1) = 0, \quad \sigma_4(p_2) = 0 \iff p_2'(1) = 0, \\
\sigma_4(p_3) = 0 &\iff p_3'(1) = 0, \quad \sigma_4(p_4) = 1 \iff p_4'(1) = 1.
\end{aligned} \tag{2.114}$$

Once again, the 16 relations (2.114) allow us to obtain the 16 coefficients of the four polynomials p_1, p_2, p_3, p_4 in the canonical basis for P_3. The result is

$$\begin{aligned}
p_1(x) &= (x-1)^2(2x+1), \quad p_2(x) = x(x-1)^2, \\
p_3(x) &= x^2(3-2x), \quad\quad\quad\; p_4(x) = (x-1)x^2.
\end{aligned} \tag{2.115}$$

2.8.2 Finite-Element Methods in Two Spatial Dimensions

Triangular Meshes

In this section we discuss finite-element methods in which the mesh element G is an arbitrary triangle with vertices M_1, M_2, and M_3 in the plane $(O; x, y)$ (see Fig. 2.3).

- **P_0 Finite-Element Analysis**

1. The approximation space $P(G) \equiv P_0$ comprises the constant functions on the triangle G. The dimension of the space P_0 is 1.

2. We consider the linear form σ defined by

Fig. 2.3 Triangular mesh element

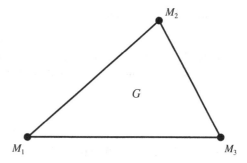

$$\sigma : p \longrightarrow \frac{1}{\text{area}(G)} \iint_G p(x, y) \, dx dy. \tag{2.116}$$

3. We now determine the basis function p of P_0 satisfying the defining property $\sigma(p) = 1$:

$$\sigma(p) = 1 \iff \frac{1}{\text{area}(G)} \iint_G p(x, y) \, dx dy = 1, \tag{2.117}$$

where $p(x, y) = \text{constant}$, for all $(x, y) \in G$. This implies that the canonical basis function p is the constant function equal to 1 everywhere on the triangle G.

- **P_1 Finite-Element Analysis**

1. The approximation space P_1 comprises the polynomial functions of degree less than or equal to one for the pair of variables (x, y). In other words, every function p of P_1 can be written in the form

$$p(x, y) = ax + by + c, \tag{2.118}$$

where (a, b, c) is an arbitrary triplet in $\mathbb{R}^3$. The previous definition implies that the dimension of the space P_1 is 3.

2. We consider the three linear forms defined by

$$\sigma_1 : p \to p(M_1), \quad \sigma_2 : p \to p(M_2) \quad \sigma_3 : p \to p(M_3). \tag{2.119}$$

3. The three functions p_1, p_2, p_3 in the canonical basis correspond to the three barycentric functions $\lambda_1, \lambda_2, \lambda_3$, whose existence is proven in [5].

Note, however, that these polynomial functions in the pair of variables (x, y), with degree less than or equal to one, satisfy the canonical property

$$\sigma_j(\lambda_i) \equiv \lambda_i(M_j) = \delta_{ij}, \tag{2.120}$$

by definition.

- *P₂ **Finite-Element Analysis***

1. The approximation space $P(G) \equiv P_2$ comprises the polynomial functions of degree less than or equal to two in the pair of variables (x, y). In other words, every function p in P_2 can be written in the form

$$p(x, y) = ax^2 + by^2 + cxy + dx + ey + f, \tag{2.121}$$

where (a, b, c, d, e, f) is an arbitrary point in $\mathbb{R}^6$. The definition (2.121) implies that the dimension of the space P_2 is 6.

2. To define the six linear forms σ_i, $i = 1, \ldots, 6$, we introduce three further nodes M_{12}, M_{13}, and M_{23} at the middle of each side of the triangle G, as shown in Fig. 2.4. The six linear forms are now defined as follows:

$$\sigma_1 : p \longrightarrow p(M_1), \qquad \sigma_2 : p \longrightarrow p(M_2), \tag{2.122}$$

$$\sigma_3 : p \longrightarrow p(M_3), \qquad \sigma_4 : p \longrightarrow p(M_{12}), \tag{2.123}$$

$$\sigma_5 : p \longrightarrow p(M_{13}), \qquad \sigma_6 : p \longrightarrow p(M_{23}). \tag{2.124}$$

3. The functions $p_1, p_2, p_3, p_4, p_5, p_6$ in the canonical basis are constructed as follows.

Take the example of the function p_1. This second-degree polynomial in the pair (x, y) must vanish at the points M_2, M_3, M_{12}, M_{13}, and M_{23}.

- On the segment $M_2 M_3$, the polynomial p_1, whose trace is a trinomial of second degree in the variable parameterizing the segment $M_2 M_3$, is identically zero, since it vanishes at the three points M_2, M_3, and M_{23}.

But since the segment $M_2 M_3$ is characterized by the equation $\lambda_1 = 0$, this means that the barycentric function λ_1 is a factor in the expression for the polynomial p_1.

- Similarly, the polynomial p_1 is zero at the nodes M_{13} and M_{12}. Since the barycentric functions λ_i are affine in x and in y, at these two nodes λ_1 is equal to $1/2$ on the segment $M_{13} M_{12}$.

In other words, factoring the quantity $\lambda_1 - 1/2$ from p_1, we ensure that p_1 is indeed zero at the nodes M_{13} and M_{12}.

The polynomial structure of the function p_1 is thus

$$p_1(M) = \alpha \lambda_1(M) \left[\lambda_1(M) - \frac{1}{2} \right], \tag{2.125}$$

where α is a constant to be determined such that the polynomial p_1 takes the value 1 at its characteristic node, i.e., at the node M_1.

Fig. 2.4 Triangular element for the P_2 finite-element analysis

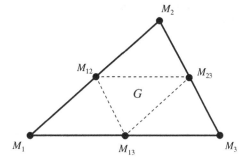

Note also that the expression (2.125) indeed gives p_1 the structure of a second-degree polynomial in the pair of variables (x, y), since the polynomial λ_1 is of first degree in the pair (x, y).

We can thus write

$$p_1(M_1) \equiv \alpha\,\lambda_1(M_1)\left[\lambda(M_1) - \frac{1}{2}\right] = \frac{1}{2}\alpha, \qquad (2.126)$$

which ensures that $p_1(M_1) = 1$, whereupon we deduce that $\alpha = 2$.

Finally, the polynomial p_1 takes the form

$$p_1(M) = \lambda_1(M)\big[2\lambda_1(M) - 1\big]. \qquad (2.127)$$

The other polynomials in the canonical basis are obtained by the same procedure.

The results are as follows:

$$p_1(M) = \lambda_1(M)(2\lambda_1(M) - 1), \qquad p_2(M) = \lambda_2(M)\big[2\lambda_2(M) - 1\big], \quad (2.128)$$

$$p_3(M) = \lambda_3(M)\big[2\lambda_3(M) - 1\big], \qquad p_{12}(M) = 4\lambda_1\lambda_2(M), \qquad (2.129)$$

$$p_{13}(M) = 4\lambda_1\lambda_3(M), \qquad\qquad p_{23}(M) = 4\lambda_2\lambda_3(M). \qquad (2.130)$$

● **P_3 Finite-Element Analysis**

1. The approximation space $P(G) \equiv P_3$ comprises the polynomial functions of degree less than or equal to three for the pair of variables (x, y).

In other words, every function p in P_3 can be written in the form

$$p(x, y) = ax^3 + by^3 + cx^2y + dxy^2 + ex^2 + fy^2 + gxy + hx + iy + j, \quad (2.131)$$

where $(a, b, c, d, e, f, g, h, i, j)$ is an arbitrary point in $\mathbb{R}^{10}$.

Fig. 2.5 Triangular element
for the P_3 finite-element
analysis

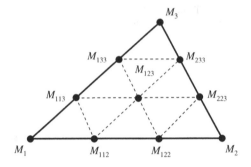

The definition (2.131) implies that the dimension of the space P_3 is 10.

2. To define the 10 linear forms σ_i, $(i = 1, \ldots, 10)$, we introduce seven further nodes M_{112}, M_{122}, M_{113}, M_{133}, M_{223}, M_{233}, and M_{123} a third of the way along each side of the triangle G, as shown in Fig. 2.5.

The 10 linear forms are then defined as follows:

$$\sigma_1 : p \to p(M_1), \qquad \sigma_2 : p \to p(M_2), \tag{2.132}$$

$$\sigma_3 : p \to p(M_3), \qquad \sigma_4 : p \to p(M_{112}), \tag{2.133}$$

$$\sigma_5 : p \to p(M_{122}), \qquad \sigma_6 : p \to p(M_{223}), \tag{2.134}$$

$$\sigma_7 : p \to p(M_{233}), \qquad \sigma_8 : p \to p(M_{113}), \tag{2.135}$$

$$\sigma_9 : p \to p(M_{133}), \qquad \sigma_{10} : p \to p(M_{123}). \tag{2.136}$$

3. The 10 functions p_i, $(i = 1, \ldots, 10)$, of the canonical basis are determined by the same kind of arguments as were used for the P_2 triangular finite-element analysis.

Consider again the polynomial p_1 characterizing the node M_1, i.e., satisfying $p_1(M_1) = 1$. Since it must vanish at the other nine nodes, we deduce the following factors:

- λ_1 is a factor of p_1, because this polynomial must vanish at the nodes M_2, M_3, M_{223}, and M_{233}.

- $(\lambda_1 - 2/3)$ must be a factor of p_1, because it vanishes at the nodes M_{112} and M_{113}.

- $(\lambda_1 - 1/3)$ must be a factor p_1, because it vanishes at the nodes M_{122}, M_{133}, and M_{123}.

The polynomial p_1 thus has the form

$$p_1(M) = \alpha \, \lambda_1(M) \left[\lambda_1(M) - \frac{1}{3} \right] \left[\lambda_1(M) - \frac{2}{3} \right], \tag{2.137}$$

where once again, the constant α is adjusted so that the polynomial p_1 is equal to 1 at the node M_1.

Note also that the form of the polynomial p_1 defined by (2.137) is consistent with the definition (2.131) of functions belonging to P_3, since the barycentric function λ_1 is a first-degree polynomial in the pair of variables (x, y).

It is straightforward to show that $\alpha = 9/2$, and the final form of p_1 is thus

$$p_1(M) = \frac{9}{2}\lambda_1(M)\left[\lambda_1(M) - \frac{1}{3}\right]\left[\lambda_1(M) - \frac{2}{3}\right]. \tag{2.138}$$

By the obvious symmetry, the polynomials p_2 and p_3 can be deduced immediately from the expression for p_1, giving

$$p_2(M) = \frac{9}{2}\lambda_2(M)\left[\lambda_2(M) - \frac{1}{3}\right]\left[\lambda_2(M) - \frac{2}{3}\right], \tag{2.139}$$

$$p_3(M) = \frac{9}{2}\lambda_3(M)\left[\lambda_3(M) - \frac{1}{3}\right]\left[\lambda_3(M) - \frac{2}{3}\right]. \tag{2.140}$$

Now consider the polynomial p_{112}. We can immediately extract certain factors:

- λ_1 is a factor in the expression for p_{112}, because this polynomial must vanish at the nodes M_2, M_3, M_{223}, and M_{233}.

- λ_2 must be a factor of p_{112}, because it must vanish at the nodes M_1, M_3, M_{113}, and M_{133}.

- $(\lambda_1 - 1/3)$ must be a factor of p_1, because it must vanish at the nodes M_{122}, M_{133}, and M_{123}.

Therefore, p_{112} has the structure

$$p_{112}(M) = \beta\,\lambda_1(M)\lambda_2(M)\left[\lambda_1(M) - \frac{1}{3}\right], \tag{2.141}$$

where the constant β is adjusted so that p_{112} is equal to 1 at the node M_{112}.
Since $\lambda_1 = 2/3$ and $\lambda_2 = 1/3$ at the node M_{112}, we obtain

$$\beta = \frac{27}{2}. \tag{2.142}$$

Finally, the basis function p_{112} has the form

$$p_{112}(M) = \frac{27}{2}\lambda_1(M)\lambda_2(M)\left[\lambda_1(M) - \frac{1}{3}\right]. \tag{2.143}$$

Once again, by the symmetry of the situation, the other basis functions p_{ijk}, for triplets (i, j, k) differing from $(1, 2, 3)$, are obtained immediately:

$$p_{122}(M) = \frac{27}{2}\lambda_1(M)\lambda_2(M)\left[\lambda_2(M) - \frac{1}{3}\right], \qquad (2.144)$$

$$p_{113}(M) = \frac{27}{2}\lambda_1(M)\lambda_3(M)\left[\lambda_1(M) - \frac{1}{3}\right], \qquad (2.145)$$

$$p_{133}(M) = \frac{27}{2}\lambda_1(M)\lambda_3(M)\left[\lambda_3(M) - \frac{1}{3}\right], \qquad (2.146)$$

$$p_{223}(M) = \frac{27}{2}\lambda_2(M)\lambda_3(M)\left[\lambda_2(M) - \frac{1}{3}\right], \qquad (2.147)$$

$$p_{233}(M) = \frac{27}{2}\lambda_1(M)\lambda_3(M)\left[\lambda_3(M) - \frac{1}{3}\right]. \qquad (2.148)$$

We end by examining the last polynomial function in the canonical basis of P_3, viz., p_{123}.

This polynomial contains the following factors:

- λ_1 is a factor in the expression for p_{123}, because this polynomial must vanish at the nodes M_2, M_3, M_{223}, and M_{233}.

- λ_2 is a factor of p_{123}, because it must vanish at the nodes M_1, M_3, M_{113}, and M_{133}.

- λ_3 is a factor of p_{123}, because it must vanish at the nodes M_1, M_2, M_{112}, and M_{122}.

The function p_{123} thus has the following polynomial structure:

$$p_{123}(M) = \gamma\lambda_1(M)\lambda_2(M)\lambda_3(M), \qquad (2.149)$$

where the constant γ is adjusted so that the polynomial p_{123} satisfies the characteristic property at the node M_{123}, namely $p_{123}(M_{123}) = 1$.

Since the barycentric functions λ_1, λ_2, and λ_3 are all equal to 1/3 at the node M_{123}, we clearly obtain

$$\gamma = 27. \qquad (2.150)$$

Finally, the polynomial p_{123} has the form

$$p_{123}(M) = 27\lambda_1(M)\lambda_2(M)\lambda_3(M). \qquad (2.151)$$

Quadrilateral Meshes

In this section, we discuss finite-element methods in which the geometry G of the primitive mesh element is a square $[0, 1] \times [0, 1]$ with vertices M_1, M_2, M_3, and M_4 in the plane $(O; x, y)$, as shown in Fig. 2.6.

Fig. 2.6 Square primitive
element for plane finite-
element analysis

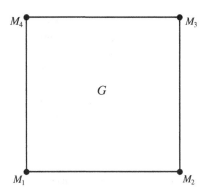

- ### Q_1 Finite-Element Analysis

1. The space $P(G) \equiv Q_1$ is defined as the set of polynomials of degree less than
   or equal to 1 in each of the two variables x and y. So every function p in Q_1 has
   the form

$$p(x,y) = axy + bx + cy + d, \tag{2.152}$$

   where (a, b, c, d) runs over $\mathbb{R}^4$. By simple inspection of the definition (2.152),
   we see that Q_1 has dimension 4.

2. We introduce the four linear forms defined by

$$\sigma_i : p \longrightarrow p(M_i), \quad \forall i = 1, \ldots, 4. \tag{2.153}$$

3. To determine the four canonical basis functions p_i, $(i = 1, \ldots, 4)$, of the space
   Q_1, we recall that these functions must satisfy the definition:

$$\sigma_j(p_i) = p_i(M_j) = \delta_{ij}$$

So each of the canonical basis functions characterizes a single vertex of the square
G, taking the value 1 at this vertex and 0 at all the others. We use this fact to
examine the factorization properties of these functions.

Consider, for example, the polynomial p_1, which has the following properties:

- Since p_1 vanishes on the segment M_2M_3 parameterized by $x = 1$, the monomial $(x - 1)$ must be a factor in the expression for p_1.

- Since p_1 vanishes on the segment M_3M_4 parameterized by $y = 1$, the monomial $(y - 1)$ must also be a factor in the expression for p_1.

Hence the canonical basis function p_1 must have the structure

$$p_1(x,y) = \alpha(x - 1)(y - 1), \tag{2.154}$$

where the constant α is determined as usual in such a way that $p_1(M_1) = 1$. We thus find that the coefficient α is equal to 1 and the function p_1 in the canonical basis is given by

$$p_1(x,y) = (x-1)(y-1). \qquad (2.155)$$

By analogous reasoning, the three other canonical basis functions of Q_1 are given by

$$p_2(x,y) = x(1-y), \qquad p_3(x,y) = xy, \qquad p_4(x,y) = y(1-x). \quad (2.156)$$

- **Q_2 Finite-Element Analysis**

1. The space $P(G) \equiv Q_2$ is the set of polynomials of degree less than or equal to two in each of the variables x and y. So every function p in Q_2 can be written in the form

$$p(x,y) = ax^2y^2 + bx^2y + cxy^2 + dx^2 + ey^2 + fxy + gx + hy + i, \quad (2.157)$$

where $(a, b, c, d, e, f, g, h, i)$ runs over $\mathbb{R}^9$. The definition (2.157) implies that Q_2 has dimension 9.

2. To define the nine linear forms σ_i, we introduce five further discretization nodes M_5, M_6, M_7, M_8, and M_9, where the first four of these lie in the middle of each side of the square G, and M_9 lies in the middle of the square, as shown in Fig. 2.7.

The nine linear forms $\sigma_i, i = 1, \ldots, 9$ are then defined by

$$\sigma_i : p \longrightarrow p(M_i), \quad \forall i = 1, \ldots, 9. \qquad (2.158)$$

3. We apply exactly the same logic as for the Q_1 quadrilateral finite-element analysis, obtaining the following nine canonical basis functions p_i, $(i = 1, \ldots, 9)$:

$$
\begin{aligned}
p_1(x, y) &= (1-x)(1-2x)(1-y)(1-2y), \\
p_2(x, y) &= x(2x-1)(1-y)(1-2y), \\
p_3(x, y) &= xy(2x-1)(2y-1), \\
p_4(x, y) &= y(1-x)(1-2x)(2y-1), \\
p_5(x, y) &= 4x(1-x)(1-y)(1-2y), \\
p_6(x, y) &= 4xy(2x-1)(1-y), \\
p_7(x, y) &= 4xy(1-x)(2y-1), \\
p_8(x, y) &= 4y(1-x)(1-2x)(1-y), \\
p_9(x, y) &= 16xy(1-x)(1-y).
\end{aligned}
\qquad (2.159)
$$

Fig. 2.7 Square mesh element
for the Q_2 finite-element
analysis

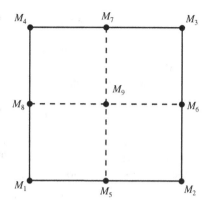

Fig. 2.8 Cubic mesh element
for finite-element analysis in
3-dimensional space

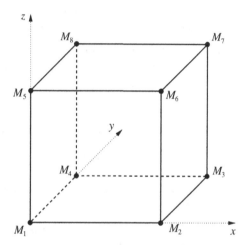

2.8.3 Finite-Element Methods in Three Spatial Dimensions

Cubic Meshes

In this last section, we discuss a finite-element analysis in which the mesh element
G is a cube $[0, 1] \times [0, 1] \times [0, 1]$ with vertices M_i, $(i = 1, \ldots, 8)$, in the space
$(O; x, y, z)$ (see Fig. 2.8).

1. The space Q_1 is defined as the set of polynomials of degree less than or equal to
   1 in each of the variables x, y, and z. So every function p in Q_1 can be expressed
   in the form

$$p(x,y,z) = axyz + bxy + cxz + dyz + ex + fy + gz + h, \qquad (2.160)$$

where (a, b, c, d, e, f, g, h) runs over $\mathbb{R}^8$. By simple inspection of the definition
(2.160), we see that Q_1 has dimension 8.

2. We introduce the eight linear forms defined by

$$\sigma_i : p \longrightarrow p(M_i), \quad \forall i = 1, \dots, 8. \qquad (2.161)$$

3. As usual, the eight canonical basis functions p_i, $(i = 1, \dots, 8)$, have to satisfy the
relations $\sigma_j(p_i) \equiv p_i(M_j) = \delta_{ij}$. To construct these, we identify the monomial
factors in the expressions for each.

Consider, for example, the polynomial p_1 characterizing the node M_1, where it
takes the value unity, and which is zero at the seven other nodes M_i, $(i = 2, \dots, 8)$.

- The monomial $(1 - x)$ must be a factor in the expression for p_1 so that it
  vanishes at the nodes M_2, M_3, M_6, and M_7.

- The monomial $(1 - y)$ must be a factor in the expression for p_1 so that it
  vanishes at the nodes M_3, M_4, M_7, and M_8.

- The monomial $(1 - z)$ must be a factor in the expression for p_1 so that it
  vanishes at the nodes M_5, M_6, M_7, and M_8.

The basic function p_1 thus has the structure

$$p_1(x,y,z) = \alpha(1 - x)(1 - y)(1 - z), \qquad (2.162)$$

where the constant α is adjusted so that p_1 takes the value unity at its characteristic
node M_1.

Now at the node M_1, $x = y = z = 0$, which implies that α is equal to 1 and the
function p_1 has the final form

$$p(x,y,z) = (1 - x)(1 - y)(1 - z). \qquad (2.163)$$

By analogous reasoning, we obtain each polynomial in the canonical basis of Q_1:

$$p_1(x, y, z) = (1 - x)(1 - y)(1 - z), \quad p_2(x, y, z) = x(1 - y)(1 - z),$$
$$p_3(x, y, z) = xy(1 - z), \qquad\qquad p_4(x, y, z) = (1 - x)y(1 - z),$$
$$p_5(x, y, z) = (1 - x)(1 - y)z, \qquad p_6(x, y, z) = x(1 - y)z,$$
$$p_7(x, y, z) = xyz, \qquad\qquad\qquad p_8(x, y, z) = (1 - x)yz.$$
$$\qquad (2.164)$$

References

1. G. Duvaut, *Mécanique des milieux continus* (Dunod, Paris, 1998)
2. H. Brézis, *Analyse fonctionnelle, théorie et applications* (Masson, Paris, 1983)
3. R. Dautray, J.-L. Lions, *Analyse mathématique et calcul numérique pour les sciences et les techniques* (Masson, Paris, 1987)
4. M. Moussaoui, in *Singularities and Constructive Methods for Their Treatment*, ed. by P. Grisvard, W. Wendland, J.R. Whiteman. Sur l'approximation des solutions du problème de Dirichlet dans un ouvert avec coins. Lecture Notes in Mathematics, vol. 1121 (Springer, Berlin, 1984), p. 136
5. D. Euvrard, *Résolution des équations aux dérivées partielles de la physique, de la mécanique et des sciences de l'ingénieur* (Masson, Paris, 1994)

Part II
Worked Problems

Chapter 3
Variational Formulations of Problems with Elliptical Boundary Conditions

3.1 Heat Transfer with Mixed Boundary Conditions

3.1.1 Statement of the Problem

Under certain conditions, the heat exchange between a fluid occupying a bounded open subset Ω of $\mathbb{R}^3$ with sufficiently regular boundary $\partial\Omega$, subdivided into three distinct parts Γ_i, $(i = 1, 2, 3)$, $\partial\Omega = \Gamma_1 \cup \Gamma_2 \cup \Gamma_3$, and the surrounding medium can be described by the following mathematical formulation.

Let $u \in H^2(\Omega)$ be a real-valued function on Ω that describes the temperature field and solves the continuous problem (**CP**):

$$(\text{CP}) \begin{cases} -\Delta u = f \text{ in } \Omega, & (3.1) \\[2mm] u = 0, \text{ on } \Gamma_1, & (3.2) \\[2mm] \dfrac{\partial u}{\partial n} = g_2, \text{ on } \Gamma_2, & (3.3) \\[2mm] \dfrac{\partial u}{\partial n} + ku = g_3, \text{ on } \Gamma_3, & (3.4) \end{cases}$$

where f, g_2, and g_3 are three given functions in $L^2(\Omega)$, $L^2(\Gamma_2)$, and $L^2(\Gamma_3)$, respectively, and k is a given real positive parameter.

Recall that the space $H^2(\Omega)$ is defined by

$$H^2(\Omega) = \left\{ v : \Omega \rightarrow \mathbb{R}, \ v \in L^2(\Omega), \ \frac{\partial v}{\partial x_i} \in L^2(\Omega), \ \frac{\partial^2 v}{\partial x_i \partial x_j} \in L^2(\Omega) \right\},$$

where $(i, j) \in \{1, 2, 3\}$.

J. Chaskalovic, *Mathematical and Numerical Methods for Partial Differential Equations*, Mathematical Engineering, DOI: 10.1007/978-3-319-03563-5_3, © Springer International Publishing Switzerland 2014

▶ **Variational Formulation**

1) Show that if u solves the continuous problem **(CP)**, then u is also a solution of the variational problem **(VP)**:

$$(\textbf{VP}) \quad \begin{cases} \text{Find } u \in V \text{ solution of} \\ a(u, v) = L(v), \quad \forall v \in V, \end{cases} \tag{3.5}$$

where the space V is defined by

$$V = H^1_{\Gamma_1}(\Omega) \equiv \left\{ v : \Omega \to \mathbb{R}, \ v \in H^1(\Omega), \ v = 0 \text{ on } \Gamma_1 \right\},$$

equipped with the usual norm on the space $H^1(\Omega)$, viz.,

$$\|v\|^2_{H^1(\Omega)} = \int_\Omega v^2 \, d\Omega + \int_\Omega |\nabla v|^2 \, d\Omega. \tag{3.6}$$

The forms $a(., .)$ and $L(.)$ are defined by

$$a(u, v) = \int_\Omega \nabla u \cdot \nabla v \, d\Omega + k \int_{\Gamma_3} uv \, d\Gamma, \tag{3.7}$$

$$L(v) = \int_\Omega fv \, d\Omega + \int_{\Gamma_2} g_2 v \, d\Gamma + \int_{\Gamma_3} g_3 v \, d\Gamma. \tag{3.8}$$

Show that the integrals converge and justify the boundary condition on Γ_1 for the functions v in V.

▶ **Existence and Uniqueness of the Solution to the Variational Formulation. A Priori Estimation**

2) Show that the space $H^1_{\Gamma_1}(\Omega)$ is a Hilbert space for the norm $\| \cdot \|_{H^1(\Omega)}$.

3) Show that $a(., .)$ is a symmetric, continuous bilinear form on $V \times V$.

4) Show that the bilinear form $a(., .)$ is coercive on $V \times V$.

5) Show that the linear form $L(.)$ is continuous on V.

6) Establish existence and uniqueness of the solution to the variational problem **(VP)**.

7) Show directly, without using the Lax–Milgram theorem, that the solution to the variational problem is unique, then that this solution exists. Hint: show that $\sqrt{a(v, v)}$ is a norm equivalent to the H^1 norm on $H^1_{\Gamma_1}(\Omega)$, then use the Riesz theorem.

8) Show that the variational problem is equivalent to the following minimization problem:

Find $u \in H^1_{\Gamma_1}(\Omega)$ such that $I(u) \leq I(v), \quad \forall v \in H^1_{\Gamma_1}(\Omega)$,

where

$$I(v) = \frac{1}{2}a(v, v) - L(v).$$

9) *A priori* estimation. Show that if u solves **(VP)**, then there is a constant $C > 0$ such that

$$\|u\|_{H^1(\Omega)} \leq C\left[\|f\|_{L^2(\Omega)} + \|g_2\|_{L^2(\Gamma_2)} + \|g_3\|_{L^2(\Gamma_3)}\right].$$

10) Demonstrate equivalence of the continuous problem **(CP)** and the variational problem **(VP)**.

3.1.2 Solution

1) Let v be a real-valued test function defined on Ω and belonging to the function space V. Multiplying (3.1) by v and integrating over Ω, we obtain

$$-\int_{\Omega} \Delta u \cdot v \, d\Omega = \int_{\Omega} f v \, d\Omega, \quad \forall v \in V. \tag{3.9}$$

As usual, we use Green's formula (1.119) in order to bring in the boundary conditions (3.2)–(3.4) given in the statement of the problem **(CP)**:

$$\int_{\Omega} \nabla u \cdot \nabla v \, d\Omega - \int_{\partial\Omega} \frac{\partial u}{\partial n} v \, d\Gamma = \int_{\Omega} f v \, d\Omega, \quad \forall v \in V. \tag{3.10}$$

The surface integral over $\partial\Omega$ splits into three terms, one for each subset Γ_1, Γ_2, and Γ_3 of $\partial\Omega$, so that with the boundary conditions (3.2)–(3.4), we obtain

$$\int_{\partial\Omega} \frac{\partial u}{\partial n} v \, d\Gamma = \int_{\Gamma_1} \frac{\partial u}{\partial n} v \, d\Gamma + \int_{\Gamma_2} g_2 v \, d\Gamma + \int_{\Gamma_3} (-ku + g_3) v \, d\Gamma. \tag{3.11}$$

Hence, in order to incorporate the three boundary conditions (3.2)–(3.4), and in particular the homogeneous Dirichlet condition (3.2) on Γ_1, into the future variational formulation **(VP)**, we require the test functions v to satisfy this boundary condition:

$$v = 0 \text{ on } \Gamma_1. \tag{3.12}$$

In this case, the variational Eq. (3.10) becomes

$$\int_{\Omega} \nabla u \cdot \nabla v \, d\Omega - \int_{\Gamma_2} g_2 v \, d\Gamma - \int_{\Gamma_3} (-ku + g_3) v \, d\Gamma = \int_{\Omega} f v \, d\Omega, \quad \forall v \in V. \tag{3.13}$$

The variational problem **(VP)** then takes the following form:

$$\textbf{(VP)} \quad \begin{cases} \text{Find } u \in V \text{ solution of } a(u, v) = L(v), \quad \forall v \in V, \text{ with} \\[2mm] a(u, v) = \displaystyle\int_{\Omega} \nabla u \cdot \nabla v \, d\Omega + \int_{\Gamma_3} kuv \, d\Gamma, \\[2mm] L(v) = \displaystyle\int_{\Omega} f v \, d\Omega + \int_{\Gamma_2} g_2 v \, d\Gamma + \int_{\Gamma_3} g_3 v \, d\Gamma. \end{cases} \tag{3.14}$$

Definition of the Space V. The functional context V in which we shall seek the solution u of the variational formulation **(VP)** is defined in such a way as to guarantee convergence of the various integrals it involves:

- **Convergence of** $\displaystyle\int_{\Omega} \nabla u \cdot \nabla v \, d\Omega$.

We note that

$$\left| \int_{\Omega} \nabla u \cdot \nabla v \, d\Omega \right| \leq \|\nabla u\|_{L^2(\Omega)} \|\nabla v\|_{L^2(\Omega)} \leq \|u\|_{H^1(\Omega)} \|v\|_{H^1(\Omega)}, \qquad (3.15)$$

where we have used the Cauchy–Schwarz inequality. Hence, if u and v belong to $H^1(\Omega)$, then the integral $\displaystyle\int_{\Omega} \nabla u \cdot \nabla v \, d\Omega$ will converge.

- **Convergence of** $\displaystyle\int_{\Gamma_3} kuv \, d\Gamma$.

This time, we have

$$\left| \int_{\Gamma_3} kuv \, d\Gamma \right| \leq k\|u\|_{L^2(\Gamma_3)} \|v\|_{L^2(\Gamma_3)} \leq k\|u\|_{L^2(\partial\Omega)} \|v\|_{L^2(\partial\Omega)} \qquad (3.16)$$

$$\leq kC^2 \|u\|_{H^1(\Omega)} \|v\|_{H^1(\Omega)},$$

where the constant C is the one introduced in the trace theorem (Theorem 1.3), which is valid for all functions in $H^1(\Omega)$.

So, if u and v are two functions in $H^1(\Omega)$, then the integral $\displaystyle\int_{\Gamma_3} kuv \, d\Gamma$ will also converge.

- **Convergence of** $\displaystyle\int_{\Omega} fv \, d\Omega$.

Here we have

$$\left| \int_{\Omega} fv \, d\Omega \right| \leq \|f\|_{L^2(\Omega)} \|v\|_{L^2(\Omega)} \leq \|f\|_{L^2(\Omega)} \|v\|_{H^1(\Omega)}. \qquad (3.17)$$

Once again, it is enough to insist that v belong to $H^1(\Omega)$ in order to ensure convergence of $\displaystyle\int_{\Omega} fv \, d\Omega$.

- **Convergence of** $\displaystyle\int_{\Gamma_2} g_2 v \, d\Gamma$.

In this case, we write

$$\left| \int_{\Gamma_2} g_2 v \, d\Gamma \right| \leq \|g_2\|_{L^2(\Gamma_2)} \|v\|_{L^2(\Gamma_2)} \qquad (3.18)$$

$$\leq \|g_2\|_{L^2(\Gamma_2)} \|v\|_{L^2(\partial\Omega)} \leq C\|g_2\|_{L^2(\Gamma_2)} \|v\|_{H^1(\Omega)}.$$

The condition that v belong to $H^1(\Omega)$ implies convergence of $\int_{\Gamma_2} g_2 v \, d\Gamma$.

• **Convergence of** $\int_{\Gamma_3} g_3 v \, d\Gamma$.

The bound on this integral is established as for the last one, simply replacing g_2 by g_3 and Γ_2 by Γ_3.

In conclusion, the space $V = H^1_{\Gamma_1}(\Omega)$ defined by

$$H^1_{\Gamma_1}(\Omega) \equiv \left\{ v : \Omega \to \mathbb{R}, \ v \in H^1(\Omega), \ v = 0 \text{ on } \Gamma_1 \right\}$$

ensures the existence of all the integrals in the variational formulation **(VP)**, while including also the homogeneous Dirichlet condition on the region Γ_1 of the boundary, whence the formulation **(VP)** contains all the information present in the continuous problem **(CP)**.

2) We use Lemma 1.8 to show that the space $H^1_{\Gamma_1}(\Omega)$ is a Hilbert space for the norm induced from the Sobolev space $H^1(\Omega)$. To this end, we show that the vector subspace $H^1_{\Gamma_1}(\Omega)$ of $H^1(\Omega)$ is closed in $H^1(\Omega)$.

Therefore, consider an arbitrary sequence $(v_n)_{n \in \mathbb{N}}$ of functions in $H^1_{\Gamma_1}(\Omega)$ that converge to some element v of $H^1(\Omega)$ for the norm $\| . \|_{H^1(\Omega)}$.

We must show that v belongs to $H^1_{\Gamma_1}(\Omega)$, i.e., that the trace γ_{Γ_1} of v on Γ_1 is zero: $\gamma_{\Gamma_1}(v) = 0$.

Now,

$$\left\| \gamma_{\Gamma_1}(v) \right\|_{L^2(\partial\Omega)} = \left\| \gamma_{\Gamma_1}(v - v_n) \right\|_{L^2(\partial\Omega)}, \tag{3.19}$$

since the norm $L^2(\partial\Omega)$ of $\gamma_{\Gamma_1}(v_n)$ is zero whenever the sequence $(v_n)_{n \in \mathbb{N}}$ belongs to $H^1_{\Gamma_1}(\Omega)$. Equation (3.19) then gives the following bound on $\gamma_{\Gamma_1}(v)$ in the $L^2(\partial\Omega)$ norm:

$$\left\| \gamma_{\Gamma_1}(v) \right\|_{L^2(\partial\Omega)} \leq \| \gamma_0(v - v_n) \|_{L^2(\partial\Omega)} \tag{3.20}$$

$$\leq C \| v - v_n \|_{H^1(\Omega)},$$

where γ_0 denotes the trace map on the whole boundary $\partial\Omega$ of Ω, as introduced in Sect. 1.5.3. We also used the fact that $\gamma_{\Gamma_1}(v_n) = 0$, since v_n belongs to $H^1_{\Gamma_1}(\Omega)$, and the trace theorem (Theorem 1.3) in $H^1(\Omega)$.

Taking the limit in the inequality (3.20), we deduce that the trace of v on Γ_1 is zero, since the $L^2(\partial\Omega)$ norm of $\gamma_{\Gamma_1}(v)$ is itself zero, i.e., $\gamma_{\Gamma_1}(v) = 0$.

Hence, $H^1_{\Gamma_1}(\Omega)$ is a closed vector subspace of $H^1(\Omega)$ for the norm $\| . \|_{H^1(\Omega)}$, and this, in turn, implies that it is a Hilbert space for this same norm and its associated scalar product $(., .)_{H^1(\Omega)}$ (see Lemma 1.17).

3) In this question, we establish the properties of the form $a(.,.)$ defined by

$$a(u, v) = \int_\Omega \nabla u \cdot \nabla v \, d\Omega + \int_{\Gamma_3} kuv \, d\Gamma. \tag{3.21}$$

- From the definition of the space V, chosen to be $H^1_{\Gamma_1}(\Omega)$, $a(.,.)$ is a well-defined real-valued map on $V \times V$. In other words, the integrals in (3.21) are convergent when u and v belong to V.

- Given the structure of $a(.,.)$, it is obviously a symmetric bilinear form.

- Continuity of $a(.,.)$ is demonstrated as follows. For all $(u, v) \in V \times V$, we have

$$|a(u, v)| = \left| \int_\Omega \nabla u \cdot \nabla v \, d\Omega + \int_{\Gamma_3} kuv \, d\Gamma \right|, \tag{3.22}$$

whence

$$|a(u, v)| \leq \|\nabla u\|_{L^2(\Omega)} \|\nabla v\|_{L^2(\Omega)} + k\|u\|_{L^2(\Gamma_3)} \|v\|_{L^2(\Gamma_3)} \tag{3.23}$$

$$\leq \|\nabla u\|_{L^2(\Omega)} \|\nabla v\|_{L^2(\Omega)} + k\|u\|_{L^2(\partial\Omega)} \|v\|_{L^2(\partial\Omega)}$$

$$\leq \|u\|_{H^1(\Omega)} \|v\|_{H^1(\Omega)} + kC^2 \|u\|_{H^1(\Omega)} \|v\|_{H^1(\Omega)}$$

$$\leq (1 + kC^2) \|u\|_{H^1(\Omega)} \|v\|_{H^1(\Omega)}.$$

Here we have once again used the Cauchy–Schwarz inequality, and also the continuity of the trace map applied to a function in $H^1(\Omega)$ (see Theorem 1.3).

We conclude that there exists $M > 0$ defined by $M \equiv 1 + kC^2$ such that

$$\forall (u, v) \in V \times V, \quad |a(u, v)| \leq M\|u\|_{H^1(\Omega)} \|v\|_{H^1(\Omega)}, \tag{3.24}$$

so the form $a(.,.)$ is indeed well defined and continuous on $V \times V$.

4) We must show that the bilinear form $a(.,.)$ is coercive, i.e., that there is $\alpha_0 > 0$ such that

$$\forall v \in V, \quad a(v, v) \geq \alpha_0 \|v\|^2_{H^1(\Omega)}. \tag{3.25}$$

Now for any function v in $H^1_{\Gamma_1}(\Omega)$, we have the Poincaré inequality (see the remarks following Lemma 1.22), which gives

$$\exists C > 0 \text{ such that } \forall v \in H^1_{\Gamma_1}(\Omega), \quad \int_\Omega v^2 \, d\Omega \leq C \int_\Omega |\nabla v|^2 \, d\Omega. \tag{3.26}$$

In terms of norms, this becomes

$$\exists\, C > 0 \text{ such that } \forall\, v \in H^1_{\Gamma_1}(\Omega), \quad \|v\|^2_{L^2(\Omega)} \le C \,\|\nabla v\|^2_{L^2(\Omega)}. \tag{3.27}$$

Adding $\|\nabla v\|^2_{L^2(\Omega)}$ to both sides of (3.27), we then bring in the norm $\| \cdot \|^2_{H^1(\Omega)}$:

$$\forall\, v \in H^1_{\Gamma_1}(\Omega), \ \|v\|^2_{H^1(\Omega)} \equiv \|v\|^2_{L^2(\Omega)} + \|\nabla v\|^2_{L^2(\Omega)} \le (1 + C)\,\|\nabla v\|^2_{L^2(\Omega)}. \tag{3.28}$$

We thus obtain

$$\forall\, v \in H^1_{\Gamma_1}(\Omega), \quad \|\nabla v\|^2_{L^2(\Omega)} \ge \frac{1}{(1+C)} \|v\|^2_{H^1(\Omega)}. \tag{3.29}$$

In this case, the lower bound on $a(v, v)$ is found from

$$a(v, v) \equiv \|\nabla v\|^2_{L^2(\Omega)} + k\|v\|^2_{L^2(\Gamma_3)} \ge \|\nabla v\|^2_{L^2(\Omega)} \ge \frac{1}{(1+C)} \|v\|^2_{H^1(\Omega)}. \tag{3.30}$$

The coercivity constant α_0 is thus given by $\alpha_0 \equiv 1/(1 + C)$.

5) This question deals with the properties of the form $L(.)$ defined by

$$L(v) = \int_\Omega f v \, d\Omega + \int_{\Gamma_2} g_2 v \, d\Gamma + \int_{\Gamma_3} g_3 v \, d\Gamma. \tag{3.31}$$

- From the investigation in the first question about the convergence of the integrals defining $L(.)$ and the choice of function space $H^1_{\Gamma_1}(\Omega)$, we know that $L(.)$ is a suitably defined form on $H^1_{\Gamma_1}(\Omega)$.

- Linearity of the form $L(.)$ is immediate from the structure of the definition.

- Continuity of the form $L(.)$ is then shown as follows:

$$|L(v)| = \left| \int_\Omega f v \, d\Omega + \int_{\Gamma_2} g_2 v \, d\Gamma + \int_{\Gamma_3} g_3 v \, d\Gamma \right| \tag{3.32}$$

$$\le \|f\|_{L^2(\Omega)} \|v\|_{L^2(\Omega)} + \|g_2\|_{L^2(\Gamma_2)} \|v\|_{L^2(\Gamma_2)} + \|g_3\|_{L^2(\Gamma_3)} \|v\|_{L^2(\Gamma_3)}$$

$$\le \|f\|_{L^2(\Omega)} \|v\|_{L^2(\Omega)} + \left[\|g_2\|_{L^2(\Gamma_2)} + \|g_3\|_{L^2(\Gamma_3)} \right] \|v\|_{L^2(\Gamma)}$$

$$\le \left\{ \|f\|_{L^2(\Omega)} + C \left[\|g_2\|_{L^2(\Gamma_2)} + \|g_3\|_{L^2(\Gamma_3)} \right] \right\} \|v\|_{H^1(\Omega)}.$$

To obtain this last bound, we used the Cauchy–Schwarz inequality and the continuity of the trace map define in Theorem 1.3 for every function in $H^1(\Omega)$.

So if we define the constant M by

$$M \equiv \|f\|_{L^2(\Omega)} + C\left[\|g_2\|_{L^2(\Gamma_2)} + \|g_3\|_{L^2(\Gamma_3)}\right],$$

the inequality (3.32) becomes

$$\forall v \in H^1_{\Gamma_1}(\Omega), \quad |L(v)| \le M\|v\|_{H^1(\Omega)}, \tag{3.33}$$

and the form $L(.)$ is indeed continuous for the norm $\| \cdot \|_{H^1(\Omega)}$.

6) All the premises of the Lax–Milgram theorem (Theorem 1.11) are satisfied, i.e.,

- $\left(H^1_{\Gamma_1}(\Omega), \| \cdot \|_{H^1(\Omega)}\right)$ is a Hilbert space,

- $a(.,.)$ is a continuous coercive bilinear form on $H^1_{\Gamma_1}(\Omega) \times H^1_{\Gamma_1}(\Omega)$,

- $L(.)$ is a continuous linear form on $H^1_{\Gamma_1}(\Omega)$.

Note that all the above properties were established for the same norm, viz., the norm $\| \cdot \|_{H^1(\Omega)}$.

There is therefore one and only one solution u of the variational problem **(VP)** in $H^1_{\Gamma_1}(\Omega)$ such that

$$a(u, v) = L(v), \quad \forall v \in H^1_{\Gamma_1}(\Omega),$$

where $a(.,.)$ and $L(.)$ are defined by (3.14).

7) The existence and uniqueness of the solution u of the variational problem **(VP)** defined by (3.14) can be established directly, without appealing to the Lax–Milgram theorem:

- **Uniqueness.** Suppose that u_1 and u_2 are two solutions to the variational problem **(VP)** specified by (3.14). Then we have

$$\begin{cases} a(u_1, v) = L(v), & \forall v \in H^1_{\Gamma_1}(\Omega), \\ a(u_2, v) = L(v), & \forall v \in H^1_{\Gamma_1}(\Omega). \end{cases} \tag{3.34}$$

Subtracting one from the other and using the bilinearity of the form $a(.,.)$ and linearity of the form $L(.)$, it follows that

$$a(u_2 - u_1, v) = 0, \quad \forall v \in H^1_{\Gamma_1}(\Omega). \tag{3.35}$$

Then if we choose v equal to $u_2 - u_1$, we have

$$a(u_2 - u_1, u_2 - u_1) = 0. \tag{3.36}$$

Furthermore, $a(.,.)$ is coercive, so

$$\exists \alpha_0 > 0, \text{ such that } \forall v \in V, \quad a(v, v) \geq \alpha_0 \|v\|^2_{H^1(\Omega)}. \tag{3.37}$$

Then choosing $v = u_2 - u_1$ in (3.37), we obtain

$$a(u_2 - u_1, u_2 - u_1) \geq \alpha_0 \|u_2 - u_1\|^2_{H^1(\Omega)}. \tag{3.38}$$

Therefore, using the property (3.36) in (3.38), it follows that

$$0 \geq \alpha_0 \|u_2 - u_1\|^2_{H^1(\Omega)}. \tag{3.39}$$

Consequently, $\|u_2 - u_1\|_{H^1(\Omega)}$ is zero, and this, in turn, implies that $u_2 - u_1$ is also zero, whence u_2 must be equal to u_1.

This proves the uniqueness of the solution u to the variational problem **(VP)** specified by (3.14).

• **Existence.** In order to establish the existence of a solution u to the variational problem **(VP)** specified by (3.14), we observe that with the continuity and the coercivity of the bilinear form $a(.,.)$, we have the double inequality

$$\exists M > 0, \ \exists \alpha_0 > 0 : \forall v \in H^1_{\Gamma_1}(\Omega), 0 \leq \alpha_0 \|v\|^2_{H^1(\Omega)} \leq a(v, v) \leq M \|v\|^2_{H^1(\Omega)}. \tag{3.40}$$

We now define the map $\| . \|_1$ by

$$\| . \|_1 : H^1_{\Gamma_1}(\Omega) \longrightarrow \mathbb{R}_+$$

$$v \longmapsto \|v\|_1 \equiv \sqrt{a(v, v)}.$$

Since $a(.,.)$ is bilinear, symmetric, and coercive, hence positive definite, it defines a scalar product $(.,.)_1$ on $H^1_{\Gamma_1}(\Omega)$, viz.,

$$\forall v \in H^1_{\Gamma_1}(\Omega), \quad \|v\|_1 = \sqrt{(v, v)_1} \equiv \sqrt{a(v, v)}, \tag{3.41}$$

and the map $\| . \|_1$ is the associated norm on $H^1_{\Gamma_1}(\Omega)$.

Finally, (3.40) implies that the new norm $\| . \|_1$ is equivalent to the norm on $H^1_{\Gamma_1}(\Omega)$ induced from $H^1(\Omega)$. It follows that the space $H^1_{\Gamma_1}(\Omega)$ equipped with the norm $\| . \|_1$ is also a Hilbert space.

Since we have also shown that the form $L(.)$ is linear and continuous on $H^1_{\Gamma_1}(\Omega)$ for the norm $\| . \|_{H^1(\Omega)}$, hence for the norm $\| . \|_1$ by equivalence, the Riesz representation theorem (Theorem 1.7) for a linear form on a Hilbert space implies

$$\exists u \in H^1_{\Gamma_1}(\Omega) \text{ such that } L(v) = (u, v)_1 \equiv a(u, v), \tag{3.42}$$

and hence the existence and uniqueness of the solution u to the variational problem (**VP**) specified by (3.14) in the space $H^1_{\Gamma_1}(\Omega)$.

8) Equivalence of the variational problem (**VP**) and the minimization problem (**MP**) can be shown by direct application of Theorem 1.13 in Sect. 1.6.

Indeed, once we have established the existence and uniqueness of the variational formulation (**VP**), in particular using the Lax–Milgram theorem, then according to Theorem 1.13, we need only check that the bilinear form $a(.,.)$ is symmetric, and this is indeed the case here, so the equivalence of the two formulations follows.

9) The aim in this question is to make an *a priori* estimate of the solution u to the variational problem (**VP**) specified by (3.14). We must therefore find an upper bound for the solution u for a suitable norm in terms of the norms of the input data, viz., f, g_2, and g_3.

In the present case, since we have shown that there exists one and only one solution u to the variational problem (**VP**) specified by (3.14) in the function space $H^1_{\Gamma_1}(\Omega)$, which we have equipped with the natural norm on the Sobolev space $H^1(\Omega)$, it makes sense to find an upper bound for the norm $\|.\|_{H^1(\Omega)}$ of the solution u.

Now, if u solves the variational problem (**VP**) specified by (3.14), we have

$$a(u, v) = L(v), \quad \forall v \in H^1_{\Gamma_1}(\Omega). \tag{3.43}$$

In the special case $v = u$, (3.43) becomes

$$a(u, u) = L(u). \tag{3.44}$$

Since the bilinear form $a(.,.)$ is coercive, we thus have the inequality

$$\exists \alpha_0 > 0, \text{ such that } a(u, u) \geq \alpha_0 \|u\|^2_{H^1(\Omega)}. \tag{3.45}$$

We now proceed to bound $L(u)$ using the Cauchy–Schwarz inequality and the continuity in $H^1(\Omega)$ of the trace map defined in Theorem 1.3, whereupon

$$|L(u)| = \left| \int_\Omega fu \, d\Omega + \int_{\Gamma_2} g_2 u \, d\Gamma + \int_{\Gamma_3} g_3 u \, d\Gamma \right| \tag{3.46}$$

$$\leq \|f\|_{L^2(\Omega)}\|u\|_{L^2(\Omega)} + \|g_2\|_{L^2(\Gamma_2)}\|u\|_{L^2(\Gamma_2)} + \|g_3\|_{L^2(\Gamma_3)}\|u\|_{L^2(\Gamma_3)}$$

$$\leq \|f\|_{L^2(\Omega)}\|u\|_{L^2(\Omega)} + \left[\|g_2\|_{L^2(\Gamma_2)} + \|g_3\|_{L^2(\Gamma_3)}\right] \|u\|_{L^2(\Gamma)}$$

$$\leq \left\{ \|f\|_{L^2(\Omega)} + D\left[\|g_2\|_{L^2(\Gamma_2)} + \|g_3\|_{L^2(\Gamma_3)}\right] \right\} \|u\|_{H^1(\Omega)}.$$

Then, using the two inequalities (3.45) and (3.46) in the Eq. (3.44) satisfied by the solution u, we obtain

$$\alpha_0 \|u\|^2_{H^1(\Omega)} \leq \left\{ \|f\|_{L^2(\Omega)} + D\left[\|g_2\|_{L^2(\Gamma_2)} + \|g_3\|_{L^2(\Gamma_3)} \right] \right\} \|u\|_{H^1(\Omega)}. \quad (3.47)$$

Dividing through by the $\|.\|_{H^1(\Omega)}$ norm of the solution u, we obtain

$$\|u\|_{H^1(\Omega)} \leq \frac{1}{\alpha_0} \left\{ \|f\|_{L^2(\Omega)} + D\left[\|g_2\|_{L^2(\Gamma_2)} + \|g_3\|_{L^2(\Gamma_3)} \right] \right\} \quad (3.48)$$

$$\leq C\left[\|f\|_{L^2(\Omega)} + \|g_2\|_{L^2(\Gamma_2)} + \|g_3\|_{L^2(\Gamma_3)} \right],$$

where the constant C can be defined by

$$C \equiv \frac{\max(1, D)}{\alpha_0}. \quad (3.49)$$

10) The question of the equivalence between the so-called strong formulation or continuous problem **(CP)** and the so-called weak formulation of the variational problem **(VP)** must be treated with the utmost care, since the finite-element method produces an approximation to the variational formulation **(VP)** that we hope then to apply to the continuous problem **(CP)**.

While the sufficiency of **(CP)** is immediate in most applications, the situation is quite different for the converse, as we shall explain in more detail when we come to that.

(CP) $\Longrightarrow$ **(VP)**. By construction of the formulation **(VP)** as specified by (3.14), if u belonging to $H^2(\Omega)$ solves the continuous problem **(CP)**, we have shown in the first question that u is also a solution of the variational problem **(VP)**.

(VP) $\Longrightarrow$ **(CP)**. The converse is a much more delicate matter, since the two formulations are defined in functional contexts with different regularities: $H^2(\Omega)$ for the continuous problem **(CP)** and $H^1(\Omega)$ for the variational problem **(VP)**.

In other words, either we establish a full converse, by first showing (if it is the case) that the solution u of the variational problem **(VP)** that we know to be in $H^1_{\Gamma_1}(\Omega)$ is also in $H^2(\Omega)$, then showing the equivalence of the two formulations, or we aim for a "partial converse," in the words of Raviart [1], by asking the following question: if u is a solution to the variational formulation **(VP)** that we assume to belong to $H^2(\Omega)$, is u also a solution to the continuous problem **(CP)**?

It is precisely this partial converse that we shall try to obtain here. Indeed, those solutions of the variational problem **(VP)** that do not belong to $H^2(\Omega)$, if there are such, will be of no interest to us in establishing the equivalence of the two formulations **(CP)** and **(VP)**, given that since the beginning of this presentation we have been looking for solutions of the continuous problem **(CP)** that belong to $H^2(\Omega)$.

Hence let u be a solution to the variational problem **(VP)** specified by (3.14), which we assume to belong to $H^2(\Omega)$. Then

$$\int_{\Omega} \nabla u \cdot \nabla v \, d\Omega + k \int_{\Gamma_3} uv \, d\Gamma = \int_{\Omega} fv \, d\Omega + \int_{\Gamma_2} g_2 v \, d\Gamma + \int_{\Gamma_3} g_3 v \, d\Gamma. \quad (3.50)$$

We use Green's formula (1.159) to transform the integral containing the gradients of u and v and bring back the Laplacian of u:

$$-\int_{\Omega} \Delta u \cdot v \, d\Omega + \int_{\partial\Omega} \frac{\partial u}{\partial n} v \, d\Gamma + k \int_{\Gamma_3} uv \, d\Gamma = \int_{\Omega} fv \, d\Omega + \int_{\Gamma_2} g_2 v \, d\Gamma + \int_{\Gamma_3} g_3 v \, d\Gamma. \quad (3.51)$$

Rewriting the integral over $\partial\Omega$ as a sum of three terms, one for each of its parts Γ_1, Γ_2, and Γ_3, noting that v vanishes on Γ_1, and gathering together similar terms, we find that for all $v \in H^1_{\Gamma_1}(\Omega)$,

$$-\int_{\Omega} (\Delta u + f)v \, d\Omega + \int_{\Gamma_2} \left(\frac{\partial u}{\partial n} - g_2\right) v \, d\Gamma + \int_{\Gamma_3} \left(\frac{\partial u}{\partial n} + ku - g_3\right) v \, d\Gamma = 0. \quad (3.52)$$

Among the functions v in $H^1_{\Gamma_1}(\Omega)$, we choose those with zero trace on Γ_2 and Γ_3, which amounts to considering functions in $H^1_0(\Omega)$, since these functions are automatically zero on Γ_1. Equation (3.52) then gives

$$\int_{\Omega} (\Delta u + f)v \, d\Omega = 0, \quad \forall v \in H^1_0(\Omega). \quad (3.53)$$

We have already treated such a problem in Sect. 2.3. In that case, we used density arguments to show that

$$-\Delta u = f \quad \text{(a.e.) in } \Omega. \quad (3.54)$$

Returning to (3.52) and taking into account (3.54), we thus have

$$\int_{\Gamma_2} \left(\frac{\partial u}{\partial n} - g_2\right) v \, d\Gamma + \int_{\Gamma_3} \left(\frac{\partial u}{\partial n} + ku - g_3\right) v \, d\Gamma = 0, \quad \forall v \in H^1_{\Gamma_1}(\Omega). \quad (3.55)$$

Now we consider the special case of functions v in $H^1_{\Gamma_1}(\Omega)$ that have zero trace on Γ_2. In this case, (3.55) becomes

$$\int_{\Gamma_3} \left(\frac{\partial u}{\partial n} + ku - g_3\right) v \, d\Gamma = 0, \; \forall v \in H^1_{\Gamma_1}(\Omega) \cap \{v/v = 0 \text{ on } \Gamma_2\}. \quad (3.56)$$

Since we are considering a solution u of the variational problem (VP) that also belongs to $H^2(\Omega)$, it follows that the quantity $\partial u/\partial n + ku - g_3$ is in $L^2(\Gamma_3)$.

So, if the space of traces on Γ_3 of functions belonging to $H^1_{\Gamma_1}(\Omega) \cap \{v : v = 0 \text{ on } \Gamma_2\}$ is dense in $L^2(\Gamma_3)$, we thus establish that

$$\frac{\partial u}{\partial n} + ku - g_3 = 0, \quad \text{(a.e.) on } \Gamma_3. \quad (3.57)$$

Using this, (3.55) becomes

$$\int_{\Gamma_2} \left(\frac{\partial u}{\partial n} - g_2 \right) v \, d\Gamma = 0, \quad \forall v \in H^1_{\Gamma_1}(\Omega). \qquad (3.58)$$

Using the density argument again, we then have

$$\frac{\partial u}{\partial n} - g_2 = 0, \quad \text{(a.e.) on } \Gamma_2. \qquad (3.59)$$

This ends the demonstration of the partial converse, whence the solution u of the variational formulation (**VP**) specified by (3.14) is also a solution of the continuous problem (**CP**) defined by (3.1)–(3.4), and *vice versa*.

———————————————

3.2 Heat Conduction

3.2.1 Statement of the Problem

Consider a container $\Omega \in \mathbb{R}^3$ in contact with an external medium at constant temperature u_0. The wall $\Gamma = \partial \Omega$ separating the inside of Ω from this external medium is permeable to heat, and there is a flow of heat through Γ that is proportional to the temperature. There is no source of heat inside Ω.

For a system like this, the equilibrium temperature u is governed by the equations of the continuous problem $(\mathbf{CP})_1$ defined as follows:

Find $u \in H^2$ solution of

$$(\mathbf{CP})_1 \quad \begin{cases} \Delta u = 0 \text{ in } \Omega, \\ -\dfrac{\partial u}{\partial n} = k(u - u_0) \text{ on } \Gamma, \end{cases} \tag{3.60}$$

where k is a positive constant and $\mathbf{n}$ is the outward-pointing unit normal to Γ.

▶ **Part I**

1) Establish a variational formulation $(\mathbf{VP})_1$ associated with the problem $(\mathbf{CP})_1$ in the following form:

$$(\mathbf{PV})_1 \quad \begin{cases} \text{Find } u \in V \text{ solution of} \\ a(u, v) = L(v), \quad \forall v \in V. \end{cases} \tag{3.61}$$

Define the quantities $a(.,.)$ and $L(.)$, and specify the function space V.

2) Show that $a(.,.)$ is a symmetric, continuous bilinear form on $V \times V$.

3) Show that the bilinear form $a(.,.)$ is coercive on $V \times V$.

4) Show that the linear form $L(.)$ is continuous on V.

5) Establish the existence and uniqueness of the solution of the variational problem $(\mathbf{VP})_1$. Express the solution u to the problem.

6) Prove the equivalence of the continuous problem $(\mathbf{CP})_1$ and the variational formulation $(\mathbf{VP})_1$.

▶ **Part II**

We wish to maintain a constant average temperature u_1 in Ω that differs from the external temperature u_0. For this purpose, we place a temperature control at the center of Ω, occupying the region $\omega_c \subset \Omega$. This temperature control can supply or remove an amount F of heat, where F is a positive or negative constant depending on the situation.

The function u represents the temperature field in Ω at equilibrium. Under such conditions, it can be shown that the system is governed by the following equations:

$$
\textbf{(CP)}_2 \quad
\begin{cases}
-\Delta u = F & \text{in } \omega_c, \\[4pt]
-\Delta u = 0 & \text{in } \Omega \backslash \omega_c, \\[4pt]
-\dfrac{\partial u}{\partial n} = k(u - u_0) & \text{on } \Gamma, \\[4pt]
[[u]] = 0 & \text{on } \partial \omega_c, \\[4pt]
\left[\left[\dfrac{\partial u}{\partial n}\right]\right] = 0 & \text{on } \partial \omega_c,
\end{cases}
\qquad (3.62)
$$

where $[[v]]$ denotes the discontinuity in any function v in crossing the boundary $\partial \omega_c$:

$$
[[v]] \equiv v^+ - v^-, \qquad (3.63)
$$

with v^+ the trace of v on $\partial \omega_c$ as we approach from the outside of ω_c, i.e., $\Omega \backslash \omega_c$, and v^- the trace of v on $\partial \omega_c$ as we approach from the inside of ω_c.

7) Establish the variational formulation $\textbf{(VP)}_2$ associated with the continuous problem $\textbf{(CP)}_2$.

8) Show that there is one and only one solution to the variational problem $\textbf{(VP)}_2$.

9) Prove the equivalence of the continuous problem $\textbf{(CP)}_2$ and the variational formulation $\textbf{(VP)}_2$.

10) Make an *a priori* estimate of the solution u of the variational formulation $\textbf{(VP)}_2$.

Show that there exists a positive constant C depending on the problem data (F, k, u_0) such that

$$
\|u\|_{H^1(\Omega)} \leq C. \qquad (3.64)
$$

3.2.2 Solution

▶ **Part I**

1) In order to associate a variational formulation $(\mathbf{VP})_1$ with the continuous problem $(\mathbf{CP})_1$, we multiply the partial differential equation of the continuous problem $(\mathbf{CP})_1$, as defined by (3.60), by a test function v belonging to some function space that remains to be specified, and we integrate over the domain Ω:

$$\int_\Omega -\Delta u \cdot v \, d\Omega = 0, \quad \forall v \in V. \tag{3.65}$$

Applying Green's formula in a formal sense, we obtain

$$\int_\Omega \Delta u \cdot v \, d\Omega = -\int_\Omega \nabla u \cdot \nabla v \, d\Omega + \int_{\partial\Omega} \frac{\partial u}{\partial n} v \, d\Gamma, \quad \text{with} \quad \frac{\partial u}{\partial n} = \nabla u \cdot \mathbf{n}, \tag{3.66}$$

where $\mathbf{n}$ is the outward-pointing unit normal to the domain Ω. We now use the boundary condition on the boundary Γ, viz.,

$$\frac{\partial u}{\partial n} = -k(u - u_0),$$

whereupon (3.66) becomes

$$\int_\Omega \Delta u \cdot v \, d\Omega = -\int_\Omega \nabla u \cdot \nabla v \, d\Omega - \int_\Gamma k(u - u_0) v \, d\Gamma = 0, \quad \forall v \in V. \tag{3.67}$$

We thus arrive at the formal variational problem $(\mathbf{VP})_1$:

Find $u \in V$ that solves

$$\int_\Omega \nabla u \cdot \nabla v \, d\Omega + k \int_\Gamma u v \, d\Gamma = k \int_\Gamma u_0 v \, d\Gamma, \quad \forall v \in V. \tag{3.68}$$

In order to specify a suitable functional context guaranteeing convergence of the integrals appearing in (3.68), we note the following convergence properties:

● **Convergence of** $\displaystyle\int_\Omega \nabla u \cdot \nabla v \, d\Omega$.

We have

$$\left| \int_\Omega \nabla u \cdot \nabla v \, d\Omega \right| \leq \|\nabla u\|_{L^2(\Omega)} \|\nabla v\|_{L^2(\Omega)} \leq \|u\|_{H^1(\Omega)} \|v\|_{H^1(\Omega)}, \tag{3.69}$$

where we have applied the Cauchy–Schwarz inequality. So if u and v belong to $H^1(\Omega)$, the integral $\int_\Omega \nabla u \cdot \nabla v \, d\Omega$ is convergent.

- **Convergence of** $\int_\Gamma uv \, d\Gamma$.

Here we have

$$\left| \int_\Gamma uv \, d\Gamma \right| \le \|u\|_{L^2(\Gamma)} \|v\|_{L^2(\Gamma)} \le \|u\|_{L^2(\partial\Omega)} \|v\|_{L^2(\partial\Omega)} \tag{3.70}$$

$$\le C^2 \|u\|_{H^1(\Omega)} \|v\|_{H^1(\Omega)},$$

where we have applied the trace theorem (Theorem 1.3) for functions in $H^1(\Omega)$.

Once again, if u and v belong to $H^1(\Omega)$, then the integral $\int_\Gamma uv \, d\Gamma$ converges.

- **Convergence of** $\int_\Gamma u_0 v \, d\Gamma$.

This time, we have

$$\left| \int_{\Gamma_2} u_0 v \, d\Gamma \right| \le \|u_0\|_{L^2(\Gamma)} \|v\|_{L^2(\Gamma)} \le \|u_0\|_{L^2(\Gamma)} \|v\|_{L^2(\partial\Omega)} \tag{3.71}$$

$$\le C \|u_0\|_{L^2(\Gamma)} \|v\|_{H^1(\Omega)}.$$

Note that the norm $\|.\|_{L^2(\Gamma)}$ of the external temperature u_0 is finite, since it is assumed to be constant.

In conclusion, if the variational formulation $(\mathbf{VP})_1$ specified by (3.68) is to be valid, the function space V should be taken as the Sobolev space $H^1(\Omega)$.

We then introduce the forms $a(.,.)$ and $L(.)$ defined by

$$a : H^1(\Omega) \times H^1(\Omega) \longrightarrow \mathbb{R}, \tag{3.72}$$

$$(u, v) \longmapsto a(u, v) = \int_\Omega \nabla u \cdot \nabla v \, d\Omega + k \int_\Gamma uv \, d\Gamma,$$

$$L : H^1(\Omega) \longrightarrow \mathbb{R}, \tag{3.73}$$

$$v \longmapsto L(v) \equiv k \int_\Gamma u_0 v \, d\Gamma.$$

With these definitions, the variational formulation $(\mathbf{VP})_1$ associated with the problem $(\mathbf{CP})_1$ can indeed be written in the form

$$\textbf{(VP)}_1 \quad \begin{cases} \text{Find } u \in H^1(\Omega) \text{ solution of} \\[4pt] a(u,v) = L(v), \quad \forall v \in H^1(\Omega). \end{cases} \tag{3.74}$$

2) The form $a(.,.)$ defined by (3.72) is trivially bilinear by inspection, due to the linearity of the operator ∇ and the operation of integration.

We now prove that the form $a(.,.)$ is continuous on $H^1(\Omega) \times H^1(\Omega)$. To do this, we consider an arbitrary pair of functions (u,v) in $H^1(\Omega) \times H^1(\Omega)$.

Then

$$|a(u,v)| \leq \left| \int_\Omega \nabla u \cdot \nabla v \, d\Omega + k \int_\Gamma uv \, d\Gamma \right| \leq \left| \int_\Omega \nabla u \cdot \nabla v \, d\Omega \right| + \left| k \int_\Gamma uv \, d\Gamma \right| \tag{3.75}$$

$$\leq \int_\Omega |\nabla u \cdot \nabla v| \, d\Omega + k \int_\Gamma |uv| \, d\Gamma.$$

Using the Cauchy–Schwarz inequality, it then follows that

$$|a(u,v)| \leq \|\nabla u\|_{L^2(\Omega)} \|\nabla v\|_{L^2(\Omega)} + k \|u\|_{L^2(\Gamma)} \|v\|_{L^2(\Gamma)}, \tag{3.76}$$

whence, from the definition of the norm $\|.\|_{H^1(\Omega)}$,

$$|a(u,v)| \leq \|u\|_{H^1(\Omega)} \|v\|_{H^1(\Omega)} + k \|u\|_{L^2(\Gamma)} \|v\|_{L^2(\Gamma)}. \tag{3.77}$$

The trace theorem (Theorem 1.3) ensures that the map γ_0, which associates with any function v of $H^1(\Omega)$ its trace $\gamma_0(v)$ in $L^2(\partial\Omega)$, is linear and continuous:

$$\exists C > 0 \text{ such that } \|v\|_{L^2(\Gamma)} \leq C \|v\|_{H^1(\Omega)}, \quad \forall v \in H^1(\Omega). \tag{3.78}$$

Then using this result in the inequality (3.77), it follows that

$$|a(u,v)| \leq (1 + kC^2) \|u\|_{H^1(\Omega)} \|v\|_{H^1(\Omega)}, \tag{3.79}$$

which proves the continuity of the bilinear form $a(.,.)$.

3) Regarding the coercivity of the bilinear form $a(.,.)$, we use a result already mentioned in the notes for Definition 1.21 on the Sobolev space $H_0^1(\Omega)$ and relating to the equivalence of the norms $H_0^1(\Omega)$ extended to $H^1(\Omega)$.

In particular, it was noted that in $H^1(\Omega)$, the norm $\|v\|_{H^1(\Omega)}$ is equivalent to the norm $\left[\|v\|_{L^2(\Gamma)}^2 + \|\nabla v\|_{L^2(\Omega)}^2 \right]^{1/2}$.

Put another way, there exist two real numbers α and β such that

$$\alpha \|v\|_{H^1(\Omega)} \leq \left[\|v\|_{L^2(\Gamma)}^2 + \|\nabla v\|_{L^2(\Omega)}^2 \right]^{1/2} \leq \beta \|v\|_{H^1(\Omega)}. \tag{3.80}$$

So, in order to establish the coercivity of the bilinear form $a(., .)$ defined by (3.72), we must find a constant $\alpha_0 > 0$ such that

$$\forall v \in H^1(\Omega), \quad |a(v, v)| \geq \alpha_0 \|v\|_{H^1(\Omega)}^2. \tag{3.81}$$

To do this, we consider two cases:

- *First case*: $k \geq 1$. For values of k greater than or equal to 1, we have

$$k \int_\Gamma v^2 \, d\Gamma \geq \int_\Gamma v^2 \, d\Gamma. \tag{3.82}$$

Hence there exists α defined by (3.80) such that

$$|a(v, v)| \geq \|v\|_{L^2(\Gamma)}^2 + \|\nabla v\|_{L^2(\Omega)}^2 \geq \alpha^2 \|v\|_{H^1(\Omega)}^2. \tag{3.83}$$

As a consequence, the value of α_0 required in (3.81) corresponds to $\alpha_0 \equiv \alpha^2$.

- *Second case*: $k < 1$. For these values of k, we may write

$$\int_\Omega |\nabla v|^2 \, d\Omega + k \int_\Gamma v^2 \, d\Gamma \geq k \left[\int_\Omega |\nabla v|^2 \, d\Omega + \int_\Gamma v^2 \, d\Gamma \right]. \tag{3.84}$$

It then follows that

$$|a(v, v)| \geq k \left(\|v\|_{L^2(\Gamma)}^2 + \|\nabla v\|_{L^2(\Omega)}^2 \right) \geq k\alpha^2 \|v\|_{H^1(\Omega)}^2. \tag{3.85}$$

Hence we may take the constant α_0 for this second case to be $\alpha_0 \equiv k\alpha^2$.

4) The map L defined by (3.73) is obviously linear, given its structure. Regarding its continuity, we apply the Cauchy–Schwarz inequality once again to obtain

$$|L(v)| = \left| k \int_\Gamma u_0 v \, d\Gamma \right| \leq k \|u_0\|_{L^2(\Gamma)} \|v\|_{L^2(\Gamma)}. \tag{3.86}$$

We then use the continuity of the trace map γ_0 for every function in $H^1(\Omega)$ (see Theorem 1.3) to obtain finally

$$|L(v)| \leq C_L \|v\|_{H^1(\Omega)}, \tag{3.87}$$

where the continuity constant C_L of the form L is given by

$$C_L \equiv Ck\|u_0\|_{L^2(\Gamma)}, \tag{3.88}$$

and C corresponds to the constant arising in the trace theorem (Theorem 1.3).

5) To prove the existence and uniqueness of the solution to the variational problem **(VP)**$_1$, we need only apply the Lax–Milgram theorem (Theorem 1.11), provided we first check that its premises are satisfied:

1. $H^1(\Omega)$ with the scalar product $((u, v))_{H^1(\Omega)} = ((u, v))_{L^2(\Omega)} + ((\nabla u, \nabla v))_{L^2(\Omega)}$ is a Hilbert space.

2. The other premises concern the bilinear form $a(., .)$, which must be continuous and coercive on $H^1(\Omega) \times H^1(\Omega)$, and the linear form $L(.)$, which must also be continuous on $H^1(\Omega)$. These properties have been checked in the previous questions.

There is then a unique solution u to the variational problem **(VP)**$_1$ belonging to $H^1(\Omega)$, i.e., such that

$$a(u, v) = L(v), \quad \forall v \in H^1(\Omega).$$

Note that the trivial solution to the variational problem **(VP)** is given by

$$u = u_0, \tag{3.89}$$

where u_0 is the constant temperature of the external medium.

It is straightforward to check that $u = u_0$ is also a solution to the continuous problem **(CP)**, which should come as no surprise, once we have established the equivalence of the continuous problem **(CP)** and the variational formulation **(VP)**. And this is the subject of the following question.

6) We prove the equivalence of the continuous problem **(CP)**$_1$ and the formulation **(VP)**$_1$ in two steps.

We have effectively proven sufficiency by the very construction of the variational formulation **(VP)**$_1$, which makes it clear that every solution u of the problem **(CP)**$_1$ belonging to $H^2(\Omega)$ will solve the variational problem **(VP)**$_1$.

So the real subject of this question is to prove the converse.

As already mentioned on several occasions, in this book we shall only consider a partial converse defined as follows: if u solves the variational problem **(VP)**$_1$ and in addition belongs to $H^2(\Omega)$, it is then a solution to the continuous problem **(CP)**$_1$.

Now in this case, using Green's formula, we have

$$-\int_\Omega \Delta u \cdot v \, d\Omega + \int_\Gamma \frac{\partial u}{\partial n} v \, d\Gamma + \int_\Gamma kuv \, d\Gamma = k \int_\Gamma u_0 v \, d\Gamma, \quad \forall v \in H^1(\Omega). \tag{3.90}$$

Collecting the various integrals on the boundary Γ of Ω yields

$$-\int_\Omega \Delta u \cdot v \, d\Omega + \int_\Gamma \left[\frac{\partial u}{\partial n} + k(u - u_0)\right] v \, d\Gamma = 0, \quad \forall v \in H^1(\Omega). \quad (3.91)$$

Among the functions v in $H^1(\Omega)$, we choose those that are in $\mathcal{D}(\Omega)$. In this case, (3.91) becomes

$$\int_\Omega \Delta u \cdot v \, d\Omega = 0, \quad \forall v \in \mathcal{D}(\Omega), \quad (3.92)$$

because functions in $\mathcal{D}(\Omega)$ are zero on the boundary Γ of Ω. Now (3.92) is to be interpreted in the sense of distributions:

$$\Delta u = 0 \text{ in } \mathcal{D}'(\Omega). \quad (3.93)$$

Even more precisely, since the solution u of the variational formulation $(\mathbf{VP})_1$ is also considered to be in $H^2(\Omega)$ for the purposes of this partial converse, we deduce that Δu belongs to $L^2(\Omega)$ and (3.92) can be interpreted in the sense of distributions by saying that

$$\langle T_{\Delta u}, v \rangle = 0, \quad \forall v \in \mathcal{D}(\Omega), \quad (3.94)$$

where $T_{\Delta u}$ is the regular distribution associated with Δu by Definition 1.12. Under these conditions, (3.94) leads to

$$T_{\Delta u} = 0 \text{ in } \mathcal{D}'(\Omega). \quad (3.95)$$

It is now easy to conclude, since according to Lemma 1.12, the map J from $L^2(\Omega)$ into $\mathcal{D}'(\Omega)$ is injective. In other words, the kernel of J is the zero space, whence

$$T_f = 0 \text{ in } \mathcal{D}'(\Omega) \implies f = 0 \text{ in } L^2(\Omega). \quad (3.96)$$

Therefore, (3.95) implies that

$$\Delta u = 0 \text{ in } L^2(\Omega), \quad (3.97)$$

and hence finally,

$$\Delta u = 0 \text{ (a.e.) in } \Omega, \quad (3.98)$$

where (a.e.) denotes as usual "almost everywhere," i.e., the equality (3.98) holds everywhere in Ω, except possibly on a set of measure zero.

Taking into account (3.98), the relation (3.91) can now be written

$$\int_\Gamma \left[\frac{\partial u}{\partial n} + k(u - u_0)\right] v \, d\Gamma = 0, \quad \forall v \in H^1(\Omega). \quad (3.99)$$

But according to Theorem 1.6, $\partial u/\partial n$ is in $L^2(\Gamma)$, since the solution u is assumed to be in $H^2(\Omega)$. It follows that the quantity $\partial u/\partial n + k(u - u_0)$ is also in $L^2(\Gamma)$.

Using density arguments of the kind already discussed in detail in Sect. 2.3 and assuming that the space of traces on Γ of the functions in $H^1(\Omega)$ is dense in $L^2(\Gamma)$, we see that the relation (3.99) implies

$$\int_\Gamma \left[\frac{\partial u}{\partial n} + k(u - u_0) \right] v\, d\Gamma = 0, \quad \forall v \in L^2(\Gamma). \tag{3.100}$$

As a consequence,

$$\frac{\partial u}{\partial n} + k(u - u_0) = 0, \quad \text{(a.e.) on } \Gamma. \tag{3.101}$$

▶ **Part II**

7) Let v be a test function defined on the whole of Ω. Further, let v^+ be the trace of v on $\partial \omega_c$ as we approach from the outside $\Omega \backslash \omega_c$ and let v^- be the trace of v on $\partial \omega_c$ as we approach from the inside ω_c.

Each of the two partial differential equations in the statement of the problem **(CP)** in (3.62) is now multiplied by v, and the two resulting equations are added to yield

$$-\int_{\Omega \backslash \omega_c} \Delta u \cdot v\, d\Omega - \int_{\omega_c} \Delta u \cdot v\, d\Omega = \int_{\omega_c} Fv\, d\Omega, \quad \forall v \in V, \tag{3.102}$$

where V is a function space to be specified, but which must contain real-valued functions on Ω.

Using Green's formula (1.159), we have

$$\int_\Omega \nabla u \cdot \nabla v\, d\Omega - \int_\Gamma \frac{\partial u}{\partial n} v\, d\Gamma - \int_{\partial \omega_c} \frac{\partial u^+}{\partial n^+} v^+\, d\Gamma - \int_{\partial \omega_c} \frac{\partial u^-}{\partial n^-} v^-\, d\Gamma = \int_{\omega_c} Fv\, d\Omega, \tag{3.103}$$

where n^+ is the inward-pointing unit normal on the boundary of ω_c and n^- the outward-pointing unit normal.

It follows that if we set $n \equiv n^+ = -n^-$, (3.103) can be rewritten in the form

$$\int_\Omega \nabla u \cdot \nabla v\, d\Omega - \int_\Gamma \frac{\partial u}{\partial n} v\, d\Gamma - \int_{\partial \omega_c} \left[\left[\frac{\partial u}{\partial n} v \right]\right] d\Gamma = \int_{\omega_c} Fv\, d\Omega, \quad \forall v \in V, \tag{3.104}$$

where we use the notation $[[.]]$ defined by (3.63) in the statement of the problem to denote the discontinuity in a quantity.

We now observe that the boundary condition on $\partial \omega_c$ referring to the discontinuity in u does not appear in (3.104), whereas the boundary condition on the discontinuity in its normal derivative is potentially present through the expression $\left[\left[\frac{\partial u}{\partial n} v\right]\right]$.

For this reason, we require the functions v in the space V to satisfy the condition

$$[[v]] = v^+ - v^- = 0. \tag{3.105}$$

Under such conditions, the solution u to the variational formulation **(VP)** that we seek in V will also satisfy the discontinuity condition (3.105).

Hence, using the condition (3.105), the discontinuity $\left[\left[\frac{\partial u}{\partial n} v\right]\right]$ on the boundary of ω_c can be written

$$\left[\left[\frac{\partial u}{\partial n} v\right]\right] = \left[\left[\frac{\partial u}{\partial n}\right]\right] v^+ = 0, \tag{3.106}$$

according to the second discontinuity condition on the normal derivative $\partial u / \partial n$ of the solution u.

Finally, we make use of the boundary condition on Γ to obtain

$$\int_\Omega \nabla u \cdot \nabla v \, d\Omega + k \int_\Gamma (u - u_0) v \, d\Gamma = \int_{\omega_c} F v \, d\Omega, \quad \forall v \in V, \tag{3.107}$$

or alternatively,

$$\int_\Omega \nabla u \cdot \nabla v \, d\Omega + k \int_\Gamma u v \, d\Gamma = \int_{\omega_c} F v \, d\Omega + k \int_\Gamma u_0 v \, d\Gamma, \quad \forall v \in V. \tag{3.108}$$

Definition of the Space V. To specify the function space V in which we shall seek the solution u of the variational problem **(VP)**$_2$ specified by (3.108), we must determine sufficient conditions on u and v to guarantee convergence of the various integrals appearing in it.

To do this, we note that the variational formulation **(VP)**$_2$ differs from **(VP)**$_1$ solely through the presence of the integral over ω_c, whose convergence is ensured by the following series of bounds:

$$\left| \int_{\omega_c} F v \, d\Gamma_c \right| \leq \|F\|_{L^2(\omega_c)} \|v\|_{L^2(\omega_c)} \tag{3.109}$$

$$\leq \|F\|_{L^2(\omega_c)} \|v\|_{L^2(\Omega)}$$

$$\leq \|F\|_{L^2(\omega_c)} \|v\|_{H^1(\Omega)}.$$

In conclusion, the variational formulation **(VP)**$_2$ specified by (3.108) can be written in the generic form

$$(\textbf{VP})_2 \quad \begin{cases} \text{Find } u \in V \text{ solution of} \\ a(u, v) = L(v), \quad \forall v \in V, \end{cases} \tag{3.110}$$

where we have introduced the forms $L(.)$ and $a(.,.)$ and the function space V as follows:

$$a : V \times V \longrightarrow \mathbb{R}, \tag{3.111}$$

$$(u, v) \longmapsto a(u, v) = \int_{\Omega} \nabla u \cdot \nabla v \, d\Omega + k \int_{\Gamma} uv \, d\Gamma,$$

$$L : V \longrightarrow \mathbb{R}, \tag{3.112}$$

$$v \longmapsto L(v) = \int_{\omega_c} Fv \, d\Omega + k \int_{\Gamma} u_0 v \, d\Gamma,$$

$$V \equiv \left\{ v : \Omega \to \mathbb{R}, \ v \in H^1(\Omega) \text{ such that } [[v]] = 0 \right\}. \tag{3.113}$$

8) To prove the existence and uniqueness of the solution to the variational problem **(VP)₂** specified by (3.110)–(3.113), we use the same method as for the first question, i.e., we apply the Lax–Milgram theorem (Theorem 1.11):

- We begin by noting that the space V is a vector subspace of the Sobolev space $H^1(\Omega)$. In particular, we note that if two functions v_1 and v_2 satisfy the discontinuity condition (3.105), the same will also be true for every linear combination $\alpha_1 v_1 + \alpha_2 v_2$, where (α_1, α_2) is an arbitrary pair of real numbers in $\mathbb{R}^2$:

$$[[\alpha_1 v_1 + \alpha_2 v_2]] = \alpha_1(v_1^+ - v_1^-) + \alpha_2(v_2^+ - v_2^-) = 0, \tag{3.114}$$

since v_1 and v_2 are in the space V and hence satisfy the zero discontinuity property on the boundary $\partial \omega_c$.

We thus equip the space V with the norm induced from $H^1(\Omega)$, which is given by

$$\|v\|_{H^1(\Omega)}^2 = \int_{\Omega} v^2 \, d\Omega + \int_{\Omega} |\nabla v|^2 \, d\Omega. \tag{3.115}$$

We must now check that the space V equipped with the norm $\|.\|_{H^1(\Omega)}$ is a Hilbert space.

To this end, we shall show that V is a closed subspace of $H^1(\Omega)$ and then use Property 1.8, adapting it to the present situation. We must thus show that every sequence of functions $(v_n)_{n \in \mathbb{N}}$ in V converging for the norm $H^1(\Omega)$ converges to some element v in V, and in particular that it satisfies the discontinuity condition $[[v]] = 0$.

Now the discontinuity in v can be found as follows:

$$[[v]] = v^+ - v^-$$
$$= (v^+ - v_n^+) + (v_n^+ - v_n^-) + (v_n^- - v^-) \qquad (3.116)$$
$$= (v^+ - v_n^+) + (v_n^- - v^-),$$

where we have used the fact that the discontinuity in the sequence $(v_n)_{n \in \mathbb{N}}$ on the boundary $\partial \omega_c$ is zero, since each term in the sequence belongs to V.

By the trace theorem (Theorem 1.3), we have the following bounds:

$$\exists\, C^+ > 0 \text{ such that } \left\| v_n^+ - v^+ \right\|_{L^2(\partial \omega_c)} \le \| v_n - v \|_{L^2(\partial \omega_c \cup \Gamma)} \qquad (3.117)$$
$$\le C^+ \| v_n - v \|_{H^1(\Omega \backslash \omega_c)},$$

$$\exists\, C^- > 0 \text{ such that } \left\| v_n^- - v^- \right\|_{L^2(\partial \omega_c)} \le C^- \| v_n - v \|_{H^1(\omega_c)}, \qquad (3.118)$$

whence it follows immediately that

$$\left\| v_n^+ - v^+ \right\|_{L^2(\partial \omega_c)} \le C^+ \| v_n - v \|_{H^1(\Omega)}, \qquad (3.119)$$

$$\left\| v_n^- - v^- \right\|_{L^2(\partial \omega_c)} \le C^- \| v_n - v \|_{H^1(\Omega)}. \qquad (3.120)$$

We can now evaluate the discontinuity in v (which is the limit of the sequence of the v_n) on $\partial \omega_c$ using the relation (3.116). Taking the limit in the two bounds (3.119) and (3.120), we have

$$\lim_{n \to +\infty} \left\| v_n^+ - v^+ \right\|_{L^2(\partial \omega_c)} = \lim_{n \to +\infty} \left\| v_n^- - v^- \right\|_{L^2(\partial \omega_c)} = 0, \qquad (3.121)$$

because the sequence v_n converges to v for the norm on $H^1(\Omega)$. Then (3.116) implies that

$$\left\| v^+ - v^- \right\|_{L^2(\partial \omega_c)} = 0, \qquad (3.122)$$

and hence

$$[[v]] = v^+ - v^- = 0 \quad \text{(a.e.) on } \partial \omega_c. \qquad (3.123)$$

The space V is thus closed in $H^1(\Omega)$. and it is therefore a Hilbert space for the norm induced from $H^1(\Omega)$.

- Regarding the properties of the bilinear form $a(.,.)$ specified by (3.111), since it is the same bilinear form as we considered in Question 3, and noting that the space V has been given the same norm, namely the norm $\|.\|_{H^1(\Omega)}$, the continuity and coercivity of $a(.,.)$ are immediate.

- It remains to establish the continuity of the linear form $L(.)$ defined by (3.112). We have the following argument:

$$|L(v)| \leq \|F\|_{L^2(\omega_c)}\|v\|_{L^2(\omega_c)} + k\|u_0\|_{L^2(\Gamma)}\|v\|_{L^2(\Gamma)}$$

$$\leq \|F\|_{L^2(\omega_c)}\|v\|_{L^2(\omega_c)} + k\|u_0\|_{L^2(\Gamma)}\|v\|_{L^2(\partial\omega_c\cup\Gamma)}$$

$$\leq \|F\|_{L^2(\omega_c)}\|v\|_{L^2(\Omega)} + kC^+\|u_0\|_{L^2(\Gamma)}\|v\|_{H^1(\Omega\backslash\omega_c)}$$

$$\leq \|F\|_{L^2(\omega_c)}\|v\|_{L^2(\Omega)} + kC^+\|u_0\|_{L^2(\Gamma)}\|v\|_{H^1(\Omega)}$$

$$\leq \left[\|F\|_{L^2(\omega_c)} + kC^+\|u_0\|_{L^2(\Gamma)}\right]\|v\|_{H^1(\Omega)}, \tag{3.124}$$

where the constant C^+ is the one appearing in (3.117).

The Lax–Milgram theorem (Theorem 1.1) now ensures that there exists one and only one function u in the function space V that solves the variational problem $(\mathbf{VP})_2$ defined by (3.110)–(3.113).

9) The main task here is to show that the solution to the variational problem $(\mathbf{VP})_2$ defined by (3.110)–(3.113) is also a solution to the continuous problem $(\mathbf{CP})_2$.

Indeed, the converse is obvious, following essentially from the construction of the variational formulation $(\mathbf{VP})_2$ as presented in Question 7, whenever the solution u to the continuous problem $(\mathbf{CP})_2$ belongs to $H^2(\Omega)$.

So, let us now assume that the solution u of the variational problem $(\mathbf{VP})_2$ belongs to $H^2(\Omega)$, so that u satisfies

$$\int_\Omega \nabla u \cdot \nabla v \, d\Omega + k\int_\Gamma uv \, d\Gamma = \int_{\omega_c} Fv \, d\Omega + k\int_\Gamma u_0 v \, d\Gamma, \quad \forall v \in V. \tag{3.125}$$

If we write the integral over Ω as a sum of integrals over ω_c and $\Omega\backslash\omega_c$, then use Green's formula for each integral and collect similar terms, we find that for every function v in V,

$$-\int_{\omega_c} (\Delta u + F)v \, d\Omega - \int_{\Omega\backslash\omega_c} \Delta u \cdot v \, d\Omega + \int_{\partial\omega_c} \left[\!\left[\frac{\partial u}{\partial n}v\right]\!\right] d\Gamma$$

$$+ \int_\Gamma \left[\frac{\partial u}{\partial n} + k(u - u_0)\right]v \, d\Gamma = 0. \tag{3.126}$$

We now consider particular functions v belonging to V in order to obtain the various equations of the continuous problem $(\mathbf{CP})_2$:

1. Our first choice for the functions v in V takes them to be defined by

$$v = \begin{cases} v_{\omega_c} & \text{in } \omega_c, \\ 0 & \text{otherwise,} \end{cases} \tag{3.127}$$

where we consider functions v_{ω_c} in $H_0^1(\omega_c)$, i.e., with zero trace on $\partial\omega_c$, to ensure that the discontinuity in v in crossing the boundary $\partial\omega_c$ is zero, whereupon v is indeed an element of V.

In this case, (3.126) becomes

$$-\int_{\omega_c} (\Delta u + F)v_{\omega_c}\, d\Omega = 0, \quad \forall v_{\omega_c} \in H_0^1(\omega_c). \tag{3.128}$$

As we have already shown on several occasions, either by density arguments as in Sect. 2.3 or using distributions as in Question 6, we may then conclude that

$$-\Delta u = F \ \text{(a.e.) on } \omega_c. \tag{3.129}$$

2. Our second choice for the functions v in V is defined as follows:

$$v = \begin{cases} v_{\Omega\backslash\omega_c} & \text{in } \Omega\backslash\omega_c, \\ 0, & \text{otherwise,} \end{cases} \tag{3.130}$$

where the functions $v_{\Omega\backslash\omega_c}$ belong to the space $H_*^1(\Omega\backslash\omega_c)$, defined as the set of functions in $H^1(\Omega\backslash\omega_c)$ whose trace vanishes on ω_c.

Under these conditions, the discontinuity in the functions v is zero in crossing the boundary $\partial\omega_c$, whence v is indeed an element of V.

Hence, (3.126) can be written in the form

$$-\int_{\Omega\backslash\omega_c} (\Delta u)v_{\Omega\backslash\omega_c}\, d\Omega + \int_\Gamma \left[\frac{\partial u}{\partial n} + k(u - u_0)\right]v_{\Omega\backslash\omega_c}\, d\Gamma = 0, \ \forall v_{\Omega\backslash\omega_c} \in H_*^1(\Omega\backslash\omega_c). \tag{3.131}$$

Among the functions $v_{\Omega\backslash\omega_c}$ in $H_*^1(\Omega\backslash\omega_c)$, we thus select those whose trace vanishes on Γ.

Put another way, the functions $v_{\Omega\backslash\omega_c}$ now belong to $H_0^1(\Omega\backslash\omega_c)$. In this case, (3.131) implies that

$$-\int_{\Omega\backslash\omega_c} \Delta u \cdot v_{\Omega\backslash\omega_c}\, d\Omega = 0, \quad \forall v_{\Omega\backslash\omega_c} \in H_0^1(\Omega\backslash\omega_c). \tag{3.132}$$

A density argument then shows that

$$-\Delta u = 0 \ \text{(a.e.) in } \Omega\backslash\omega_c. \tag{3.133}$$

Returning to (3.131) and taking into account (3.133), we obtain

$$\int_\Gamma \left[\frac{\partial u}{\partial n} + k(u - u_0) \right] v_{\Omega \setminus \omega_c} \, d\Gamma = 0, \quad \forall v_{\Omega \setminus \omega_c} \in H_*^1(\Omega \setminus \omega_c). \tag{3.134}$$

Assuming that the space of traces on Γ of functions in $H_*^1(\Omega \setminus \omega_c)$ is dense in $L^2(\Gamma)$, we have

$$\int_\Gamma \left[\frac{\partial u}{\partial n} + k(u - u_0) \right] v_{\Omega \setminus \omega_c} \, d\Gamma = 0, \quad \forall v_{\Omega \setminus \omega_c} \in L^2(\Gamma). \tag{3.135}$$

Since the solution u of the variational problem $(\textbf{VP})_2$ belongs to $H^2(\Omega)$, the trace theorem (Theorem 1.6) implies that

$$\frac{\partial u}{\partial n} + k(u - u_0) \in L^2(\Gamma). \tag{3.136}$$

It then follows from (3.135) that

$$\frac{\partial u}{\partial n} + k(u - u_0) = 0 \quad \text{(a.e.) on } \Gamma. \tag{3.137}$$

3. To conclude here, it remains to establish that the discontinuity in the solution u to the variational formulation $(\textbf{VP})_2$ is zero across $\partial \omega_c$.

   Given the results (3.129), (3.133), and (3.137), the variational Eq. (3.126) can be written

$$\int_{\partial \omega_c} \left[\!\!\left[\frac{\partial u}{\partial n} v \right]\!\!\right] d\Gamma = 0, \quad \forall v \in V. \tag{3.138}$$

Now every function v in V satisfies $[\![v]\!] = v^+ - v^- = 0$. If v denotes the identical values of v^+ and v^- in crossing $\partial \omega_c$, viz.,

$$v = v^+ = v^-,$$

then (3.138) can be rewritten in the form

$$\int_{\partial \omega_c} \left[\!\!\left[\frac{\partial u}{\partial n} \right]\!\!\right] v \, d\Gamma = 0, \quad \forall v \in V. \tag{3.139}$$

Assuming once again that the traces of the functions in V are dense in $L^2(\partial \omega_c)$, (3.139) becomes

$$\int_{\partial \omega_c} \left[\!\!\left[\frac{\partial u}{\partial n} \right]\!\!\right] v \, d\Gamma = 0, \quad \forall v \in L^2(\partial \omega_c), \tag{3.140}$$

and finally,

$$\left[\!\left[\frac{\partial u}{\partial n}\right]\!\right] \text{ (a.e.) on } \partial\omega_c. \tag{3.141}$$

4. Our converse is completely proven once we observe that the functions v in V have zero discontinuity across $\partial\omega_c$. In particular, the solution u of the variational problem $(\mathbf{VP})_2$ then satisfies

$$[[u]] = u^+ - u^- = 0. \tag{3.142}$$

In conclusion, every solution u of the variational formulation $(\mathbf{VP})_2$ that belongs to $H^2(\Omega)$ is also a solution to the continuous problem $(\mathbf{CP})_2$.

10) An *a priori* estimate of the solution u of the variational problem $(\mathbf{VP})_2$ specified by (3.110)–(3.113) can be obtained using the coercivity of the bilinear form $a(.,.)$ and the continuity of the linear form $L(.)$.

Indeed, according to Questions 3 and 8, there is a strictly positive real number γ defined by

$$\gamma = \begin{vmatrix} \alpha^2 & \text{if } k \geq 1, \\ k\alpha^2 & \text{if } k < 1, \end{vmatrix} \tag{3.143}$$

where the constant α is the one introduced in Question 3 to express the equivalence of the norm $\|.\|_{H^1(\Omega)}$ and the norm defined by

$$\forall v \in H^1(\Omega), \quad |v|_1 \equiv \left[\|v\|_{L^2(\Gamma)}^2 + \|\nabla v\|_{L^2(\Omega)}^2\right]^{1/2}. \tag{3.144}$$

The coercivity of the bilinear form $a(.,.)$ is expressed by

$$\forall v \in V, \quad a(v,v) \geq \gamma\|v\|_{H^1(\Omega)}^2. \tag{3.145}$$

Furthermore, if u solves the variational problem $(\mathbf{VP})_2$, we have

$$a(u,v) = L(v), \quad \forall v \in V. \tag{3.146}$$

If we now insert in (3.146) the special case in which v is equal to the solution u, we have

$$a(u,u) = L(u). \tag{3.147}$$

Therefore, combining the lower bound (3.145) on the bilinear form $a(.,)$ and the upper bound (3.124) on the linear form $L(.)$, we see that the relation (3.147) implies that

$$\gamma\|u\|_{H^1(\Omega)}^2 \leq \left[\|F\|_{L^2(\omega_c)} + kC^+\|u_0\|_{L^2(\Gamma)}\right]\|u\|_{H^1(\Omega)}. \tag{3.148}$$

Dividing through by the norm $\|.\|_{H^1(\Omega)}$ of u yields

$$\|u\|_{H^1(\Omega)} \leq \frac{1}{\gamma} \left[\|F\|_{L^2(\omega_c)} + kC^+ \|u_0\|_{L^2(\Gamma)} \right]. \tag{3.149}$$

The solution u of the variational formulation $(\mathbf{VP})_2$ is thus bounded in the norm $\|.\|_{H^1(\Omega)}$ as a function of the input data F, k, and u_0.

Regarding the constant C^+, according to Poincaré's inequality (1.130), it depends only on Ω.

3.3 Incompressible Fluid

3.3.1 Statement of the Problem

We consider a viscous fluid occupying a bounded and connected open subset Ω of $\mathbb{R}^3$, with a regular boundary $\partial\Omega$. The fluid flow is modeled by the equations for a viscous incompressible fluid, namely Stokes's equations, valid for small velocities and continuous flow.

Let $\mathbf{u}(\mathbf{x})$ be the velocity field, with components $u_i(x)$, $1 \leq i \leq 3$, and let $p(x)$ be the pressure field at each point $x = (x_1, x_2, x_3) \in \Omega$. The constant positive density of the fluid is denoted by ρ and the dynamic viscosity by μ, also constant and positive.

The fluid is subjected to an external force field $\mathbf{f}(\mathbf{x})$ with components $f_i(x)$, $1 \leq i \leq 3$.

The velocity and pressure $(\mathbf{u}, p)$ must solve the Stokes system of partial differential equations (**CP**) defined by

$$(\mathbf{CP}) \quad \begin{cases} -\mu\,\Delta\mathbf{u} + \nabla p = f, & \text{in } \Omega, & (3.150) \\[2mm] \operatorname{div}\mathbf{u} = 0, & \text{in } \Omega, & (3.151) \\[2mm] \mathbf{u} = 0, & \text{on } \partial\Omega. & (3.152) \end{cases}$$

▶ **Mathematical Preliminaries and Variational Formulation**

We define the following function spaces:

$$\left(L^2(\Omega)\right)^3 = \left\{\mathbf{v} = (v_1, v_2, v_3), \ v_i \in L^2(\Omega), \ i = 1, 2, 3\right\}, \tag{3.153}$$

$$\left(H^1(\Omega)\right)^3 = \left\{\mathbf{v} = (v_1, v_2, v_3), \ v_i \in H^1(\Omega), \ i = 1, 2, 3\right\}, \tag{3.154}$$

$$\left(H_0^1(\Omega)\right)^3 = \left\{\mathbf{v} \in (H^1(\Omega))^3, \ v_i(x) = 0, \ x \in \partial\Omega, \ i = 1, 2, 3\right\}, \tag{3.155}$$

$$\left(H^2(\Omega)\right)^3 = \left\{\mathbf{v} = (v_1, v_2, v_3), \ v_i \in H^2(\Omega), \ i = 1, 2, 3\right\}, \tag{3.156}$$

$$V = \left\{\mathbf{v} \in \left(H_0^1(\Omega)\right)^3, \ \operatorname{div}\mathbf{v} = 0 \text{ in } \Omega\right\}, \tag{3.157}$$

and the norms:

$$\|\mathbf{v}\| = \left[\sum_{i=1}^3 \|v_i\|^2_{H^1(\Omega)}\right]^{1/2}, \tag{3.158}$$

$$|\mathbf{v}| = \left[\sum_{i=1}^3 \|\nabla v_i\|^2_{L_2(\Omega)}\right]^{1/2} = \left[\sum_{i,j=1}^3 \left\|\frac{\partial v_i}{\partial x_j}\right\|^2_{L_2(\Omega)}\right]^{1/2}. \tag{3.159}$$

1) Show that the space V is a closed vector subspace of the space $\left(H_0^1(\Omega)\right)^3$ for the norm $\|\,.\,\|$ defined by (3.158).

2) Establish the equivalence of the norms $\|\,.\,\|$ et $|\,.\,|$ on V and deduce that the space V is a Hilbert space for the norm $|\,.\,|$ defined by (3.159).

3) Show that for every field $\mathbf{v}$ belonging to $\left(H^1(\Omega)\right)^3$ and with zero divergence, and for every function p belonging to $H^1(\Omega)$, we have

$$\operatorname{div}(\mathbf{v}p) = \mathbf{v}\cdot\nabla p. \tag{3.160}$$

Deduce that for every pair $(\mathbf{v}, p) \in V \times H^1(\Omega)$, we have

$$\int_\Omega \mathbf{v}\cdot\nabla p\, d\Omega = 0. \tag{3.161}$$

Hint: Use the Green–Ostrogradski formula: for a vector function $\mathbf{v} \in \left(H^1(\Omega)\right)^3$, if Ω is a sufficiently regular bounded open subset of $\mathbb{R}^3$, we have

$$\int_\Omega \operatorname{div}\mathbf{v}\, d\Omega = \int_{\partial\Omega} \mathbf{v}\cdot\mathbf{n}\, d\Gamma.$$

4) We now assume that the force field $\mathbf{f}$ is a given function in $(L^2(\Omega))^3$. Show that if $\mathbf{u} \in \left(H^2(\Omega)\right)^3$ solves the continuous problem **(CP)**, then $\mathbf{u}$ also solves the variational problem **(VP)** defined as follows:

$$\textbf{(VP)} \quad \begin{cases} \text{Find } \mathbf{u} \in V \text{ solution of} \\ a(\mathbf{u}, \mathbf{v}) = L(\mathbf{v}), \quad \forall \mathbf{v} \in V, \end{cases} \tag{3.162}$$

where $a(.,.)$ and $L(.)$ are given by

$$a(\mathbf{u}, \mathbf{v}) = \mu \int_\Omega \nabla\mathbf{u}\cdot\nabla\mathbf{v}\, d\Omega, \qquad L(\mathbf{v}) = \int_\Omega \mathbf{f}\cdot\mathbf{v}\, d\Omega, \tag{3.163}$$

with

$$\mathbf{f}\cdot\mathbf{v} = \sum_{i=1}^{3} f_i v_i, \tag{3.164}$$

and

$$\nabla\mathbf{u}\cdot\nabla\mathbf{v} = \sum_{i=1}^{3} \nabla u_i\cdot\nabla v_i = \sum_{i,j=1}^{3} \frac{\partial u_i}{\partial x_j}\frac{\partial v_i}{\partial x_j}. \tag{3.165}$$

▶ **Existence and Uniqueness of the Solution to the Variational Formulation.**
A Priori **Estimation**

5) Show that $a(.,.)$ is a continuous and coercive bilinear form on $V \times V$.

6) Show that $L(.)$ is a continuous linear form on V.

7) Deduce the existence and uniqueness of the solution to the variational problem **(VP)**.

8) Show directly, without using the Lax–Milgram theorem, that the solution to the variational problem **(VP)** is unique, then that this solution exists, by application of the Riesz representation theorem.

9) Show that the unique solution **u** to the variational problem **(VP)** is also the unique minimum of a functional $J(\mathbf{v})$ on the space V, to be specified.

10) To demonstrate the equivalence of the continuous problem **(CP)** and the variational formulation **(VP)**, the aim will be to establish a new velocity–pressure formulation **(PV*)** equivalent to **(VP)** and then show that it is also equivalent to the continuous problem **(CP)**.

To do this, assume the following result:

De Rham's Lemma: If Ω is a regular bounded subset of $\mathbb{R}^3$ and $l(\mathbf{v})$ is a continuous linear form on $\left(H_0^1(\Omega)\right)^3$, then the form $l(\mathbf{v})$ vanishes on V if and only if there exists a scalar function $q \in L^2(\Omega)$ such that

$$l(\mathbf{v}) = \int_\Omega q \operatorname{div} \mathbf{v} \, d\Omega, \quad \forall \mathbf{v} \in \left(H_0^1(\Omega)\right)^3. \qquad (3.166)$$

Show that under the hypotheses of de Rham's Lemma, the function q is defined up to an additive constant.

11) By making a particular choice for the linear form $l(.)$, use de Rham's Lemma to deduce that the variational formulation **(VP)** is equivalent to the variational formulation **(VP*)** defined as follows:

$$\textbf{(VP*)} \quad \begin{cases} \text{Find } (\mathbf{u}, p) \in V \times L^2(\Omega) \text{ solution of} \\ \mu \displaystyle\int_\Omega \nabla\mathbf{u} \cdot \nabla\mathbf{v} \, d\Omega - \int_\Omega \mathbf{f} \cdot \mathbf{v} \, d\Omega = \int_\Omega p \operatorname{div} \mathbf{v} \, d\Omega, \quad \forall \mathbf{v} \in \left(H_0^1(\Omega)\right)^3. \end{cases}$$

12) Assuming that the solution $(\mathbf{u}, p)$ of the variational problem **(VP*)** belongs to $\left(H^2(\Omega)\right)^3 \times H^1(\Omega)$, prove the equivalence of the continuous problem **(CP)** and the variational formulation **(VP*)**.

13) *A priori* estimate. Show that if **u** solves the problem (**VP**), then there is a constant $C > 0$ such that

$$|\mathbf{u}| \le C\|\mathbf{f}\|_{(L^2(\Omega))^3},$$

where $|\,.\,|$ is the norm defined by (3.159).

3.3.2 Solution

1) Note first that V is obviously a vector subspace of $\left(H_0^1(\Omega)\right)^3$. This follows from the linearity of the divergence operator and the homogeneous Dirichlet condition on $\partial\Omega$.

To show that V is closed in $\left(H_0^1(\Omega)\right)^3$ for the norm $\|\,.\,\|$ defined by (3.158), we consider a sequence of functions $(\mathbf{v}_n)_{n\in\mathbb{N}}$ in V that converges to v for the norm $\|\,.\,\|$.

It suffices then to show that the limit $\mathbf{v}$ of the sequence $(\mathbf{v}_n)_{n\in\mathbb{N}}$ is also in V, i.e., that $\mathbf{v}$ has zero divergence.

Now,

$$\text{div } \mathbf{v} = \text{div } (\mathbf{v} - \mathbf{v}_n) + \text{div } \mathbf{v}_n = \text{div } (\mathbf{v} - \mathbf{v}_n) \tag{3.167}$$

$$= \sum_{i=1}^{3} \left[\frac{\partial(v_n)_i}{\partial x_i} - \frac{\partial v_i}{\partial x_i} \right], \quad \forall n \in \mathbb{N}, \tag{3.168}$$

where we note that $\mathbf{v}_n$ has zero divergence because it is an element of V. Furthermore, $\mathbf{v}_n$ converges to $\mathbf{v}$ for the norm $\|\,.\,\|$ defined by (3.158).

It follows that

$$\forall i = 1, 2, 3, \quad \left\| \frac{\partial(v_n)_i}{\partial x_i} - \frac{\partial v_i}{\partial x_i} \right\|_{L^2(\Omega)} \longrightarrow 0. \tag{3.169}$$

Then (3.167) and (3.168) imply that for all n in $\mathbb{N}$,

$$\|\text{div } \mathbf{v}\|_{L^2(\Omega)} = \|\text{div } (\mathbf{v} - \mathbf{v}_n)\|_{L^2(\Omega)} = \left\| \sum_{i=1}^{3} \left[\frac{\partial(v_n)_i}{\partial x_i} - \frac{\partial v_i}{\partial x_i} \right] \right\|_{L^2(\Omega)}$$

$$\leq \sum_{i=1}^{3} \left\| \frac{\partial(v_n)_i}{\partial x_i} - \frac{\partial v_i}{\partial x_i} \right\|_{L^2(\Omega)}. \tag{3.170}$$

Taking the limit as n tends to $+\infty$ yields

$$\|\text{div } \mathbf{v}\|_{L^2(\Omega)} = 0, \tag{3.171}$$

and div $\mathbf{v}$ is zero almost everywhere in Ω, ensuring that $\mathbf{v}$ is indeed an element of V. Consequently, V is closed in $\left(H_0^1(\Omega)\right)^3$ for the norm $\|\,.\,\|$.

2) The equivalence of the norms $\|\,.\,\|$ and $|\,.\,|$ on V, as defined by (3.158) and (3.159), is a direct consequence of Poincaré's inequality, which is valid in V as a vector subspace of $\left(H_0^1(\Omega)\right)^3$. Indeed, we can then write

$\exists\, C > 0$ such that $\forall\, v_i \in H_0^1(\Omega)$, $\quad \|v_i\|^2_{L^2(\Omega)} \leq C\|\nabla v_i\|^2_{L^2(\Omega)}$, $\quad i \in \{1, 2, 3\}$.

$$(3.172)$$

Adding the L^2 norm of the gradient of v_i to each side of the inequality (3.172), we obtain

$$\|v_i\|^2_{H^1(\Omega)} \leq (1 + C)\|\nabla v_i\|^2_{L^2(\Omega)}, \quad i \in \{1, 2, 3\}. \tag{3.173}$$

Hence, returning to the definition (3.158) of the norm $\|\,.\,\|$, we obtain

$$\|\mathbf{v}\| = \left[\sum_{i=1}^{3} \|v_i\|^2_{H^1(\Omega)}\right]^{1/2} \leq \left[\sum_{i=1}^{3} (1 + C)\|\nabla v_i\|^2_{L^2(\Omega)}\right]^{1/2} \leq \sqrt{1 + C}\,|\mathbf{v}|. \tag{3.174}$$

Finally, there exist $\alpha = 1$ and $\beta = \sqrt{1 + C}$ such that

$$\forall\, \mathbf{v} \in V, \quad \alpha|\mathbf{v}| \leq \|\mathbf{v}\| \leq \beta|\mathbf{v}|, \tag{3.175}$$

and the norms $\|\,.\,\|$ and $|\,.\,|$ defined by (3.158) and (3.159) are indeed equivalent on V.

We deduce that V equipped with the norm $|\,.\,|$ is a Hilbert space. Indeed, in Question 1, we established that V was a closed vector subspace of $\left(H_0^1(\Omega)\right)^3$ for the norm $\|\,.\,\|$, which we have just proved to be equivalent to $|\,.\,|$.

Therefore, V is also closed for the norm $|\,.\,|$, and is thus a Hilbert space for this same norm, according to Lemma 1.8.

Note. The norm $|\,.\,|$ indeed comes from a scalar product φ on $V \times V$ defined by

$$\begin{aligned} \varphi : V \times V &\longrightarrow \mathbb{R} \\ (\mathbf{u}, \mathbf{v}) &\longmapsto \varphi(\mathbf{u}, \mathbf{v}) \equiv \int_{\Omega} \nabla \mathbf{u} \cdot \nabla \mathbf{v}\, d\Omega. \end{aligned} \tag{3.176}$$

In particular, note that $\varphi(\mathbf{v}, \mathbf{v}) = 0$ implies $\mathbf{v} = \mathbf{0}$, since the vector field $\mathbf{v}$ also belongs to $\left(H_0^1(\Omega)\right)^3$.

Finally, for all functions $\mathbf{v}$ in V, we have

$$\begin{aligned} \varphi(\mathbf{v}, \mathbf{v}) &= \int_{\Omega} \nabla \mathbf{v} \cdot \nabla \mathbf{v}\, d\Omega = \int_{\Omega} \sum_{i=1}^{3} |\nabla v_i|^2\, d\Omega \\ &= \sum_{i=1}^{3} \int_{\Omega} |\nabla v_i|^2\, d\Omega = \sum_{i=1}^{3} \|\nabla v_i\|^2_{L_2(\Omega)} \\ &= |\mathbf{v}|^2. \end{aligned}$$

3) Let $\mathbf{v}$ be a velocity field in $\left(H^1(\Omega)\right)^3$ with zero divergence and let p be a pressure field in $H^1(\Omega)$. We then have

$$\operatorname{div}(\mathbf{v}p) = p \operatorname{div}\mathbf{v} + \mathbf{v}\cdot\nabla p = \mathbf{v}\cdot\nabla p. \tag{3.177}$$

For every pair $(\mathbf{v}, p) \in V \times H^1(\Omega)$, we then deduce that

$$\int_\Omega \mathbf{v}\cdot\nabla p \, d\Omega = \int_\Omega \operatorname{div}(\mathbf{v}p) \, d\Omega = \int_{\partial\Omega} p\,\mathbf{v}\cdot\mathbf{n}\,d\Gamma, \tag{3.178}$$

by the Green–Ostrogradski formula. Now every vector field $\mathbf{v}$ of V vanishes on the boundary $\partial\Omega$ of Ω. Hence

$$\forall(\mathbf{v}, p) \in V \times H^1(\Omega), \quad \int_\Omega \mathbf{v}\cdot\nabla p \, d\Omega = 0. \tag{3.179}$$

4) Let $\mathbf{v}$ be a test vector field with components v_i in a function space to be determined. We multiply the ith Stokes Eq. (3.150), ($i = 1, 2, 3$), by v_i and integrate the resulting equation over Ω to obtain

$$-\mu \int_\Omega \Delta u_i \cdot v_i \, d\Omega + \int_\Omega \frac{\partial p}{\partial x_i} v_i \, d\Omega = \int_\Omega f_i v_i \, d\Omega, \quad \forall i \in \{1, 2, 3\}. \tag{3.180}$$

Applying Green's formula to the first integral of (3.180) leads to

$$\mu \int_\Omega \nabla u_i \cdot \nabla v_i \, d\Omega - \int_{\partial\Omega} \frac{\partial u_i}{\partial n} v_i \, d\Gamma + \int_\Omega \frac{\partial p}{\partial x_i} v_i \, d\Omega = \int_\Omega f_i v_i \, d\Omega. \tag{3.181}$$

Equation (3.181) does not directly involve the value of the solution vector field $\mathbf{u}$ on the boundary of Ω, so we require the test vector fields $\mathbf{v}$, among which we shall seek the solution $\mathbf{u}$, to vanish on $\partial\Omega$:

$$\mathbf{v} = \mathbf{0} \text{ on } \partial\Omega. \tag{3.182}$$

Consequently, (3.181) can be written

$$\mu \int_\Omega \nabla u_i \cdot \nabla v_i \, d\Omega + \int_\Omega \frac{\partial p}{\partial x_i} v_i \, d\Omega = \int_\Omega f_i v_i \, d\Omega, \quad \forall \mathbf{v} \text{ s.t. } \mathbf{v} = \mathbf{0} \text{ on } \partial\Omega. \tag{3.183}$$

Likewise, the zero divergence condition on the solution $\mathbf{u}$ of the continuous problem **(CP)** does not appear in the formulation (3.183). In addition, we thus require the test vector fields $\mathbf{v}$ to satisfy

$$\operatorname{div}\mathbf{v} = 0. \tag{3.184}$$

Finally, assuming that the vector field $\mathbf{f}$ is a given element of $\left(L^2(\Omega)\right)^3$, and given the structure of the integrals appearing in the formulation (3.183), it is natural to consider test vector fields $\mathbf{v}$ in $\left(H^1(\Omega)\right)^3$ and a pressure field p in $H^1(\Omega)$ in order to guarantee the convergence of the given integrals.

In this case, Question 3 ensures that

$$\int_\Omega \mathbf{v} \cdot \nabla p \, d\Omega = \sum_{i=1}^{3} \int_\Omega \frac{\partial p}{\partial x_i} v_i \, d\Omega = 0. \tag{3.185}$$

Having summed i from 1 to 3, we see that the formulation (3.183) implies that

$$\mu \sum_{i=1}^{3} \int_\Omega \nabla u_i \cdot \nabla v_i \, d\Omega = \sum_{i=1}^{3} \int_\Omega f_i v_i \, d\Omega, \quad \forall \mathbf{v} \in V, \tag{3.186}$$

where V is defined by bringing together all the different conditions imposed on the test vector field $\mathbf{v}$, whence

$$V \equiv \left\{ \mathbf{v} \in \left(H_0^1(\Omega) \right)^3, \quad \text{div } \mathbf{v} = 0 \text{ in } \Omega \right\}. \tag{3.187}$$

Finally, note that (3.186) can be written in vector notation as

$$\mu \int_\Omega \nabla \mathbf{u} \cdot \nabla \mathbf{v} \, d\Omega = \int_\Omega \mathbf{f} \cdot \mathbf{v} \, d\Omega, \quad \forall \mathbf{v} \in V. \tag{3.188}$$

Introducing the forms $a(.,.)$ and $L(.)$ defined by (3.163), we conclude that the solution velocity field $\mathbf{u}$ satisfies the variational formulation (**VP**) specified by the following:

$$\textbf{(VP)} \quad \begin{cases} \text{Find } \mathbf{u} \in V \text{ solution of} \\ a(\mathbf{u}, \mathbf{v}) = L(\mathbf{v}), \quad \forall \mathbf{v} \in V. \end{cases}$$

5) The following properties of the map $a(.,.)$ defined by (3.163) are clear by inspection:

- The map $a(.,.)$ is real-valued and is therefore a form.

- $a(.,.)$ is symmetric, i.e., $\forall (\mathbf{u}, \mathbf{v}) \in V \times V$, we have $a(\mathbf{u}, \mathbf{v}) = a(\mathbf{v}, \mathbf{u})$.

- $a(.,.)$ is a bilinear form. By symmetry of $a(.,.)$, it suffices to show that

$$\forall \lambda \in \mathbb{R}, \quad \forall (\mathbf{u}_1, \mathbf{u}_2) \in V \times V, \quad \forall \mathbf{v} \in V,$$

$$a(\mathbf{u}_1 + \lambda \mathbf{u}_2, \mathbf{v}) = a(\mathbf{u}_1, \mathbf{v}) + \lambda a(\mathbf{u}_2, \mathbf{v}),$$

using the linearity of the gradient and integral operators.

- $a(.,.)$ is continuous on $V \times V$. Since $\forall (\mathbf{u}, \mathbf{v}) \in V \times V$, we have

$$|a(\mathbf{u}, \mathbf{v})| = \mu \left| \sum_{i=1}^{3} \int_{\Omega} \nabla u_i \cdot \nabla v_i \, d\Omega \right| \tag{3.189}$$

$$\leq \mu \sum_{i=1}^{3} \left| \int_{\Omega} \nabla u_i \cdot \nabla v_i \, d\Omega \right| \tag{3.190}$$

$$\leq \mu \sum_{i=1}^{3} \|\nabla u_i\|_{L^2(\Omega)} \|\nabla v_i\|_{L^2(\Omega)} \tag{3.191}$$

$$\leq \mu \left[\sum_{i=1}^{3} \|\nabla u_i\|_{L^2(\Omega)}^2 \right]^{1/2} \left[\sum_{i=1}^{3} \|\nabla v_i\|_{L^2(\Omega)}^2 \right]^{1/2} \tag{3.192}$$

$$\leq \mu |\mathbf{u}| |\mathbf{v}|, \tag{3.193}$$

where we have used the triangle inequality in $\mathbb{R}^3$ in going from (3.189) to (3.190), the Cauchy–Schwarz inequality in $L^2(\Omega)$ in going from (3.190) to (3.191), the Cauchy–Schwarz inequality in $\mathbb{R}^3$ in going from (3.191) to (3.192), and the definition (3.159) of the norm $|.|$ in going from (3.192) to (3.193).

The bilinear form $a(.,.)$ is thus continuous on $V \times V$ for the norm $|.|$ defined by (3.159):

$$\exists \, M \equiv \mu > 0 \text{ such that } \forall \, (\mathbf{u}, \mathbf{v}) \in V \times V, \quad |a(\mathbf{u}, \mathbf{v})| \leq M |\mathbf{u}| \, |\mathbf{v}|. \tag{3.194}$$

• The bilinear form $a(.,.)$ is coercive. This result is trivial in our case. It follows from the choice of norm $|.|$. Indeed, we have

$$\forall \mathbf{v} \in V, \quad a(\mathbf{v}, \mathbf{v}) = \mu \varphi(\mathbf{v}, \mathbf{v}) = \mu |\mathbf{v}|^2 \geq \mu |\mathbf{v}|^2, \tag{3.195}$$

where φ denotes the scalar product (3.176) that yields the norm $|.|$. The coercivity of the form $a(.,.)$ then follows, since

$$\exists \, \alpha \equiv \mu > 0 \text{ such that } \forall \mathbf{v} \in V, \quad a(\mathbf{v}, \mathbf{v}) \geq \alpha |\mathbf{v}|^2. \tag{3.196}$$

6) The map $L(.)$ defined by (3.163) trivially satisfies the following properties:

• The map $L(.)$ is real-valued and is thus a form.

• The linearity of $L(.)$ follows from the linearity of the integral operator.

• $L(.)$ is continuous on V, since for every $\mathbf{v}$ in V, we have

$$|L(\mathbf{v})| = \left| \int_{\Omega} \mathbf{f} \cdot \mathbf{v} \, d\Omega \right| \tag{3.197}$$

$$\leq \|\mathbf{f}\|_{[L^2(\Omega)]^3} \|\mathbf{v}\|_{[L^2(\Omega)]^3} \tag{3.198}$$

$$\leq \|\mathbf{f}\|_{[L^2(\Omega)]^3} \|\mathbf{v}\|, \tag{3.199}$$

using the Cauchy–Schwarz inequality in $(L^2(\Omega))^3$ and the definition (3.158) of the norm $\|\cdot\|$. Furthermore, in Question 2, we established that the norm $\|\cdot\|$ is equivalent to $|\cdot|$ [see (3.175)]:

$$\forall \mathbf{v} \in V, \quad |\mathbf{v}| \leq \|\mathbf{v}\| \leq \sqrt{1 + C} \, |\mathbf{v}|. \tag{3.200}$$

The inequality (3.199) thus implies that

$$|L(\mathbf{v})| \leq \sqrt{1 + C} \, \|\mathbf{f}\|_{[L^2(\Omega)]^3} |\mathbf{v}|. \tag{3.201}$$

We thereby establish that $L(.)$ is continuous for the norm $|\cdot|$ defined by (3.159), since

$$\exists \, M \equiv \sqrt{1 + C} \, \|\mathbf{f}\|_{[L^2(\Omega)]^3} > 0 \ \text{ such that } \ \forall \mathbf{v} \in V, \quad |L(\mathbf{v})| \leq M|\mathbf{v}|. \tag{3.202}$$

7) All the conditions of the Lax–Milgram theorem have been shown to hold in the previous questions. Note in particular that we have established the Hilbert space structure of V, the continuity and coercivity of the bilinear form $a(.,.)$, and the continuity of the linear form $L(.)$, and all these properties have been demonstrated for *one and the same* norm $|\cdot|$.

It follows that there exists a unique solution $\mathbf{u}$ in V satisfying the variational problem **(VP)** specified by the following:

$$\textbf{(VP)} \quad \begin{cases} \text{Find } \mathbf{u} \in V \text{ solution of} \\ a(\mathbf{u}, \mathbf{v}) = L(\mathbf{v}), \quad \forall \mathbf{v} \in V, \end{cases} \tag{3.203}$$

where $a(.,.)$ and $L(.)$ are defined by (3.163).

8) The existence and uniqueness of the solution $\mathbf{u}$ of the variational problem **(VP)** can be established directly without appealing to the Lax–Milgram theorem:

• **Uniqueness.** Let $\mathbf{u}_1$ and $\mathbf{u}_2$ be two solutions to the variational problem **(VP)**. Then

$$\begin{cases} a(\mathbf{u}_1, \mathbf{v}) = L(\mathbf{v}), \quad \forall \mathbf{v} \in V, \\ a(\mathbf{u}_2, \mathbf{v}) = L(\mathbf{v}), \quad \forall \mathbf{v} \in V. \end{cases} \tag{3.204}$$

By subtracting one from the other and exploiting the bilinearity of the form $a(.,.)$ and the linearity of the form $L(.)$, it follows that

$$a(\mathbf{u}_2 - \mathbf{u}_1, \mathbf{v}) = 0, \quad \forall \mathbf{v} \in V. \tag{3.205}$$

In particular, if we choose $\mathbf{v} = \mathbf{u}_2 - \mathbf{u}_1$, we have

$$a(\mathbf{u}_2 - \mathbf{u}_1, \mathbf{u}_2 - \mathbf{u}_1) = 0. \tag{3.206}$$

Since the bilinear form $a(., .)$ is coercive for the norm $|.|$, we have

$$\exists \alpha > 0 \text{ such that } \forall \mathbf{v} \in V, \quad a(\mathbf{v}, \mathbf{v}) \geq \alpha |\mathbf{v}|^2. \tag{3.207}$$

Choosing $\mathbf{v} = \mathbf{u}_2 - \mathbf{u}_1$, we then obtain

$$a(\mathbf{u}_2 - \mathbf{u}_1, \mathbf{u}_2 - \mathbf{u}_1) \geq \alpha |\mathbf{u}_2 - \mathbf{u}_1|^2. \tag{3.208}$$

Using the property (3.206) in (3.208), it follows that

$$0 \geq \alpha |\mathbf{u}_2 - \mathbf{u}_1|^2. \tag{3.209}$$

As a consequence, $|\mathbf{u}_2 - \mathbf{u}_1|$ is zero, implying that the difference $\mathbf{u}_2 - \mathbf{u}_1$ is also zero, and hence that $\mathbf{u}_2$ is equal to $\mathbf{u}_1$. This proves the uniqueness of the solution $\mathbf{u}$ of the variational problem (**VP**).

- **Existence.** To prove the existence of the solution $\mathbf{u}$ to the variational problem (**VP**), we note that the bilinear form $a(., .)$ defined by (3.163) differs from the scalar product φ defined by (3.176) by the multiplicative factor μ.

  This means that $a(., .)$ is also a scalar product, denoted by $(., .)_1$, on $V \times V$, associated with a norm $|.|_1$, which is nothing but the norm $|.|$ multiplied by μ.

  Put another way, if we introduce the map $|.|_1$ defined by

$$|.|_1 : V \longrightarrow \mathbb{R}_+$$

$$\mathbf{v} \longmapsto |\mathbf{v}|_1 \equiv \sqrt{a(\mathbf{v}, \mathbf{v})},$$

we have

$$\forall \mathbf{v} \in V, \quad |\mathbf{v}|_1 = \sqrt{(\mathbf{v}, \mathbf{v})_1} \equiv \sqrt{a(\mathbf{v}, \mathbf{v})}. \tag{3.210}$$

Since the space V is also a Hilbert space for this norm $|.|_1$, and since the form $L(.)$ is linear and continuous on V for this same norm, the Riesz representation theorem (Theorem 1.7) for linear forms implies that

$$\exists! \, \mathbf{u} \in V \text{ such that } L(\mathbf{v}) = (\mathbf{u}, \mathbf{v})_1 \equiv a(\mathbf{u}, \mathbf{v}). \tag{3.211}$$

This proves the existence and uniqueness in the space V of the solution $\mathbf{u}$ to the variational problem (**VP**).

9) As we saw in Question 1, the space V is a closed subspace of the Sobolev space $\left(H_0^1(\Omega)\right)^3$ for the norm $\|\,.\,\|$. Since this norm is equivalent to the norm $|\,.\,|$ defined by (3.159) (see Question 2), all the properties we have demonstrated for the bilinear form $a(.,.)$ and the linear form $L(.)$ for the norm $|\,.\,|$ will also be valid for the norm $\|\,.\,\|$.

In addition, since the bilinear form $a(.,.)$ defined by (3.163) is symmetric, it is a direct consequence of Stampacchia's theorem (see Theorem 1.10 in Sect. 1.6) that the variational problem **(VP)** is equivalent to the minimization problem specified as follows:

$$
\textbf{(MP)} \quad
\begin{cases}
\text{Find } \mathbf{u} \in V \text{ solution of} \\[4pt]
J(\mathbf{u}) \equiv \min_{\mathbf{v} \in V} J(\mathbf{v}), \\[8pt]
\text{where } J(\mathbf{v}) = \dfrac{1}{2}\, a(\mathbf{v}, \mathbf{v}) - L(\mathbf{v}).
\end{cases}
$$

Note. The equivalence between the variational formulation **(VP)** and the minimization problem **(MP)** can be established directly by applying Theorem 1.13, also presented in Sect. 1.6.

10) Let $l(\mathbf{v})$ be a continuous linear form on $\left(H_0^1(\Omega)\right)^3$, vanishing on the subspace V of $\left(H_0^1(\Omega)\right)^3$. By de Rham's Lemma, there is q in $L^2(\Omega)$ such that

$$
l(\mathbf{v}) = \int_\Omega q \, \mathrm{div}\, \mathbf{v} \, \mathrm{d}\Omega, \quad \forall \mathbf{v} \in (H_0^1(\Omega))^3. \tag{3.212}
$$

Regarding the uniqueness of q, suppose there are two solutions q_1 and q_2 of (3.212). Then

$$
l(\mathbf{v}) = \int_\Omega q_1 \, \mathrm{div}\, \mathbf{v} \, \mathrm{d}\Omega = \int_\Omega q_2 \, \mathrm{div}\, \mathbf{v} \, \mathrm{d}\Omega, \quad \forall \mathbf{v} \in (H_0^1(\Omega))^3, \tag{3.213}
$$

and hence

$$
\int_\Omega (q_2 - q_1) \mathrm{div}\, \mathbf{v} \, \mathrm{d}\Omega = 0, \quad \forall \mathbf{v} \in (H_0^1(\Omega))^3. \tag{3.214}
$$

Moreover, since $\left(C_0^\infty(\Omega)\right)^3$ is contained in $\left(H_0^1(\Omega)\right)^3$, we may consider in (3.214) the special case that the vector fields $\mathbf{v}$ belong to the space $\left(C_0^\infty(\Omega)\right)^3$. We then have

$$
\int_\Omega (q_2 - q_1) \mathrm{div}\, \mathbf{v} \, \mathrm{d}\Omega = 0, \quad \forall \mathbf{v} \in \left(C_0^\infty(\Omega)\right)^3. \tag{3.215}
$$

Since we only have L^2 regularity for q_1 and q_2, Eq. (3.215) can be transformed only using the derivative in the sense of distributions (see Sect. 1.4):

$$\int_{\Omega} (q_2 - q_1) \operatorname{div} \mathbf{v} \, d\Omega = \sum_{i=1}^{3} \int_{\Omega} (q_2 - q_1) \frac{\partial v_i}{\partial x_i} \, d\Omega = \sum_{i=1}^{3} \left\langle q_2 - q_1, \frac{\partial v_i}{\partial x_i} \right\rangle \quad (3.216)$$

$$= -\sum_{i=1}^{3} \left\langle \frac{\partial}{\partial x_i} (q_2 - q_1), v_i \right\rangle = 0, \quad \forall \mathbf{v} \in \left(C_0^{\infty}(\Omega) \right)^3.$$

By considering successively vector fields $\mathbf{v}$ in which two out of the three components are zero, we must have

$$\frac{\partial}{\partial x_i} (q_2 - q_1) = 0 \quad \text{in } \mathscr{D}'(\Omega), \quad \forall i \in \{1, 2, 3\}. \quad (3.217)$$

Since the open set Ω is assumed to be connected, we deduce that

$$q_2 - q_1 = \text{constant (a.e.) in } \Omega. \quad (3.218)$$

In other words, the functions q in $L^2(\Omega)$ that solve (3.212) are defined up to an additive constant.

11) Let $(\mathbf{u}, p)$ be a solution of the variational problem **(VP*)** defined by

(VP*) $\left\{ \begin{array}{l} \text{Find } (\mathbf{u}, p) \in V \times L^2(\Omega) \text{ solution of} \\[2mm] \mu \int_{\Omega} \nabla \mathbf{u} \cdot \nabla \mathbf{v} \, d\Omega - \int_{\Omega} \mathbf{f} \cdot \mathbf{v} \, d\Omega = \int_{\Omega} p \operatorname{div} \mathbf{v} \, d\Omega, \quad \forall \mathbf{v} \in \left(H_0^1(\Omega) \right)^3, \end{array} \right.$

$$(3.219)$$

Choosing zero-divergence vector fields $\mathbf{v}$ in (3.219), we immediately find that $\mathbf{u}$ solves the variational problem **(VP)**.

Conversely, let $\mathbf{u}$ in V be a solution of the variational problem **(VP)**. We introduce the linear form $l(.)$ defined by

$$l : V \longrightarrow \mathbb{R}$$

$$\mathbf{v} \longmapsto l(\mathbf{v}) \equiv \mu \int_{\Omega} \nabla \mathbf{u} \cdot \nabla \mathbf{v} \, d\Omega - \int_{\Omega} \mathbf{f} \cdot \mathbf{v} \, d\Omega.$$

The continuity of the bilinear form $a(., .)$ and the linear form $L(.)$ for the norm $| \, . \, |$ ensure that the linear form $l(.)$ is also continuous on $(H_0^1(\Omega))^3$ for the same norm. Furthermore, $l(.)$ vanishes on V if $\mathbf{u}$ solves the variational problem **(VP)**.

By de Rham's Lemma, there exists a function p in $L^2(\Omega)$, determined up to an additive constant, such that

$$l(\mathbf{v}) = \int_{\Omega} p \operatorname{div} \mathbf{v} \, d\Omega, \quad \forall \mathbf{v} \in \left(H_0^1(\Omega) \right)^3. \quad (3.220)$$

It follows that $(\mathbf{u}, p)$ solves the variational problem (**VP***).

12) The fact that a solution to the continuous problem (**CP**) is a solution to the variational problem (**VP**) follows from construction of the latter, which we know to be equivalent to the formulation (**VP***), according to the last question.

Let us thus show that every solution $(\mathbf{u}, p)$ of the variational formulation (**VP***) is also a solution of the continuous problem (**CP**).

To this end, in the formulation (3.219), we consider vector fields $\mathbf{v}$ belonging to $(\mathscr{D}(\Omega))^3$, which we know to be dense in $(H_0^1(\Omega))^3$ by the very definition of $(H_0^1(\Omega))^3$.

The solution $(\mathbf{u}, p)$ is also a solution of the following problem:

$$\begin{cases} \text{Find } (\mathbf{u}, p) \in V \times L^2(\Omega) \text{ solution of} \\ \mu \int_\Omega \nabla \mathbf{u} \cdot \nabla \mathbf{v} \, d\Omega - \int_\Omega \mathbf{f} \cdot \mathbf{v} \, d\Omega = \int_\Omega p \operatorname{div} \mathbf{v} \, d\Omega, \quad \forall \mathbf{v} \in (\mathscr{D}(\Omega))^3. \end{cases} \quad (3.221)$$

We now rewrite (3.221) by expanding out the scalar products in order to bring in the components of the vectors in each of the integrals:

$$\begin{cases} \text{Find } (\mathbf{u}, p) \in V \times L^2(\Omega) \text{ solution of} \\ \displaystyle\sum_{i=1}^3 \left(\int_\Omega \mu \nabla u_i \cdot \nabla v_i \, d\Omega - \int_\Omega f_i v_i \, d\Omega \right) = \sum_{i=1}^3 \int_\Omega p \frac{\partial v_i}{\partial x_i} \, d\Omega, \ \forall \mathbf{v} \in (\mathscr{D}(\Omega))^3. \end{cases}$$
$$(3.222)$$

If we also assume that the solution $(\mathbf{u}, p)$ belongs to $\left((H^2(\Omega))^3, H^1(\Omega) \right)$, we may apply Green's formula to obtain

$$\sum_{i=1}^3 \int_\Omega \left[-\mu (\Delta u_i) v_i \, d\Omega - f_i v_i \right] d\Omega = -\sum_{i=1}^3 \int_\Omega \frac{\partial p}{\partial x_i} v_i \, d\Omega, \quad \forall \mathbf{v} \in (\mathscr{D}(\Omega))^3,$$
$$(3.223)$$

taking into account the fact that the test vectors $\mathbf{v}$ vanish on the boundary $\partial\Omega$, since they belong to $(\mathscr{D}(\Omega))^3$.

Equation (3.223) can then be written in vector form:

$$\int_\Omega (\mu \Delta \mathbf{u} - \nabla p + \mathbf{f}) \cdot \mathbf{v} \, d\Omega = 0, \quad \forall \mathbf{v} \in (\mathscr{D}(\Omega))^3. \quad (3.224)$$

Noting that $(\mu \Delta \mathbf{u} - \nabla p + \mathbf{f})$ is in $(L^2(\Omega))^3$, since we assume that the solution $(\mathbf{u}, p)$ belongs to $\left((H^2(\Omega))^3, H^1(\Omega) \right)$, we can either use a density argument, appealing to the fact that $\left(\mathscr{D}(\Omega)\right)^3$ is dense in $\left(L^2(\Omega)\right)^3$, or apply the theory of distributions, to conclude that

$$\mu \Delta \mathbf{u} - \nabla p + \mathbf{f} = 0 \quad \text{(a.e.) in } \Omega. \tag{3.225}$$

Since we also know that $\mathbf{u}$ belongs to $V \cap \left(H^2(\Omega)\right)^3$, it follows by definition of the space V that

$$\text{div } \mathbf{u} = 0 \text{ in } \Omega \text{ and } \mathbf{u} = \mathbf{0} \text{ on } \partial\Omega. \tag{3.226}$$

Finally, $(\mathbf{u}, p)$ is indeed a solution of the continuous problem **(CP)** specified by (3.150–3.152).

13) Let $\mathbf{u}$ be a solution of the following variational problem **(VP)**:

$$\textbf{(VP)} \quad \begin{cases} \text{Find } \mathbf{u} \in V \text{ solution of} \\ a(\mathbf{u}, \mathbf{v}) = L(\mathbf{v}), \quad \forall \mathbf{v} \in V, \end{cases} \tag{3.227}$$

where $a(., .)$ and $L(.)$ are the bilinear and linear forms introduced in (3.163). In the formulation (3.227), we choose the special case that the test vector field $\mathbf{v}$ is equal to the solution $\mathbf{u}$. This then implies that

$$a(\mathbf{u}, \mathbf{u}) = L(\mathbf{u}). \tag{3.228}$$

Recalling the relation between the bilinear form $a(., .)$ and the norm $|.|$ defined in (3.159), Eq. (3.228) can be written

$$\mu |\mathbf{u}|^2 = a(\mathbf{u}, \mathbf{u}) = L(\mathbf{u}). \tag{3.229}$$

Furthermore, in discussing the continuity of the linear form $L(.)$, it was established that [see (3.201)]

$$\forall \mathbf{v} \in V, \quad |L(\mathbf{v})| \leq \sqrt{1 + C} \, \|\mathbf{f}\|_{[L^2(\Omega)]^3} |\mathbf{v}|. \tag{3.230}$$

Once again, if we choose $\mathbf{v} = \mathbf{u}$ in the inequality (3.230), Eq. (3.229) leads to the following upper bound:

$$\mu |\mathbf{u}|^2 \leq \sqrt{1 + C} \, \|\mathbf{f}\|_{[L^2(\Omega)]^3} |\mathbf{u}|, \tag{3.231}$$

or dividing both sides by the norm $|.|$ of $\mathbf{u}$,

$$|\mathbf{u}| \leq \frac{\sqrt{1 + C}}{\mu} \|\mathbf{f}\|_{[L^2(\Omega)]^3}. \tag{3.232}$$

Put another way, the a priori estimate (3.232) expresses the upper bound on the norm $|\,.\,|$ of the solution $\mathbf{u}$ as a function of the input data, viz., $\mathbf{f}$ and μ and the domain Ω through the constant C arising in Poincaré's inequality.

Reference

1. P.A. Raviart, J.M. Thomas, *Introduction à l'analyse numérique des équations aux dérivées partielles*, (Masson, 1988)

Chapter 4
Finite-Element Methods and Standard Differential Problems

The aim in this chapter will be to set up and discuss, in a one-dimensional context, variational formulations and the implementation of a P_1 Lagrange finite-element analysis.

More precisely, depending on the complexity of the problem under investigation, we shall apply the following analysis programme, common to all partial differential equations for which we seek an approximate solution by means of the finite-element method:

- **Stage A.** We obtain the variational (or weak) formulation **(VP)** of a given continuous problem **(CP)** (or strong formulation).

- **Stage B.** We investigate the existence and uniqueness of weak solutions to the problem **(VP)**, mainly by application of the Lax–Milgram theorem.

- **Stage C.** We analyse the regularity of the weak solutions to the variational problem **(VP)**.

- **Stage D.** We consider the equivalence of the weak and strong formulations. In particular, we show that a weak solution of the variational problem **(VP)** which also exhibits the regularity obtained in Stage C is then a strong solution of the continuous problem **(CP)**.

- **Stage E.** We write down the node equations which allow us to obtain an approximation to the weak solution of the variational problem **(VP)**. In addition, we compare the resulting node equations with those obtained for the continuous problem **(CP)** using the finite-difference method.

We may thus distinguish two parts which can be treated separately. The first part is theoretical and deals with the existence, uniqueness, and regularity of the solutions to variational formulations, together with the equivalence of the continuous and variational problems (Stages A–D).

The second part is quite distinct. It describes the approximation, mainly using a P_1 Lagrange finite-element method, of the different node equations and analysis of certain finite-difference schemes.

J. Chaskalovic, *Mathematical and Numerical Methods for Partial Differential Equations*, Mathematical Engineering, DOI: 10.1007/978-3-319-03563-5_4, © Springer International Publishing Switzerland 2014

The reader is thus free to study the whole problem, both theoretical and numerical, or to focus on one or other of the two parts. However, anyone mainly concerned with the numerical part would do well to consult the theoretical part, or at least the first question, in order to get a sufficient grasp of the variational formulation to be able to handle the numerical implementation of the finite-element method.

4.1 Dirichlet Problem

4.1.1 Statement of the Problem

The aim here is to propose a mathematical and numerical investigation of the solution to a linear differential problem subject to Dirichlet boundary conditions. Therefore, let u be a real-valued function of the real variable x, defined on $[0, 1]$. We are interested in solving the continuous problem (CP) specified as follows:

Find $u \in H^2(0, 1)$ solution of

$$\textbf{(CP)} \quad \begin{cases} -u''(x) + u(x) = f(x), & 0 \le x \le 1, \\ u(0) = u(1) = 0, \end{cases} \tag{4.1}$$

where f is a given function in $L^2(0, 1)$.

Recall also that the Sobolev space $H^2(0, 1)$ is defined by

$$H^2(0, 1) = \left\{ v : (0, 1) \to \mathbb{R}, \quad \frac{d^k v}{dx^k} \in L^2(0, 1), \quad \forall k = 0, 1, 2 \right\}. \tag{4.2}$$

▶ **Theoretical Part: Variational Formulation**

1) Let v be a real-valued test function defined on $[0, 1]$ and belonging to a function space V whose characteristics will be determined later.

Show that the continuous problem (CP) can be reformulated as a variational problem (VP) of the form:

$$\textbf{(VP)} \quad \begin{cases} \text{Find } u \in V \text{ solution of} \\ a(u, v) = L(v), \quad \forall v \in V. \end{cases} \tag{4.3}$$

Specify the bilinear form $a(., .)$, the linear form $L(.)$, and the function space V.

2) Prove the existence and uniqueness of the weak solution to the variational problem (VP) in the space $H_0^1(0, 1)$ defined by

$$H_0^1(0, 1) = \left\{ v : (0, 1) \to \mathbb{R}, \quad v \in L^2(0, 1), \quad v' \in L^2(0, 1), \quad v(0) = v(1) = 0 \right\}.$$
(4.4)

3) Show that any weak solution of the variational problem **(VP)** also belongs to $H^2(0, 1)$.

4) Deduce the equivalence of the strong formulation **(CP)** set in $H^2(0, 1)$ and the weak formulation **(VP)** considered in $H_0^1(0, 1) \cap H^2(0, 1)$.

▶ **Numerical Part: P_1 Lagrange Finite-Element Analysis**

5) We approximate the variational problem **(VP)** using a P_1 Lagrange finite-element method. To this end, we introduce a regular mesh on the interval $[0, 1]$, with constant spacing h, such that

$$\begin{cases} x_0 = 0, \quad x_{N+1} = 1, \\ x_{i+1} = x_i + h, \quad i = 0, \ldots, N. \end{cases}$$
(4.5)

We now define the approximation space $\tilde{V}$ by

$$\tilde{V} = \left\{ \tilde{v} : [0, 1] \longrightarrow \mathbb{R}, \quad \tilde{v} \in C^0([0, 1]), \quad \tilde{v}|_{[x_i, x_{i+1}]} \in P_1([x_i, x_{i+1}]) \right.$$
$$\left. \tilde{v}(0) = \tilde{v}(1) = 0 \right\},$$
(4.6)

where $P_1([x_i, x_{i+1}])$ is the space of polynomials of degree less than or equal to one defined on $[x_i, x_{i+1}]$.

– What is the dimension of $\tilde{V}$?

6) Show that the best approximation to the solution u of the variational formulation **(VP)** in $\tilde{V}$, in the sense of the norm $|.|_1$ defined by $|v|_1 \equiv \sqrt{a(v, v)}$, is the function $\tilde{u}$ in $\tilde{V}$ which solves

$$a(\tilde{u}, \tilde{v}) = L(\tilde{v}), \quad \forall \tilde{v} \in \tilde{V}.$$
(4.7)

7) Let φ_i, $(i = 1, \ldots, \dim \tilde{V})$, be the canonical basis of $\tilde{V}$ satisfying $\varphi_i(x_j) = \delta_{ij}$, where δ_{ij} is the Kronecker symbol defined by

$$\delta_{ij} = \begin{cases} 1 & \text{if } i = j, \\ 0 & \text{if } i \neq j. \end{cases}$$
(4.8)

Having written down the approximate variational formulation $\widetilde{(\text{VP})}$, with solution $\tilde{u}$, associated with the variational problem **(VP)**, show that, by choosing

$$\tilde{v}(x) = \varphi_i(x), \quad (i = 1, \ldots, \dim \widetilde{V}), \quad \tilde{u}(x) = \sum_{j=1,\ldots,\dim \widetilde{V}} \tilde{u}_j \varphi_j, \qquad (4.9)$$

we obtain the following system $\widetilde{(VP)}$:

$$\widetilde{(VP)} \quad \sum_{j=1,\ldots,\dim \widetilde{V}} A_{ij}\tilde{u}_j = b_i, \quad \forall i \in \{1, \ldots, \dim \widetilde{V}\}, \qquad (4.10)$$

where we have defined

$$A_{ij} = \int_0^1 (\varphi_i'\varphi_j' + \varphi_i\varphi_j)\,dx, \qquad b_i = \int_0^1 f\varphi_i\,dx. \qquad (4.11)$$

▶ **Function φ_i Characterising a Node Strictly in the Interior of $[0, 1]$**

8) Given the regular spacing of the mesh, the generic node equation of the system $\widetilde{(VP)}$, associated with any basis function φ_i characterising a node lying in the interior of $[0, 1]$, can be written

$$\widetilde{(VP_{int})} \quad A_{i,i-1}\tilde{u}_{i-1} + A_{i,i}\tilde{u}_i + A_{i,i+1}\tilde{u}_{i+1} = b_i, \quad \forall i = 1, \ldots, \dim \widetilde{V}. \quad (4.12)$$

Using the trapezoidal rule, calculate the four coefficients (A_{ij}, b_i).

9) Put together the above results to write down the corresponding node equation.

10) Show that this yields the same result as the central difference method applied to the differential equation of the continuous problem **(CP)**. What is the level of accuracy?

Recall that the trapezoidal rule is

$$\int_a^b \xi(s)\,ds \approx \frac{b-a}{2}\big[\xi(a) + \xi(b)\big].$$

4.1.2 Solution to Theoretical Part

▶ **Theoretical Part: Variational Formulation**

1) Let v be a sufficiently regular real-valued test function on $[0, 1]$. As always when we seek a variational formulation, the regularity of the functions v will be specified *a posteriori*.

We multiply the differential equation of the continuous problem **(CP)** by v and integrate over $[0, 1]$ to obtain

$$-\int_0^1 u''v\,dx + \int_0^1 uv\,dx = \int_0^1 fv\,dx, \quad \forall v \in V. \tag{4.13}$$

Integrating by parts,

$$\int_0^1 u'v'\,dx + u'(0)v(0) - u'(1)v(1) + \int_0^1 uv\,dx = \int_0^1 fv\,dx, \quad \forall v \in V. \tag{4.14}$$

We now note that the homogeneous boundary conditions on u, viz., $u(0) = u(1) = 0$, do not appear in the integral formulation (4.14).

Therefore, in order to preserve all the information of the continuous problem **(CP)** in the future variational formulation **(VP)**, we require the test functions v to satisfy the boundary conditions

$$v(0) = v(1) = 0. \tag{4.15}$$

This is how we ensure that the solution u, as one of the functions v in the variational search space V, will have the right properties at the boundary of the integration interval $[0, 1]$.

We thus arrive at the following formal variational formulation:

Find $u \in V$ solution of

$$\int_0^1 (u'v' + uv)\,dx = \int_0^1 fv\,dx, \quad \forall v \text{ such that } v(0) = v(1) = 0. \tag{4.16}$$

This is indeed a formal statement, since it remains to specify the regularity of the test functions that will allow us to attribute meaning to (4.16), in particular regarding the convergence of the integrals featuring in it.

This amounts to defining the function space V in which we shall seek the solution u of the integral formulation (4.16).

To achieve this, we use the Cauchy–Schwarz inequality to obtain the following upper bounds:

$$\left| \int_0^1 u'v' \, dx \right| \leq \int_0^1 |u'v'| \, dx \leq \left(\int_0^1 |u'|^2 \, dx \right)^{1/2} \left(\int_0^1 |v'|^2 \, dx \right)^{1/2}, \qquad (4.17)$$

$$\left| \int_0^1 fv \, dx \right| \leq \int_0^1 |fv| \, dx \leq \left(\int_0^1 |f|^2 \, dx \right)^{1/2} \left(\int_0^1 |v|^2 \, dx \right)^{1/2}, \qquad (4.18)$$

noting that the inequality (4.17) can be rewritten with u' replaced by u and v' by v, in order to use the same method to treat the integral of the product uv.

So if we construct the variational space V as a set of functions v in $L^2(0, 1)$ whose first derivatives also belong to $L^2(0, 1)$, the variational Eq. (4.16) will be suitably well defined.

In conclusion, not forgetting the homogeneous Dirichlet boundary conditions (4.15), the variational space V in which we seek the solution u of the variational formulation (**VP**) is precisely the Sobolev space $H_0^1(0, 1)$ defined by (4.4).

So finally, the variational formulation (**VP**) can be expressed in the form:

$$(\mathbf{VP}) \quad \begin{cases} \text{Find } u \in V \text{ solution of: } a(u, v) = L(v), \quad \forall v \in V, \\[2mm] \text{where } a(u, v) \equiv \int_0^1 \left[u'(x)v'(x) + u(x)v(x) \right] dx, \\[2mm] L(v) \equiv \int_0^1 f(x)v(x) \, dx, \\[2mm] V \equiv H_0^1(0, 1). \end{cases} \qquad (4.19)$$

2) To prove the existence and uniqueness of the solution to the variational problem (**VP**) specified by (4.19), we implement the Lax–Milgram theorem (Theorem 1.11), and this requires a suitable choice of norm on the function space $H_0^1(0, 1)$.

Now since $H_0^1(0, 1) \subset H^1(0, 1)$, it seems natural to measure the size of functions in $H_0^1(0, 1)$ using the norm induced from $H^1(0, 1)$.

In other words, we set

$$\forall v \in H_0^1(0, 1), \quad \|v\|_{H^1}^2 \equiv \int_0^1 v'^2(x) \, dx + \int_0^1 v(x)^2 \, dx \equiv \|v'\|_{L^2}^2 + \|v\|_{L^2}^2. \quad (4.20)$$

We then note that the bilinear form $a(., .)$ is none other than the scalar product which gives rise to the norm $\| . \|_{H^1}$ defined by (4.20):

$$\forall v \in H^1(0, 1), \quad a(v, v) \equiv (v, v)_{H^1} = \|v\|_{H^1}^2,$$

where $(., .)_{H^1}$ denotes the scalar product in $H^1(0, 1)$.

Under these conditions, the Sobolev spaces $H^1(0, 1)$ and $H_0^1(0, 1)$ are Hilbert spaces for the norm given by (4.20), since they are closed subspaces of $H^1(0, 1)$ (see Sect. 1.5).

Furthermore, the continuity constant for the bilinear form $a(.,.)$ is then easily obtained, since we need only apply the Cauchy–Schwarz inequality for the scalar product $(.,.)_{H^1}$.

This yields

$$|a(u, v)| \equiv |(u, v)_{H^1}| \leq (u, u)_{H^1}^{1/2} (v, v)_{H^1}^{1/2} = \|u\|_{H^1} \|v\|_{H^1}. \qquad (4.21)$$

Put another way, the continuity constant for the form $a(.,.)$ is 1.

The continuity of the linear form L is shown by interpreting the upper bound (4.18) in terms of the H^1 norm:

$$|L(v)| \leq \int_0^1 |f(x)v(x)| \, dx \leq \|f\|_{L^2} \|v\|_{L^2} \leq \|f\|_{L^2} \|v\|_{H^1}. \qquad (4.22)$$

The continuity constant for the linear form L is thus equal to the $L^2(0, 1)$ norm of f.

Finally, coercivity of the bilinear form $a(.,.)$ is immediate, since we have

$$a(v, v) = \|v'\|_{L^2}^2 + \|v\|_{L^2}^2 = \|v\|_{H^1}^2 \geq \|v\|_{H^1}^2, \qquad (4.23)$$

which means that the coercivity constant is also equal to 1.

The Lax–Milgram theorem then implies that there exists one and only one function in $H_0^1(0, 1)$ which solves the variational problem **(VP)** specified by (4.16).

Note. The Lax–Milgram theorem could easily have been implemented with a different choice of norm, particularly for the variational problem **(VP)** expressed in $H_0^1(0, 1)$, where there is a more precise norm describing the elements of this Sobolev space.

Indeed, if we change the norm by setting

$$\forall v \in H_0^1(0, 1), \quad \|v\|_{H_0^1} \equiv \left[\int_0^1 v'(x)^2 \, dx \right]^{1/2}, \qquad (4.24)$$

it is easy to show that (4.24) is a norm on $H_0^1(0, 1)$.

In particular, the first property of norms is satisfied, given that any function v in $H_0^1(0, 1)$ vanishes on the boundary of its interval of definition.

Hence, if v is a function in $H_0^1(0, 1)$ such that $\|v\|_{H_0^1} = 0$, v' therefore vanishes on $[0, 1]$. This means that v is a constant on the whole interval $[0, 1]$ and can be evaluated in particular at $x = 0$. This then implies that v is identically zero on $[0, 1]$, as required. The other properties of the $H_0^1(0, 1)$ norm are immediate.

3) It is often difficult to obtain supplementary regularity results for weak solutions to a variational problem **(VP)**, and this can therefore require rather sophisticated mathematical tools. We thus introduce the following lemma, a proof of which can be found in [1]:

> **Lemma 4.1** *Let* $g \in L_{loc}^1(I)$. *For fixed* x_0 *in* I, *we set*
>
> $$G(x) = \int_{x_0}^x g(t)\,dt, \quad \forall x \in I. \tag{4.25}$$
>
> *Then* $G \in C(I)$ *[if* I *is a bounded interval,* G *is in* $H^1(I)$*] and*
>
> $$\int_I G\varphi' = -\int_I g\varphi, \quad \forall \varphi \in C_0^1(I). \tag{4.26}$$

Now if u in $H_0^1(0, 1)$ is a solution to the variational problem **(VP)**, we have

$$\int_0^1 u'v'\,dx = \int_0^1 (f - u)v\,dx, \quad \forall v \in H_0^1(0, 1). \tag{4.27}$$

We introduce the function space $C_0^1((0, 1))$ defined by

$$C_0^1(]0, 1[) \equiv \left\{ v :]0, 1[\to \mathbb{R}, \ v \in C^1(]0, 1[), \ \operatorname{supp} v \subset]0, 1[\right\}, \tag{4.28}$$

where supp v is the support of the function v.

In (4.27) we can then choose a function v in $C_0^1(]0, 1[)$, since this relation is in fact a collection of variational equations that are valid for any function v in $H_0^1(0, 1)$, which contains $C_0^1(]0, 1[)$.

Furthermore, if we introduce the function g defined by $g = f - u$, we have

$$g \in L^2(0, 1) \subset L^1(0, 1) \subset L_{loc}^1(0, 1), \tag{4.29}$$

where the espace $L_{loc}^1(0, 1)$ is defined as follows. If K is an arbitrary closed set strictly contained within $]0, 1[$, then

$$v \in L_{loc}^1(0, 1) \implies v \in L^1(K). \tag{4.30}$$

According to Lemma 4.1, the family of variational Eq. (4.27) can then be written in $C_0^1(]0, 1[)$ in the form

$$\int_0^1 u'v' \, dx = \int_0^1 gv \, dx = -\int_0^1 Gv' \, dx \,, \quad \forall v \in C_0^1(]0, 1[), \tag{4.31}$$

where G is a primitive of g.

We then write (4.31) in the form

$$\int_0^1 (u' + G)v' \, dx = 0 \,, \quad \forall v \in C_0^1(]0, 1[). \tag{4.32}$$

Given that $u' + G \in L^2(0, 1)$, since $u \in H^1(0, 1)$ and $G \in C^0(]0, 1[) \cap H^1(0, 1)$, Lemma 4.1 applies once again to show that $u' + G$ belongs to $L_{\mathrm{loc}}^1(0, 1)$.

To conclude here, we state another result whose proof can be found in [1]:

Lemma 4.2 *If I is an open interval of $\mathbb{R}$ and f a function in $L_{\mathrm{loc}}^1(I)$ satisfying*

$$\int_I f(x)\varphi'(x) \, dx = 0 \,, \quad \forall \varphi \in C_0^1(I), \tag{4.33}$$

then $f = constant$ almost everywhere.

Applying this lemma to the variational Eq. (4.32), we find that

$$u' + G = \text{constant}, \tag{4.34}$$

whereupon u' is a function in $H^1(0, 1)$, being the difference between a constant and the function G. Finally, we deduce that u is in $H^2(0, 1)$.

4) The aim here is to establish the equivalence of the solution to the continuous problem **(CP)** and the solution to the variational problem **(VP)**.

Clearly, if u in $H^2(0, 1)$ solves the continuous problem **(CP)**, then u is a weak solution to the variational problem **(VP)**. To see this, it suffices to reconsider the construction of the variational formulation **(VP)** and observe that the formula for integration by parts is valid whenever u belongs to $H^2(0, 1)$ and v belongs to $H_0^1(0, 1)$.

We thus turn to the converse. Suppose that u is the solution to the variational problem **(VP)** belonging to $H_0^1(0, 1) \cap H^2(0, 1)$. We now use the formula for integration by parts, but in the opposite direction to when we were constructing the variational formulation **(VP)**.

This yields

$$\int_0^1 (-u'' + u - f)v \, dx = 0 \,, \quad \forall v \in H_0^1(0, 1). \tag{4.35}$$

In the variational Eq. (4.35), we then choose functions v belonging to $\mathscr{D}(0, 1)$, the set of functions v in $C^{\infty}(]0, 1[)$ with support strictly included in the interval $(0, 1)$. This is justified because $\mathscr{D}(0, 1) \subset H_0^1(0, 1)$.

We then use the density theorem (Theorem 1.1) in the following way. We rewrite (4.35) in the space $\mathscr{D}(0, 1)$:

$$\int_0^1 (-u'' + u - f)v \, dx = 0, \quad \forall v \in \mathscr{D}(0, 1). \tag{4.36}$$

Clearly, it would have been better to have this last family of equalities (4.36) for all functions v belonging to $L^2(0, 1)$, so that we could select $v = -u'' + u - f$ as a function in $L^2(0, 1)$ and conclude.

As things are, let φ be an arbitrary function in $L^2(0, 1)$. According to Theorem 1.1, $\mathscr{D}(0, 1)$ is dense in $L^2(0, 1)$, so there exists a sequence of functions φ_n in $\mathscr{D}(0, 1)$ which converges to φ in the L^2 norm, i.e.,

$$\lim_{n \to \infty} \int_0^1 |\varphi_n - \varphi|^2 \, dx = 0. \tag{4.37}$$

Now for each function φ_n in $\mathscr{D}(0, 1)$, we have (4.36) with $v = \varphi_n$, so

$$\int_0^1 (-u'' + u - f)\varphi_n \, dx = 0, \quad \forall n \in \mathbb{N}. \tag{4.38}$$

We can then obtain the same property for the functions φ of $L^2(0, 1)$ by the following argument:

$$\left| \int_0^1 \psi\varphi \, dx \right| = \left| \int_0^1 \psi(\varphi - \varphi_n) \, dx \right| \leq \left(\int_\Omega |\psi|^2 \, dx \right)^{1/2} \left(\int_0^1 |\varphi_n - \varphi|^2 \, dx \right)^{1/2}, \tag{4.39}$$

where $\psi = -u'' + u - f$. We now let n tend to $+\infty$ in the inequality (4.39), whereupon we deduce that

$$\int_0^1 \psi\varphi \, dx = 0, \quad \forall \varphi \in L^2(0, 1). \tag{4.40}$$

The proof is completed by choosing, among all the functions φ in $L^2(0, 1)$, the function $\varphi^* = -u'' + u - f$ as one of the particular functions in $L^2(0, 1)$.

We may then deduce that

$$-u'' + u - f = 0 \quad \text{in } L^2(0, 1). \tag{4.41}$$

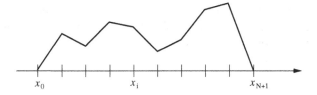

Fig. 4.1 Piecewise affine function

Note. If f also belongs to $L^2(0, 1) \cap C^0(]0, 1[)$, then the differential equation is satisfied for all x in $(0, 1)$ and the solution u is the classical solution of the continuous problem **(CP)** which belongs to $C^2(]0, 1[)$.

4.1.3 Solution to Numerical Part

▶ **Numerical Part:** P_1 **Lagrange Finite-Element Analysis**

5) The dimension of the approximation space $\tilde{V}$ can be determined in several ways. The simplest and most intuitive is to observe that the functions $\tilde{v}$ in $\tilde{V}$ are essentially piecewise continuous curves made up of straight line segments.

In fact they are affine on each mesh element $[x_i, x_{i+1}]$ and vanish at $x = 0$ and $x = 1$, like the example shown in Fig. 4.1.

Since there are $N+2$ discretisation points for the whole mesh covering the interval $[0, 1]$, any two functions in $\tilde{V}$ can be distinguished by the different values they may have at the N interior points $(x_1, \ldots, x_N)$, bearing in mind that any function $\tilde{v}$ in V must satisfy $\tilde{v}_0 = \tilde{v}_{N+1} = 0$.

Put another way, a function $\tilde{v}$ belonging to $\tilde{V}$ is completely determined by the N-uplet $(\tilde{v}_1, \ldots, \tilde{v}_N)$, whence the space $\tilde{V}$ is isomorphic to $\mathbb{R}^N$ and we deduce that $\tilde{V}$ has dimension N.

6) Recall first from Question 2 that the bilinear form $a(.,.)$ can be used to define a norm $|.|_1$, derived from a scalar product $(.,.)_1$ which is none other than the bilinear form $a(.,.)$ itself, defined by

$$\forall v \in H_0^1(0, 1), \quad |v|_1 \equiv \sqrt{a(v, v)} \equiv \sqrt{(v, v)_1}. \tag{4.42}$$

Furthermore, given the continuity and coercivity of the bilinear form $a(.,.)$, we have

$$\exists (\alpha_0, C) \in \mathbb{R}^2 \text{ such that } \forall v \in H_0^1(0, 1), \quad \alpha_0 \|v\|_{H^1}^2 \leq a(v, v) \leq C \|v\|_{H^1}^2. \tag{4.43}$$

Put another way, the norms $\|.\|_{H^1}$ and $|.|_1$ are equivalent and $H_0^1(0, 1)$ is also a Hilbert space for the norm $|.|_1$.

The best approximation to the solution u of the variational formulation in $\widetilde{V}$, in the sense of the norm $|\,.\,|_1$, is the element $\tilde{u}$ of $\widetilde{V}$ that minimises the distance between u and $\widetilde{V}$ as measured by the norm $|\,.\,|_1$, i.e.,

$$|u - \tilde{u}|_1 \leq |u - \tilde{v}|_1, \quad \forall \tilde{v} \in \widetilde{V}. \tag{4.44}$$

It is thus the projection, in the sense of the norm $|\,.\,|_1$, of u on $\widetilde{V}$, which is a closed vector subspace of $H_0^1(0, 1)$.

Since $H_0^1(0, 1)$ is a Hilbert space for the norm $|\,.\,|_1$, according to Lemma 1.10, $\tilde{u}$ is therefore characterised by

$$(u - \tilde{u}, \tilde{v})_1 = 0, \quad \forall \tilde{v} \in \widetilde{V}. \tag{4.45}$$

Expanding out, this becomes

$$(u, \tilde{v})_1 = (\tilde{u}, \tilde{v})_1, \quad \forall \tilde{v} \in \widetilde{V}, \tag{4.46}$$

and finally, reading (4.46) from right to left:

$$a(\tilde{u}, \tilde{v}) = (u, \tilde{v})_1 = L(\tilde{v}), \quad \forall \tilde{v} \in \widetilde{V}. \tag{4.47}$$

7) The approximate variational formulation is obtained by substituting the approximation functions $(\tilde{u}, \tilde{v})$ for the functions (u, v) in the variational formulation (**VP**). In addition, using the expressions in (4.9), we arrive at the following problem:

Determine the numerical sequence $(\tilde{u}_j)$, $(j = 1, \ldots, N)$, solving

$$\sum_{j=1,\ldots,N} \left[\int_0^1 \left(\varphi_i' \varphi_j' + \varphi_i \varphi_j \right) dx \right] \tilde{u}_j = \int_0^1 f \varphi_i \, dx, \quad \forall i = 1, \ldots, N. \tag{4.48}$$

We then identify the expressions for A_{ij} and b_j corresponding to (4.11).

▶ Function φ_i Characterising a Node Strictly in the Interior of $[0, 1]$

8) We now consider the basis functions φ_i characterising nodes strictly in the interior of the integration interval $[0, 1]$. The only nonzero terms in the generic equation of the system (4.48) are *a priori* those corresponding to functions φ_j whose support intersects the support of the function φ_i under consideration (see Fig. 4.2).

The relevant basis functions are thus φ_{i-1}, φ_i, and φ_{i+1}. This is why the equation expressing $(\widehat{\mathbf{VP}}_{\mathbf{int}})$ contains only the terms $A_{i,i-1}$, $A_{i,i}$, and $A_{i,i+1}$, and can thus be written as in (4.12).

Approximate Calculation of the Coefficients A_{ij}, $j = i - 1, i, i + 1$.

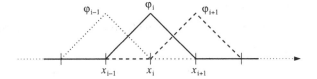

Fig. 4.2 Supports of basis functions φ_{i-1}, φ_i, and φ_i

1. *Approximation of the coefficient A_{ii}.*

   We have

$$
\begin{aligned}
A_{ii} &= \int_0^1 \left(\varphi_i'^2 + \varphi_i^2 \right) \, \mathrm{d}x = \int_{\mathrm{supp}\,\varphi_i} \left(\varphi_i'^2 + \varphi_i^2 \right) \, \mathrm{d}x \\
&= \int_{x_{i-1}}^{x_i} \left(\varphi_i'^2 + \varphi_i^2 \right) \, \mathrm{d}x + \int_{x_i}^{x_{i+1}} \left(\varphi_i'^2 + \varphi_i^2 \right) \, \mathrm{d}x \\
&\approx \left(\frac{1}{h^2} \times h \right) + \frac{h}{2}(0+1) + \left(\frac{1}{h^2} \times h \right) + \frac{h}{2}(1+0) \\
&= \frac{2}{h} + h.
\end{aligned}
\tag{4.49}
$$

   Here we have used the fact that the basis functions φ_i of $\tilde{V}$ are piecewise affine. The derivative φ_i' is constant on each element of the form $[x_i, x_{i+1}]$.

   The integrals of the derivative of φ can then be evaluated, either exactly or using the trapezoidal rule, which is exact for constant functions.

2. *Approximation of the coefficient $A_{i,i-1}$.*

   We have

$$
\begin{aligned}
A_{i,i-1} &= \int_0^1 \left(\varphi_i' \varphi_{i-1}' + \varphi_i \varphi_{i-1} \right) \, \mathrm{d}x \\
&= \int_{\mathrm{supp}\,\varphi_{i-1} \cap \mathrm{supp}\,\varphi_i} \left(\varphi_i' \varphi_{i-1}' + \varphi_i \varphi_{i-1} \right) \, \mathrm{d}x \\
&\approx \left(-\frac{1}{h^2} \times h \right) + \frac{h}{2}[(0 \times 1) + (1 \times 0)] \\
&= -\frac{1}{h}.
\end{aligned}
\tag{4.50}
$$

3. *Approximation of the coefficient $A_{i,i+1}$.*

   This can be obtained directly by noting the following symmetry properties:

   - The matrix A with coefficient A_{ij} is symmetric, i.e., $A_{ij} = A_{ji}$.

   - The mesh on the interval $[0, L]$ is invariant under translation, as a consequence of the uniform discretisation, with constant spacing h.

We then have

$$A_{i,i-1} \overset{\text{symmetry}}{=} A_{i-1,i} \overset{\text{invariance}}{=} A_{i,i+1} = -\frac{1}{h}. \tag{4.51}$$

Estimation of b_i.

   This is done by noting that the support of any basis function φ_i characterising a node strictly in the interior is a union of intervals $[x_{i-1}, x_i]$ and $[x_i, x_{i+1}]$ (see Fig. 4.2).

   It then follows that

$$b_i = \int_0^1 f\varphi_i \, dx = \int_{x_{i-1}}^{x_i} f\varphi_i \, dx + \int_{x_i}^{x_{i+1}} f\varphi_i \, dx$$

$$\approx \frac{h}{2}(0 + f_i) + \frac{h}{2}(f_i + 0)$$

$$= hf_i. \tag{4.52}$$

9) We bring together all the results (4.49–4.52) to obtain the corresponding node equation:

$$-\frac{\tilde{u}_{i-1} - 2\tilde{u}_i + \tilde{u}_{i+1}}{h^2} + \tilde{u}_i = f_i, \quad i = 1, \dots, N. \tag{4.53}$$

10) The discretisation by finite differences of the second order differential equation of the continuous problem **(CP)** is a classic result. The idea is to use Taylor expansions, once the differential equation has been expressed at the discretisation point x_i:

$$-u''(x_i) + u(x_i) = f(x_i), \quad i = 1, \dots, N. \tag{4.54}$$

With the help of the Taylor expansion, one can replace the second derivative u'' at the point x_i by a linear combination involving different values of the unknown u at various neighbouring points of the mesh.

   To obtain an order consistent with the finite-element analysis, we use the following Taylor expansions:

$$u(x_{i+1}) = u(x_i) + hu'(x_i) + \frac{h^2}{2}u''(x_i) + \frac{h^3}{6}u'''(x_i) + O(h^4), \tag{4.55}$$

$$u(x_{i-1}) = u(x_i) - hu'(x_i) + \frac{h^2}{2}u''(x_i) - \frac{h^3}{6}u'''(x_i) + O(h^4). \tag{4.56}$$

We then add these together to obtain

$$u''(x_i) = \frac{u(x_{i+1}) - 2u(x_i) + u(x_{i-1})}{h^2} + O(h^2). \tag{4.57}$$

Substituting the expression (4.57) for the second derivative of u at the point x_i into the differential Eq. (4.54), we obtain

$$-\frac{u(x_{i+1}) - 2u(x_i) + u(x_{i-1})}{h^2} + u(x_i) = f(x_i) + O(h^2), \ i = 1, \ldots, N. \tag{4.58}$$

We then replace the traces $u_i \equiv u(x_i)$ of u at the nodes x_i by the approximations $\tilde{u}_i \approx u_i$, in order to preserve the equality of each side of (4.58) when we drop the infinitely small $O(h^2)$.

This substitution leads to a situation where the resulting finite-difference analysis corresponds exactly to the node Eq. (4.53). It is then clear that the finite-difference scheme (4.53) is second order, given the approximation process we have just presented.

Indeed, if we were to substitute the values $u(x_i)$ for $\tilde{u}_i$, we would obtain the differential Eq. (4.54) to order $O(h^2)$. This is why the finite-difference scheme (4.53) is order two.

———————————

4.2 Neumann Problem

4.2.1 Statement of the Problem

The aim in this section is to propose a mathematical and numerical investigation of the solution to a linear differential problem subject to Neumann boundary conditions.

Therefore, let u be a real-valued function of the real variable $x \in [0, 1]$. We shall be concerned with the continuous problem **(CP)** specified as follows:

Find $u \in H^2(0, 1)$ solution of

$$
\textbf{(CP)} \quad \begin{cases} -u''(x) + u(x) = f(x), & 0 \le x \le 1, \\ u'(0) = u'(1) = 0, \end{cases} \tag{4.59}
$$

where f is a given function in $L^2(0, 1)$.

▶ Theoretical Part: Variational Formulation

1) Let v be a real-valued test function on $[0, 1]$, belonging to the variational space V. Show that the continuous problem **(CP)** can be transposed to the following variational formulation **(VP)**:

$$
\textbf{(VP)} \quad \begin{cases} \text{Find } u \in V \text{ solution of} \\ a(u, v) = L(v), \quad \forall v \in V. \end{cases} \tag{4.60}
$$

Specify the bilinear form $a(., .)$, the linear form $L(.)$, and the function space V.

2) Establish the existence and uniqueness of the weak solution to the variational problem **(VP)** in $H^1(0, 1)$.

3) Show that any weak solution of the variational problem **(VP)** also belongs to $H^2(0, 1)$.

4) Deduce the equivalence of the strong formulation of the continuous problem **(CP)** expressed in $H^2(0, 1)$ and the weak formulation of the variational problem **(VP)**, considered in $H^1(0, 1) \cap H^2(0, 1)$.

▶ Numerical Part: P_1 Lagrange Finite-Element Analysis

5) We approximate the variational problem **(VP)** using a P_1 Lagrange finite-element method. To do this, introduce a regular mesh on the interval $[0, 1]$, with constant spacing h, such that

$$
\begin{cases} x_0 = 0, \quad x_{N+1} = 1, \\ x_{i+1} = x_i + h, \quad i = 1, \ldots, N. \end{cases} \tag{4.61}
$$

We now define the approximation space $\tilde{V}$ by

$$\tilde{V} = \left\{ \tilde{v} : [0, 1] \longrightarrow \mathbb{R}, \ \tilde{v} \in C^0([0, 1]), \ \tilde{v}|_{[x_i, x_{i+1}]} \in P_1([x_i, x_{i+1}]) \right\}, \quad (4.62)$$

where $P_1([x_i, x_{i+1}])$ is the space of polynomials on $[x_i, x_{i+1}]$ of degree less than or equal to one.

– What is the dimension of $\tilde{V}$?

6) Let φ_i, $(i = 1, \ldots, \dim \tilde{V})$, be the canonical basis of $\tilde{V}$ satisfying $\varphi_i(x_j) = \delta_{ij}$, where δ_{ij} is the Kronecker symbol.

After writing down the approximate variational formulation $\widetilde{(VP)}$ with solution $\tilde{u}$, associated with the variational problem **(VP)**, show that, by choosing

$$\tilde{v}(x) = \varphi_i(x), \quad i = 1, \ldots, \dim \tilde{V}, \quad \tilde{u}(x) = \sum_{j=1,\ldots,\dim \tilde{V}} \tilde{u}_j \varphi_j, \quad (4.63)$$

we obtain the following system $\widetilde{(VP)}$:

$$\widetilde{(VP)} \qquad \sum_{j=1,\ldots,\dim \tilde{V}} A_{ij} \tilde{u}_j = b_i, \quad \forall i \in \{1, \ldots, \dim \tilde{V}\}, \quad (4.64)$$

where

$$A_{ij} = \int_0^1 (\varphi_i' \varphi_j' + \varphi_i \varphi_j) \, dx, \quad b_i = \int_0^1 f \varphi_i \, dx. \quad (4.65)$$

▶ **Function φ_i Characterising a Node Strictly in the Interior of $[0, 1]$**

7) Given the regularity of the mesh, the generic node equation of the system $\widetilde{(VP)}$, associated with any basis function φ_i, $(i = 1, \ldots, \dim \tilde{V} - 2)$, characterising an interior node of $[0, 1]$, can be written

$$\widetilde{(VP_{int})} \quad A_{i,i-1} \tilde{u}_{i-1} + A_{i,i} \tilde{u}_i + A_{i,i+1} \tilde{u}_{i+1} = b_i, \quad \forall i = 1, \ldots, \dim \tilde{V} - 2. \quad (4.66)$$

Using the trapezoidal rule, calculate the four coefficients (A_{ij}, b_i).

8) Bring together the above results to write down the corresponding node equation.

9) Show that we retrieve the centred finite-difference scheme associated with the differential equation of the continuous problem **(CP)**. What is the order of accuracy?

Recall that the trapezoidal rule is

$$\int_a^b \xi(s) ds \approx \frac{b-a}{2} [\xi(a) + \xi(b)].$$

▶ **Function φ_0 Characterising the Node $x_0 = 0$**

10) Considering the basis function φ_0 that characterises the node at $x_0 = 0$, show that the corresponding node equation of the system $\widetilde{(\text{VP})}$ is

$$\widetilde{(\text{VP}_0)} \qquad A_{0,0}\,\tilde{u}_0 + A_{0,1}\,\tilde{u}_1 = b_0. \tag{4.67}$$

Using the trapezoidal rule, calculate the three coefficients $A_{0,0}$, $A_{0,1}$, and b_0.

11) Write down the corresponding node equation.

12) Show that we recover the finite-difference scheme associated with the Neumann condition at $x = 0$. What is the order of accuracy?

▶ **Function φ_{N+1} Characterising the Node $x_{N+1} = 1$**

13) We now consider the basis function φ_{N+1} characterising the node $x_{N+1} = 1$. Show that the associated node equation of the system $\widetilde{(\text{VP})}$ can be written

$$\widetilde{(\text{VP}_{\text{N+1}})} \qquad A_{N+1,N}\,\tilde{u}_N + A_{N+1,N+1}\,\tilde{u}_{N+1} = b_{N+1}. \tag{4.68}$$

Using the trapezoidal rule, calculate the three coefficients $A_{N+1,N+1}$, $A_{N+1,N}$, and b_{N+1}.

14) Write down the corresponding node equation.

15) Show that we retrieve the finite-difference scheme associated with the Neumann condition $x = 1$. What is the order of accuracy?

4.2.2 Solution to Theoretical Part

▶ **Theoretical Part: Variational Formulation**

1) Let v be a sufficiently regular real-valued test function on $[0, 1]$. As already mentioned when discussing the Dirichlet problem in Sect. 4.1, the regularity of the functions v will be specified *a posteriori*, in order to give meaning to the variational formulation when it is established.

We now multiply the differential equation of the continuous problem **(CP)** by v and integrate over the interval $[0, 1]$:

$$- \int_0^1 u'' v \, dx + \int_0^1 u v \, dx = \int_0^1 f v \, dx , \quad \forall v \in V. \tag{4.69}$$

Integrating by parts,

$$\int_0^1 u' v' \, dx + u'(0)v(0) - u'(1)v(1) + \int_0^1 u v \, dx = \int_0^1 f v \, dx , \quad \forall v \in V. \tag{4.70}$$

We now observe that the homogeneous Neumann boundary conditions specified in the continuous problem **(CP)**, viz., $u'(0) = u'(1) = 0$, appear in the integral formulation (4.70).

Taking these two boundary conditions into account, this brings us to the following formulation:

Find u in V solution of

$$\int_0^1 (u' v' + u v) \, dx = \int_0^1 f v \, dx , \quad \forall v \in V. \tag{4.71}$$

At this point, the formulation (4.71) is formal in the sense that there is still no particular reason why the integrals featuring in it should be convergent.

We then note that this variational formulation is strictly analogous to the one obtained for the Dirichlet problem [see (4.16) in Sect. 4.1], except for the boundary conditions which we no longer need to impose on the test functions v in the context of the Neumann problem treated here.

This is why, exploiting the functional analysis developed in Sect. 4.1, a sufficient condition guaranteeing convergence of the integrals in the variational formulation (4.71) consists in defining the variational space V as follows:

$$V \equiv H^1(0, 1) \equiv \left\{ v : [0, 1] \to \mathbb{R}, \ v \in L^2(0, 1), \ v' \in L^2(0, 1) \right\}. \tag{4.72}$$

Finally, the variational formulation (**VP**) takes the form:

$$
\textbf{(VP)} \quad
\begin{cases}
\text{Find } u \text{ in } V \text{ solution of } a(u, v) = L(v), \quad \forall v \in V, \\[2mm]
\text{where } \ a(u, v) \equiv \displaystyle\int_0^1 (u'v' + uv)\, dx, \\[2mm]
L(v) \equiv \displaystyle\int_0^1 fv\, dx, \\[2mm]
V \equiv H^1(0, 1).
\end{cases}
\tag{4.73}
$$

2) The existence and uniqueness of the solution to the variational problem (**VP**) (4.73) are proven by applying the Lax–Milgram theorem (Theorem 1.11).

For this purpose, a crucial step is the choice of norm on the function space $H^1(0, 1)$. We thus choose to measure the size of functions v in $H^1(0, 1)$ using the natural norm defined by

$$
\forall v \in H^1(0, 1), \quad \|v\|_{H^1}^2 \equiv \int_0^1 v'(x)^2\, dx + \int_0^1 v(x)^2\, dx \equiv \|v'\|_{L^2}^2 + \|v\|_{L^2}^2. \tag{4.74}
$$

Having chosen the norm, the process of checking through the premises of the Lax–Milgram theorem is strictly the same as for the Dirichlet problem (see Sect. 4.1).

Let us summarise the different points step by step:

- The space $H^1(0, 1)$ is a Hilbert space for the norm (4.74), the scalar product giving rise to this norm coinciding exactly with the bilinear form $a(., .)$ defined by (4.73).

- The bilinear form $a(., .)$ is continuous on $H^1(0, 1) \times H^1(0, 1)$ with continuity constant equal to unity.

- The linear form $L(.)$ defined by (4.73) is continuous on $H^1(0, 1)$ with continuity constant equal to the $L^2(0, 1)$ norm of f.

- The form $a(., .)$ is H^1-elliptic (or coercive) and the ellipticity constant is equal to unity.

The Lax–Milgram theorem then implies that there exists one and only one function in $H^1(0, 1)$ that solves the variational problem (**VP**) specified by (4.73).

3) Once again, in order to answer this question, we refer back to the discussion of the Dirichlet problem in Sect. 4.1. Indeed, as already mentioned, quite sophisticated mathematical tools may be needed to obtain supplementary regularity results for weak solutions to the variational problem (**VP**).

This is why for the present study we refer the reader to Lemmas 4.1 and 4.2 quoted in our analysis of the solutions to the Dirichlet problem in Sect. 4.1.

Hence, if u is an element of $H^1(0, 1)$ that solves the variational problem **(VP)**, we have

$$\int_0^1 u'v' \, dx = \int_0^1 (f - u)v \, dx \,, \quad \forall v \in H^1(0, 1). \tag{4.75}$$

Among the functions v in $H^1(0, 1)$, we choose those belonging to $C_0^1(]0, 1[)$ which is included in $H^1(0, 1)$.

Furthermore, introducing the function g defined by $g = f - u$, we have

$$g \in L^2(0, 1) \subset L^1(0, 1) \subset L^1_{loc}(0, 1), \tag{4.76}$$

where the space $L^1_{loc}(0, 1)$ is defined as follows.

If K is any closed set strictly contained in $[0, 1]$, then

$$v \in L^1_{loc}(0, 1) \implies v \in L^1(K). \tag{4.77}$$

The family of variational Eq. (4.75) can thus be written in $C_0^1(]0, 1[)$ in the form

$$\int_0^1 u'v' \, dx = \int_0^1 gv \, dx = - \int_0^1 Gv' \, dx \,, \quad \forall v \in C_0^1(]0, 1[), \tag{4.78}$$

where G is a primitive of g. Here we have used Lemma 4.1.

We can then write (4.78) in the form

$$\int_0^1 (u' + G)v' \, dx = 0 \,, \quad \forall v \in C_0^1(]0, 1[). \tag{4.79}$$

We now have $u' + G \in L^2(0, 1)$, since $u \in H^1(0, 1)$, on the one hand, and G belongs to $C^0(]0, 1[) \cap H^1(0, 1)$, on the other, and a further application of Lemma 4.1 shows that $u' + G$ belongs to $L^1_{loc}(0, 1)$.

We can then apply Lemma 4.2 to the variational Eq. (4.79), and this implies

$$u' + G = \text{constant}. \tag{4.80}$$

We thus find that u' belongs to $H^1(0, 1)$, as the difference between a constant and the function G.

Finally, we deduce that u belongs to $H^2(0, 1)$.

4) This last question in the theoretical part deals with the equivalence of the solution to the continuous problem **(CP)** and the solution to the variational problem **(VP)**.

Now if $u \in H^2(0, 1)$ solves **(CP)**, then u is a weak solution to the variational problem **(VP)**.

To see this, we just examine the process used to establish the variational problem (**VP**) and observe that this formulation, and in particular the use of the formula for integration by parts, has a meaning whenever u is in $H^2(0, 1)$ and v in $H^1(0, 1)$.

We thus move on to the converse. Suppose therefore that u in $H^2(0, 1)$ solves the variational problem (**VP**) - according to the last question, any solution u to the variational problem belongs to $H^2(0, 1)$.

We then apply the formula for integration by parts in the opposite direction to the one used to obtain the variational formulation. This leads to

$$\int_0^1 (-u'' + u - f)v \, dx = 0, \quad \forall v \in H^1(0, 1). \tag{4.81}$$

We observe that the formulation (4.81) is identical to the one obtained for the Dirichlet problem in Sect. 4.1, apart from the functional context, viz., H_0^1 for the Dirichlet problem and H^1 here for the Neumann problem.

But now we only need note the trivial inclusion $H_0^1(0, 1) \subset H^1(0, 1)$ in order to be able to apply everything presented in Question 4 of Sect. 4.1.

Here we recall the main points of the demonstration:

- In (4.81), we choose the functions v which belong to $\mathscr{D}(0, 1)$, since we know that $\mathscr{D}(0, 1) \subset H^1(0, 1)$.

- We use the density theorem (Theorem 1.1), which shows that $\mathscr{D}(0, 1)$ is dense in $L^2(0, 1)$.

- We then show that (4.81) holds, not only in $H^1(0, 1)$, but also in a bigger space, viz., $L^2(0, 1)$.

- This means that we can choose, among all the functions v in $L^2(0, 1)$ appearing in (4.81), precisely the function

$$v^* = -u'' + u - f.$$

If in addition f belongs to $L^2(0, 1) \cap C^0(]0, 1[)$, then the differential equation is satisfied for all $x \in (0, 1)$ and the solution u is the classic solution to the continuous problem (**CP**) belonging to $C^2(]0, 1[)$.

4.2.3 Solution to Numerical Part

▶ **Numerical Part: P_1 Lagrange Finite-Element Analysis**

5) To find the dimension of the space $\tilde{V}$, we note that the definition (4.62) of the approximation space $\tilde{V}$ is very similar to the one considered in the Dirichlet problem [see Question 5 of Sect. 4.1, and in particular (4.6)].

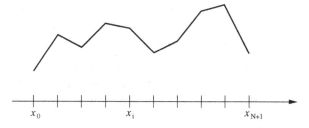

Fig. 4.3 Piecewise affine function

Referring once again to the discussion for the Dirichlet problem, it suffices to observe that, in the space $\tilde{V}$ defined by (4.62), we must add two to the dimension found there, due to the two values of the function $\tilde{v} \in \tilde{V}$ at $x = 0$ and $x = 1$.

Put another way, any function $\tilde{v} \in \tilde{V}$ is known if we specify its trace

$$(\tilde{v}_0, \tilde{v}_1, \ldots, \tilde{v}_N, \tilde{v}_{N+1})$$

at the $N + 2$ discretisation points of the mesh on $[0, 1]$, viz., $x_0, x_1, \ldots, x_N, x_{N+1}$, which uniquely define $\tilde{v}$ (see Fig. 4.3).

This therefore implies that the space $\tilde{V}$ is isomorphic to $\mathbb{R}^{N+2}$ and that the dimension of $\tilde{V}$ is thus $N + 2$.

6) The approximate variational formulation $\widetilde{(VP)}$ is obtained in the usual way by substituting the approximation functions $(\tilde{u}, \tilde{v})$ for the functions (u, v) in the variational formulation **(VP)**.

In addition, we use the expressions given in (4.63). In this way, the approximate variational formulation $\widetilde{(VP)}$ can be expressed as follows:

$$\left[\begin{array}{l} \text{Find the numerical sequence } (\tilde{u}_j), \, (j = 0, \ldots, N + 1) \text{ solution of} \\[2mm] \sum_{j=0}^{N+1} \left[\int_0^1 \left(\varphi_i' \varphi_j' + \varphi_i \varphi_j \right) dx \right] \tilde{u}_j = \int_0^1 f \varphi_i(x) \, dx \,, \quad \forall i = 1, \ldots, N + 1. \end{array} \right.$$

(4.82)

We can then identify the expressions for A_{ij} and b_j corresponding to (4.65).

▶ **Function φ_i Characterising a Node Strictly in the Interior of $[0, 1]$**

7) Looking at the variational formulation $\widetilde{(VP)}$ specified by (4.82), we note that, among the $N + 2$ equations constituting the linear system, the N equations corresponding to the interior values of j from 1 to N are exactly the same as those we found for the Dirichlet problem (see Question 7 of Sect. 4.1).

This is because the formalism of the variational formulations was exactly the same for the Dirichlet and Neumann problems, at least for the basis functions φ_i characterising nodes strictly in the interior of the interval $[0, 1]$.

As mentioned in the theoretical part, only the functional context differs between the two formulations in order to account for the different boundary conditions. There should thus be no surprise to find the same approximation described by the node equations associated with the basis functions φ_i characterising the nodes lying strictly inside the interval $[0, 1]$.

Put another way, the coefficients of the nodal Eq. (4.66) are given by

$$A_{ii} \equiv \frac{2}{h} + h, \quad A_{i,i-1} = A_{i,i+1} \equiv -\frac{1}{h}, \quad b_i \equiv h f_i. \tag{4.83}$$

8) The node equation of the approximate variational problem $\widetilde{(\mathbf{VP})}$ corresponding to a basis function φ_i characterising a node at an interior point x_i is

$$-\frac{\tilde{u}_{i+1} - 2\tilde{u}_i + \tilde{u}_{i-1}}{h^2} + \tilde{u}_i = f_i, \quad i = 1, \dots, N. \tag{4.84}$$

9) For the same reasons as in the previous questions, the analogy with the Dirichlet problem also delivers the results for the finite-difference scheme, which are obtained in the same way in the present case.

Hence, the finite-difference scheme discretising the differential equation of the continuous problem **(CP)** to second order corresponds precisely to the node Eq. (4.84).

▶ **Function φ_0 Characterising the Node $x_0 = 0$**

10) The generic Eq. (4.82) of the approximate variational problem $\widetilde{(\mathbf{VP})}$ corresponding to the basis function φ_0 characterising the node x_0 is

$$\widetilde{(\mathbf{VP_0})} \quad A_{00}\,\tilde{u}_0 + A_{01}\,\tilde{u}_1 = b_0. \tag{4.85}$$

This follows from the fact that, when we consider the approximate variational formulation $\widetilde{(\mathbf{VP})}$ specified by (4.82) for the basis function φ_0, the sum over the other basis functions φ_j leads only to zero contributions, with the exception of the basis functions φ_0 and φ_1.

This is once again due to the relative positions of the supports of each of the basis functions φ_j with respect to the support of the basis function φ_0 (see Fig. 4.4).

Fig. 4.4 Supports of the basis functions φ_0 and φ_1

Approximate Calculation of the Coefficients A_{00} and A_{01}

1. *Approximation of the coefficient A_{00}.*

   This coefficient is estimated in an analogous way to the calculation for the coefficient A_{ii} in the answer to Question 7. However, there is nevertheless a difference, since the support of the basis function φ_0 comprises only the mesh element $[x_0, x_1]$, whereas the supports of the other basis functions φ_i, $(i = 1, \ldots, N)$, all contain two mesh elements $[x_{i-1}, x_i]$ and $[x_i, x_{i+1}]$.

   The coefficient A_{00} is thus estimated as follows:

   $$A_{00} = \int_0^1 \left(\varphi_0'^2 + \varphi_0^2 \right) dx = \int_{\text{supp } \varphi_0} \left(\varphi_0'^2 + \varphi_0^2 \right) dx = \int_{x_0}^{x_1} \left(\varphi_0'^2 + \varphi_0^2 \right) dx$$

   $$\approx \left(\frac{1}{h^2} \times h \right) + \frac{h}{2} \left(0 + 1 \right), \tag{4.86}$$

   whence, finally,

   $$A_{00} \approx \frac{1}{h} + \frac{h}{2}. \tag{4.87}$$

2. *Approximation of the coefficient A_{01}.*

   Here we have

   $$A_{01} = \int_0^1 \left(\varphi_0' \varphi_1' + \varphi_0 \varphi_1 \right) dx = \int_{\text{supp } \varphi_0 \cap \text{supp } \varphi_1} \left(\varphi_0' \varphi_1' + \varphi_0 \varphi_1 \right) dx$$

   $$\approx -\frac{1}{h^2} \times h + \frac{h}{2} \left[(0 \times 1) + (1 \times 0) \right], \tag{4.88}$$

   whence

   $$A_{01} \approx -\frac{1}{h}. \tag{4.89}$$

   Note that the approximation of the coefficient A_{01} is a special case of the generic calculation presented in (4.50) of Sect. 4.1, because the mesh spacing h is constant. Thus the integral over the interval $[x_0, x_1]$ of the basis function φ_0 multiplied by the basis function φ_1 is completely equivalent to the integral of φ_i multiplied by φ_{i+1}.

To see this, it suffices to make a suitable change of variable mapping the interval $[x_i, x_{i+1}]$ to the interval $[x_0, x_1]$ and then note the fact that the coefficients $A_{i,i+1}$ and A_{01} are equal.

Estimation of b_0.

This is done using the same procedure as for b_i in (4.52) of Sect. 4.1. We thus have

$$b_0 = \int_0^1 f \varphi_0 \, dx = \int_{x_0}^{x_1} f \varphi_0 \, dx \approx \frac{h}{2}(0 + f_0) = \frac{h}{2} f_0. \qquad (4.90)$$

11) Bringing together the results (4.86–4.90), we can now obtain the corresponding node equation associated with the function φ_0 that characterises the node x_0:

$$\left(\frac{1}{h} + \frac{h}{2}\right) \tilde{u}_0 - \frac{1}{h}\tilde{u}_1 = \frac{h}{2} f_0. \qquad (4.91)$$

12) To determine the finite-difference scheme discretising the Neumann condition at $x_0 = 0$, where $u'(x_0) = 0$, note that the Taylor expansion must be taken to order 3, if we wish to maintain the order two discretisation obtained for the finite-difference scheme (4.85), associated with the differential equation of the continuous problem **(CP)** in the interior of the interval $[0, 1]$.

We thus write down the Taylor expansion

$$u(x_1) = u(x_0) + hu'(x_0) + \frac{h^2}{2}u''(x_0) + O(h^3). \qquad (4.92)$$

We can then replace the value of the first derivative which is zero at $x_0 = 0$ (this being the Neumann condition), but as usual when applying this kind of method, this brings in the second derivative of u at the point x_0.

We thus assume that we can write down the differential equation of the continuous problem **(CP)** up to the boundary of the interval, i.e., at $x_0 = 0$:

$$-u''(x_0) + u(x_0) = f(x_0). \qquad (4.93)$$

Equation (4.93) can be used to express the second derivative u'' at the point x_0 and insert it in the Taylor expansion (4.92), whereupon we obtain

$$u(x_1) = u(x_0) + hu'(x_0) + \frac{h^2}{2}[u(x_0) - f(x_0)] + O(h^3). \qquad (4.94)$$

We then pass to the approximations $\tilde{u}_i$, neglecting the terms $O(h^3)$ in (4.94). The discretised equation at $x_0 = 0$ is

$$\tilde{u}_1 = \tilde{u}_0 + \frac{h^2}{2}(\tilde{u}_0 - f_0). \qquad (4.95)$$

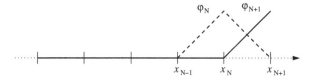

Fig. 4.5 Supports of the basis functions φ_N and φ_{N+1}

We do indeed retrieve the node Eq. (4.91) corresponding to the basis function φ_0 characterising the node x_0 of the discretisation.

▶ **Function φ_{N+1} Characterising the Node $x_{N+1} = 1$**

13) Regarding the node equation at $x_{N+1} = 1$, we can apply similar considerations to those just presented to deal with the node equation at $x_0 = 0$. We simply note that the situation for the basis functions φ_N and φ_{N+1} is symmetric with respect to the situation for the basis functions φ_0 and φ_1 (see Fig. 4.5).

For this reason, the equation of the approximate variational formulation $\widetilde{(\text{VP})}$ corresponding to the basis function φ_{N+1} has the form

$$\widetilde{(\text{VP}_{N+1})} \qquad A_{N+1,\,N}\,\tilde{u}_N + A_{N+1,\,N+1}\,\tilde{u}_{N+1} = b_{N+1}. \qquad (4.96)$$

In the same way, provided that we take care to replace the integration interval $[x_0, x_1]$ by the one corresponding to the support of the function φ_{N+1}, viz., $[x_N, x_{N+1}]$, we obtain the following results for the coefficients $A_{N+1,N}$, $A_{N+1,N+1}$, and b_{N+1}.

Approximate Calculation of the Coefficients $A_{N+1,N}$ and $A_{N+1,N+1}$

1. *Approximation of the coefficient $A_{N+1,N+1}$.*

   Here we have

$$A_{N+1,\,N+1} = \int_0^1 \left(\varphi'^2_{N+1} + \varphi^2_{N+1} \right) \, dx = \int_{\text{supp } \varphi_{N+1}} \left(\varphi'^2_{N+1} + \varphi^2_{N+1} \right) \, dx$$

$$= \int_{x_N}^{x_{N+1}} \left(\varphi'^2_{N+1} + \varphi^2_{N+1} \right) \, dx$$

$$\approx \left(\frac{1}{h^2} \times h \right) + \frac{h}{2}(0 + 1)$$

$$= \frac{1}{h} + \frac{h}{2}. \qquad (4.97)$$

2. *Approximation of the coefficient $A_{N+1,N}$.*

In this case,

$$A_{N+1, N} = \int_0^1 \left(\varphi_N' \varphi_{N+1}' + \varphi_N \varphi_{N+1} \right) dx$$

$$= \int_{\text{supp } \varphi_N \cap \text{supp } \varphi_{N+1}} \left(\varphi_N' \varphi_{N+1}' + \varphi_N \varphi_{N+1} \right) dx$$

$$\approx \left(-\frac{1}{h^2} \times h \right) + \frac{h}{2} [(0 \times 1) + (1 \times 0)] = -\frac{1}{h}. \tag{4.98}$$

Estimation of b_{N+1}.

This is done using the trapezoidal rule:

$$b_{N+1} = \int_0^1 f \varphi_{N+1} \, dx = \int_{x_N}^{x_{N+1}} f \varphi_{N+1} \, dx \approx \frac{h}{2}(0 + f_{N+1}) = \frac{h}{2} f_{N+1}. \tag{4.99}$$

14) We obtain the node equation associated with the basis function φ_{N+1} by putting together the results (4.97–4.99). This equation is perfectly symmetric with respect to the node equation associated with the basis function φ_0:

$$-\frac{1}{h} \tilde{u}_N + \left(\frac{1}{h} + \frac{h}{2} \right) \tilde{u}_{N+1} = \frac{h}{2} f_{N+1}. \tag{4.100}$$

15) Given the symmetry just mentioned in the last question, we may wonder whether we can retrieve the node Eq. (4.100) by the finite-difference method. Indeed, at the point $x_{N+1} = 1$, we must now consider a 'backward' Taylor expansion (whereas we considered a forward expansion at $x_0 = 0$):

$$u(x_N) = u(x_{N+1}) - hu'(x_{N+1}) + \frac{h^2}{2} u''(x_{N+1}) + O(h^3). \tag{4.101}$$

We then replace the second derivative at x_{N+1}, writing down the differential equation of the continuous problem **(CP)** at the point x_{N+1}:

$$u(x_N) = u(x_{N+1}) - hu'(x_{N+1}) + \frac{h^2}{2} [u(x_{N+1}) - f(x_{N+1})] + O(h^3). \tag{4.102}$$

Now using the information relating to the homogeneous Neumann condition at x_{N+1}, we obtain the counterpart of the discrete Eq. (4.95) as soon as we replace the values $u(x_i)$ by the corresponding approximations $\tilde{u}_i$:

$$\tilde{u}_N = \tilde{u}_{N+1} + \frac{h^2}{2} (\tilde{u}_{N+1} - f_{N+1}). \tag{4.103}$$

4.3 Fourier–Dirichlet Problem

4.3.1 Statement of the Problem

The aim here is to carry out a mathematical and numerical investigation of the solution to a second order linear differential problem subject to Fourier–Dirichlet boundary conditions. Let u a real-valued function of the real variable $x \in [0, 1]$. The continuous problem (**CP**) is defined by:

Find $u \in H^2(0, 1)$ solution of

$$(\textbf{CP}) \quad \begin{cases} -u''(x) + u(x) = f(x), & 0 \le x \le 1, \\ u(0) = 0, & u'(1) + ku(1) = 1, \end{cases} \tag{4.104}$$

where f is a given function in $L^2(0, 1)$ and k a given non-negative real parameter.

▶ **Theoretical Part: Variational Formulation**

1) Let v be a real-valued test function defined on $[0, 1]$ and belonging to a variational space V. Show that the continuous problem (**CP**) can be cast in a variational form (**VP**) as follows:

$$(\textbf{VP}) \quad \begin{cases} \text{Find } u \in V \text{ solution of} \\ a(u, v) = L(v), \quad \forall v \in V. \end{cases} \tag{4.105}$$

Specify the bilinear form $a(., .)$, the linear form $L(.)$, and the function space V.

2) Establish the existence and uniqueness of a weak solution to the variational problem (**VP**) in $H_*^1(0, 1)$ defined by

$$H_*^1(0, 1) = \left\{ v : (0, 1) \to \mathbb{R}, \ v \in L^2(0, 1), \ v' \in L^2(0, 1), \ v(0) = 0 \right\}. \tag{4.106}$$

3) Show that any weak solution to the variational problem (**VP**) in $H_*^1(0, 1)$ must also belong to $H^2(0, 1)$.

4) Deduce the equivalence of the strong formulation set in $H^2(0, 1)$ and the weak formulation considered in $H_*^1(0, 1) \cap H^2(0, 1)$.

▶ **Numerical Part: P_1 Lagrange Finite-Element Analysis**

5) The variational problem (**VP**) is approximated using a P_1 Lagrange finite-element method. To do this, we introduce a regular mesh on the interval $[0, 1]$, with constant spacing h, such that

$$\begin{cases} x_0 = 0, \quad x_{N+1} = 1, \\ x_{i+1} = x_i + h, \quad i = 0, \dots, N. \end{cases} \tag{4.107}$$

We now define the approximation space $\tilde{V}$ by

$$\tilde{V} = \left\{ \tilde{v} : [0, 1] \longrightarrow \mathbb{R}, \ \tilde{v} \in C^0([0, 1]), \ \tilde{v}|_{[x_i, x_{i+1}]} \in P_1([x_i, x_{i+1}]), \ \tilde{v}(0) = 0 \right\}, \tag{4.108}$$

where $P_1([x_i, x_{i+1}])$ is the space of polynomials of degree less than or equal to one on $[x_i, x_{i+1}]$.

– What is the dimension of $\tilde{V}$?

6) Let φ_i, $(i = 1, \dots, \dim \tilde{V})$, be the canonical basis of $\tilde{V}$ satisfying $\varphi_i(x_j) = \delta_{ij}$. After writing down the approximate variational formulation $\widetilde{(\text{VP})}$ with solution $\tilde{u}$, associated with the variational problem **(VP)**, show that, by choosing

$$\tilde{v}(x) = \varphi_i(x), \quad i = 1, \dots, \dim \tilde{V}, \qquad \tilde{u}(x) = \sum_{j=1,\dots,\dim \tilde{V}} \tilde{u}_j \varphi_j, \tag{4.109}$$

we obtain the following system $\widetilde{(\text{VP})}$:

$$\widetilde{(\text{VP})} \qquad \sum_{j=1,\dots,\dim \tilde{V}} A_{ij} \tilde{u}_j = b_i, \quad \forall i \in \{1, \dots, \dim \tilde{V}\}, \tag{4.110}$$

where

$$A_{ij} = \int_0^1 (\varphi_i' \varphi_j' + \varphi_i \varphi_j) \, dx + k \varphi_i(1) \varphi_j(1), \quad b_i = \int_0^1 f \varphi_i \, dx + \varphi_i(1). \tag{4.111}$$

▶ **Function φ_i Characterising a Node Strictly in the Interior of [0, 1]**

7) Given the regular spacing of the mesh, the generic node equation of the system $\widetilde{(\text{VP})}$ associated with any basis function φ_i, $(i = 1, \dots, \dim \tilde{V} - 1)$, characterising an interior node of the interval [0, 1], can be written

$$\widetilde{(\text{VP}_{\text{int}})} \qquad A_{i,i-1} \tilde{u}_{i-1} + A_{i,i} \tilde{u}_i + A_{i,i+1} \tilde{u}_{i+1} = b_i, \quad i = 1, \dots, \dim \tilde{V} - 1. \tag{4.112}$$

Using the trapezoidal rule, calculate the four coefficients (A_{ij}, b_i). Recall that the trapezoidal rule is

$$\int_a^b \xi(s)\,ds \approx \frac{b-a}{2}\big[\xi(a) + \xi(b)\big].$$

8) Bring together the above results to write down the corresponding node equation.

9) Show that we recover the centred finite-difference scheme associated with the differential equations of the continuous problem **(CP)**.

– What is the order of accuracy?

▶ **Function φ_{N+1} Characterising the Node $x_{N+1} = 1$**

10) Now consider the basis function φ_{N+1} characterising the node at $x_{N+1} = 1$. Show that the associated node equation of the system $\widetilde{(\mathbf{VP})}$ is

$$\widetilde{(\mathbf{VP_{N+1}})} \qquad A_{N+1,N}\,\tilde{u}_N + A_{N+1,N+1}\,\tilde{u}_{N+1} = b_{N+1}. \qquad (4.113)$$

Using the trapezoidal rule, calculate the three coefficients $A_{N+1,N}$, $A_{N+1,N+1}$, and b_{N+1}.

11) Write down the corresponding node equation.

12) Show that we recover the finite-difference scheme associated with the Fourier boundary condition at $x = 1$. What is the level of accuracy?

4.3.2 Solution to Theoretical Part

▶ **Theoretical Part: Variational Formulation**

1) Let v be a sufficiently regular real-valued test function on $[0, 1]$. As discussed in the presentation of the Dirichlet problem in Sect. 4.1, the regularity of the functions v will be specified *a posteriori* to ensure that the variational formulation makes good sense when it is established.

We multiply the differential equation of the continuous problem **(CP)** by v and integrate over the interval $[0, 1]$ to obtain

$$- \int_0^1 u''(x)v(x)\,\mathrm{d}x + \int_0^1 u(x)v(x)\,\mathrm{d}x = \int_0^1 f(x)v(x)\,\mathrm{d}x, \quad \forall v \in V. \quad (4.114)$$

Integrating by parts

$$\int_0^1 u'(x)v'(x)\,\mathrm{d}x + u'(0)v(0) - u'(1)v(1) + \int_0^1 u(x)v(x)\,\mathrm{d}x = \int_0^1 f(x)v(x)\,\mathrm{d}x, \quad \forall v \in V. \quad (4.115)$$

We now note that the Fourier boundary condition $u'(1) + ku(1) = 1$, as specified in the continuous problem **(CP)**, appears in the integral formulation (4.115).

Indeed, to see this, we can rewrite the Fourier condition in the form

$$u'(1) = 1 - ku(1), \quad (4.116)$$

in order to replace the first derivative of the solution u at the point $x = 1$ in (4.115).

In addition, as already noted for this type of second order differential equation, the homogeneous Dirichlet condition at $x = 0$ cannot be taken into account directly in the formulation (4.115). This is why we impose the homogeneous Dirichlet condition on the test functions v, of which the solution u constitutes a special case.

In this way we ensure that the future variational formulation **(VP)** retains all the information contained in the continuous problem **(CP)**.

These two boundary conditions, the first affecting u through the relation (4.116) and the second concerning the test functions v vanishing at $x = 0$, lead to the following variational formulation:

$$\left[\begin{array}{l} \text{Find } u \text{ in } V \text{ solution of} \\[2mm] \displaystyle\int_0^1 (u'v' + uv)\,\mathrm{d}x + ku(1)v(1) = \int_0^1 fv\,\mathrm{d}x + v(1), \quad \forall v \in V. \end{array} \right. \quad (4.117)$$

At this point in the analysis, the space V comprises functions v subject to the homogeneous Dirichlet condition at $x = 0$, i.e., $v(0) = 0$. However, the formulation (4.117) is still only formal, since there is still no guarantee that the integrals featuring in it will converge.

We note here that this variational formulation is structurally analogous to the one obtained in the context of the Dirichlet problem [see (4.16) in Sect. 4.1], apart from the boundary conditions which have to be modified for the test functions v in the context of the Fourier–Dirichlet problem we are dealing with here.

So considering the functional analysis already presented in Sect. 4.1, a sufficient condition for convergence of the integrals in the variational formulation (4.117) is provided by the following:

$$V \equiv H^1(0, 1) \equiv \left\{ v : [0, 1] \to \mathbb{R}, \ v \in L^2(0, 1), \ v' \in L^2(0, 1) \right\}, \qquad (4.118)$$

observing that $H^1(0, 1)$ is included in $C^0([0, 1])$, which ensures that the values of v are finite at any point x of the interval $[0, 1]$, and this concerns in particular $v(1)$ which appears in the formulation (4.117).

To this function space, we add the homogeneous Dirichlet condition at the point $x = 0$ to obtain, finally, the following variational formulation **(VP)**:

(VP)
$$
\begin{cases}
\text{Find } u \text{ in } V \text{ solution of } a(u, v) = L(v), \quad \forall v \in V, \\[2mm]
a(u, v) \equiv \displaystyle\int_0^1 (u'v' + uv)\, dx + ku(1)v(1), \\[3mm]
L(v) \equiv \displaystyle\int_0^1 fv\, dx + v(1), \\[3mm]
V \equiv H^1_*(0, 1),
\end{cases}
\qquad (4.119)
$$

where the space $H^1_*(0, 1)$ is defined by (4.106).

2) The existence and uniqueness of the solution to the variational problem **(VP)** defined by (4.119) are established using the Lax–Milgram theorem (Theorem 1.11).

For this purpose, we return to the $H^1(0, 1)$ norm (4.74) of Sect. 4.1, also defined in (4.120) below, thus electing to measure the size of functions in $H^1_*(0, 1)$ as we would any function in $H^1(0, 1)$, since $H^1_*(0, 1) \subset H^1(0, 1)$.

The definition of the natural norm on $H^1(0, 1)$ is

$$\forall v \in H^1(0, 1), \quad \|v\|^2_{H^1} \equiv \int_0^1 v'(x)^2\, dx + \int_0^1 v(x)^2\, dx \equiv \|v'\|^2_{L^2} + \|v\|^2_{L^2}.$$
$$(4.120)$$

Having chosen this norm, we now discuss the different premises of the Lax–Milgram theorem that must be checked before we can apply the theorem:

- The space $H_*^1(0, 1)$ is a Hilbert space for the norm defined by (4.120). To check this, we can show, for example, that $H_*^1(0, 1)$ is a closed vector subspace of $H^1(0, 1)$ for the norm (4.120) (see Sect. 1.5.3).

- The bilinear form $a(., .)$ is continuous on $H_*^1(0, 1) \times H_*^1(0, 1)$ for the norm (4.120). Indeed, if $(u, v) \in H_*^1(0, 1) \times H_*^1(0, 1)$, we have

$$|a(u, v)| \leq |(u, v)_{H^1}| + k|u(1)v(1)| \leq (u, u)_{H^1}^{1/2}(v, v)_{H^1}^{1/2} + k|u(1)v(1)|$$

$$\leq \|u\|_{H^1}\|v\|_{H^1} + k|u(1)v(1)|, \tag{4.121}$$

where $(., .)_{H^1}$ is the scalar product on H^1 from which the norm (4.120) derives.

Furthermore, whatever v is taken in $H^1(0, 1)$, we have

$$\forall x \in [0, 1], \quad v(x) = v(0) + \int_0^x v'(t)\, dt. \tag{4.122}$$

So for the special case of functions v in $H_*^1(0, 1)$, we have $v(0) = 0$. It then follows that

$$\forall x \in [0, 1], \quad v(x) = \int_0^x v'(t)\, dt. \tag{4.123}$$

We then write down (4.123) at $x = 1$ and use the Cauchy–Schwarz inequality to obtain the following upper bound:

$$|v(1)| \leq \left[\int_0^1 v'^2(t)\, dt\right]^{1/2} \equiv \|v'\|_{L^2}. \tag{4.124}$$

Going back to the inequality (4.121), we have

$$|a(u, v)| \leq \|u\|_{H^1}\|v\|_{H^1} + k\|u'\|_{L^2}\|v'\|_{L^2} \leq (1 + k)\|u\|_{H^1}\|v\|_{H^1}. \tag{4.125}$$

The bilinear form $a(., .)$ is thus continuous for the H^1 norm defined by (4.120), with continuity constant $(1 + k)$.

- The linear form $L(.)$ defined by (4.119) is continuous on $H_*^1(0, 1)$ for the norm (4.120). We proceed here as for the bilinear form $a(., .)$, noting first that

$$|L(v)| \leq \int_0^1 |fv|\, dx + |v(1)| \leq \|f\|_{L^2}\|v\|_{L^2} + |v(1)|, \tag{4.126}$$

then using the upper bound (4.124), which is valid for any function v in $H_*^1(0, 1)$, to obtain

$$|L(v)| \leq \|f\|_{L^2}\|v\|_{L^2} + \|v'\|_{L^2} \leq \left(1 + \|f\|_{L^2}\right)\|v\|_{H^1}. \qquad (4.127)$$

The linear form $L(.)$ is thus continuous on the space $H^1_*(0, 1)$ equipped with the norm (4.120) and the continuity constant is $1 + \|f\|_{L^2}$.

- The form $a(., .)$ is H^1-elliptic and the ellipticity constant is unity. Indeed, this is immediate if we note that

$$a(v, v) = \|v'\|^2_{L^2} + \|v\|^2_{L^2} + kv(1)^2 = \|v\|^2_{H^1} + kv(1)^2 \geq \|v\|^2_{H^1}. \qquad (4.128)$$

Here we used the fact that the parameter k is a given positive real number.

The Lax–Milgram theorem (Theorem 1.11) then implies that there is one and only one function in $H^1_*(0, 1)$ solving the variational problem **(VP)** specified by (4.119).

3) To show that any weak solution in $H^1_*(0, 1)$ is also a function in $H^2(0, 1)$, we use the method presented for the Dirichlet problem in Sect. 4.1. So if u is an element of $H^1_*(0, 1)$ solving the variational problem **(VP)**, we have

$$\int_0^1 u'v' \, dx = \int_0^1 (f - u)v \, dx + \left[1 - ku(1)\right]v(1), \quad \forall v \in H^1_*(0, 1). \qquad (4.129)$$

Among the functions v in $H^1_*(0, 1)$, we choose those in $C^1_0(]0, 1[) \subset H^1_*(0, 1)$. This choice implies in particular that the functions v vanish at $x = 0$ and $x = 1$.

In addition, we introduce the function g defined by $g = f - u$, whence

$$g \in L^2(0, 1) \subset L^1(0, 1) \subset L^1_{loc}(0, 1), \qquad (4.130)$$

where the space $L^1_{loc}(0, 1)$ is defined as follows: for any closed set strictly contained within $(0, 1)$,

$$v \in L^1_{loc}(0, 1) \implies v \in L^1(K). \qquad (4.131)$$

Then the family of variational Eq. (4.129) can be expressed in $C^1_0(]0, 1[)$ in the form

$$\int_0^1 u'(x)v'(x) \, dx = \int_0^1 g(x)v(x) \, dx = -\int_0^1 G(x)v'(x) \, dx, \quad \forall v \in C^1_0(]0, 1[), \qquad (4.132)$$

where G is a primitive of g. Here we have used Lemma 4.1.

Then (4.132) can be written in the form

$$\int_0^1 \left[u'(x) + G(x)\right]v'(x) \, dx = 0, \quad \forall v \in C^1_0(]0, 1[). \qquad (4.133)$$

In this form, the family of variational Eq. (4.133) is strictly the same as the one we obtained for the Dirichlet problem in (4.32) of Sect. 4.1.

The rest of the analysis follows from there and we find, by the same arguments, that any solution u of the variational problem in $H^1_*(0, 1)$ also belongs to the Sobolev space $H^2(0, 1)$.

4) This last question of the theoretical part deals with the equivalence of the solution to the continuous problem **(CP)** and the solution to the variational problem **(VP)**.

As always, it is straightforward to see this equivalence in one direction. Indeed, the construction of the solution to the variational problem **(VP)** from a given solution to the continuous problem **(CP)** is valid whenever the solution u to the continuous problem **(CP)** belongs to $H^2(0, 1)$ and the test functions v used in the variational formulation **(VP)** belong to $H^1_*(0, 1)$.

Note in particular that, if u solves the continuous problem **(CP)** and satisfies the homogeneous Dirichlet boundary condition at $x = 0$, then u, viewed now as a solution to the variational problem **(VP)**, automatically belongs to $H^1_*(0, 1)$.

So let us turn to the converse. Suppose that $u \in H^2(0, 1) \cap H^1_*(0, 1)$ solves the variational problem **(VP)** - according to the last question, we know that any solution u to the variational problem **(VP)** in $H^1_*(0, 1)$ must also belong to $H^2(0, 1)$.

Using the formula for integration by parts in the opposite direction to when we were constructing the variational formulation **(VP)**, we have

$$\int_0^1 (-u'' + u - f)v \, dx + \left[u'(1) + ku(1) - 1\right]v(1) = 0, \quad \forall v \in H^1_*(0, 1). \quad (4.134)$$

We then consider the particular case of functions v in (4.134) which belong to $H^1_0(0, 1)$, i.e., functions v vanishing at $x = 0$ and $x = 1$.

In this case, (4.134) can be written

$$\int_0^1 (-u'' + u - f)v \, dx = 0, \quad \forall v \in H^1_0(0, 1). \quad (4.135)$$

We thus observe that the formulation (4.135) is the same as the one considered in the discussion of the Dirichlet problem in Sect. 4.1. The rest of the present argument is therefore analogous.

We list the main points:

- In (4.135), we choose functions v belonging to $\mathscr{D}(0, 1)$, since $\mathscr{D}(0, 1) \subset H^1_0(0, 1)$.

- We use the density theorem (Theorem 1.1), which tells us that $\mathscr{D}(0, 1)$ is dense in $L^2(0, 1)$.

- We show that (4.135) holds, not just in $H^1_0(0, 1)$, but also in the bigger space $L^2(0, 1)$.

- Among all the functions v in $L^2(0, 1)$, we may now choose precisely the function $v^{**} = -u'' + u - f$.

If in addition the function f belongs to $L^2(0, 1) \cap C^0(]0, 1[)$, the second order differential equation is satisfied for any x in $(0, 1)$ and the solution u is the classic solution to the continuous problem (CP), i.e., belonging to $C^2(]0, 1[)$.

Having thus established that the differential equation in the problem (CP) is satisfied by the solution to the variational problem (VP), the family of Eq. (4.134) is reduced to

$$\left[u'(1) + ku(1) - 1 \right] v(1) = 0, \quad \forall v \in H^1_*(0, 1). \tag{4.136}$$

We make one final choice in (4.136) by considering the special case of the function v^* defined by $v^*(x) = x, \forall x \in [0, 1]$. Naturally, it is straightforward to check that v^* belongs to $H^1_*(0, 1)$.

In this case, (4.136) implies that u satisfies the Fourier boundary condition

$$u'(1) + ku(1) = 1. \tag{4.137}$$

This ends the proof of the converse, whereupon any solution to the variational problem (VP) belonging to $H^2(0, 1) \cap H^1_*(0, 1)$ also solves the continous problem (CP).

4.3.3 Solution to Numerical Part

▶ **Numerical Part: P_1 Lagrange Finite-Element Analysis**

5) To find the dimension of the space $\tilde{V}$, we make the following observation. Once again, the definition (4.108) of the approximation space $\tilde{V}$ makes it very similar to the one we considered for the Dirichlet problem [see Question 5, and in particular (4.6), in Sect. 4.1].

So considering the discussion there, it suffices to note that the space $\tilde{V}$ defined by (4.108) has an extra degree of freedom due to the freedom in the value of any function $\tilde{v}$ in $\tilde{V}$ at $x = 1$.

This implies that the dimension of the space $\tilde{V}$ defined by (4.6) is one greater than the dimension found for the space used in the Dirichlet problem. In other words, any function $\tilde{v}$ in $\tilde{V}$ is uniquely determined by specifying its values $(\tilde{v}_1, \ldots, \tilde{v}_{N+1})$ at the $N + 1$ discretisation points of the mesh on the interval $[0, 1]$, i.e., at the points $(x_1, \ldots, x_{N+1})$.

Therefore, the space $\tilde{V}$ is isomorphic to $\mathbb{R}^{N+1}$ and $\tilde{V}$ has dimension $N + 1$.

6) To establish the approximate variational formulation $\widetilde{(VP)}$, we replace the functions (u, v) in the variational formulation (VP) by the approximation functions $(\tilde{u}, \tilde{v})$. We also use the expressions given in (4.109). The approximate formulation $\widetilde{(VP)}$ then takes the following form:

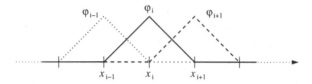

Fig. 4.6 Supports of the basis functions φ_{i-1}, φ_i, and φ_i

Determine the numerical sequence $(\tilde{u}_j)$, $(j = 1, \ldots, N + 1)$, solution of

$$\sum_{j=1}^{N+1} \left[\int_0^1 \left(\varphi_i' \varphi_j' + \varphi_i \varphi_j \right) dx + k \varphi_i(1) \varphi_j(1) \right] \tilde{u}_j$$

$$= \int_0^1 f \varphi_i(x) \, dx + \varphi_i(1). \; \forall \, i = 1, \ldots, N + 1, \qquad (4.138)$$

We can then identify the expressions for A_{ij} and b_j corresponding to the results (4.111).

▶ **Function φ_i Characterising a Node Strictly in the Interior of [0, 1]**

7) We now consider the basis functions φ_i characterising the nodes of the mesh lying strictly in the interior of the integration interval $[0, 1]$. The generic equation of the system (4.138) is exactly the same as the one obtained in (4.48) for the Dirichlet problem in Sect. 4.1.

To see this, we simply note that, for any basis function φ_i characterising a node strictly in the interior of the interval $[0, 1]$, we have $\varphi_i(1) = 0$ (see Fig. 4.6).

In this case, the node Eq. (4.138) can be written

$$\sum_{j=1}^{N+1} \left[\int_0^1 \left(\varphi_j' \varphi_i' + \varphi_j \varphi_i \right) dx \right] \tilde{u}_j = \int_0^1 f \varphi_i(x) \, dx, \quad \forall \, i = 1, \ldots, N + 1,$$

$$(4.139)$$

which thus corresponds precisely to the node Eq. (4.48) of the Dirichlet problem. We may therefore use the results proven for the Dirichlet problem directly in the present Fourier–Dirichlet problem.

Approximate Calculation of the Coefficients A_{ij}, $j = i - 1, i, i + 1$.

1. *Approximation of the coefficient A_{ii}.*

   We have

$$A_{ii} \approx \frac{2}{h} + h. \qquad (4.140)$$

2. *Approximation of the coefficient $A_{i,i-1}$.*

Here,

$$A_{i,i-1} = A_{i,i+1} \approx -\frac{1}{h}. \tag{4.141}$$

Estimation of b_i. As for the Dirichlet problem discussed in Sect. 4.1, or the Neumann problem in Sect. 4.2, we obtain

$$b_i \approx h f_i. \tag{4.142}$$

8) The node equation associated with any function φ_i characterising a strictly interior node is obtained by bringing together all the results (4.140–4.142):

$$-\frac{\tilde{u}_{i-1} - 2\tilde{u}_i + \tilde{u}_{i+1}}{h^2} + \tilde{u}_i = f_i, \quad i = 1, \ldots, N. \tag{4.143}$$

9) The finite-difference discretisation of the second order differential equation in the continuous problem **(CP)** is exactly the same as the one discussed for the Dirichlet problem. This discretisation takes the form

$$-\frac{u(x_{i+1}) - 2u(x_i) + u(x_{i-1})}{h^2} + u(x_i) = f(x_i) + O(h^2), \ i = 1, \ldots, N. \tag{4.144}$$

We then replace the traces $u_i \equiv u(x_i)$ of the function u at the nodes x_i by the corresponding approximations $\tilde{u}_i \approx u_i$, in such a way as to maintain the equality between the two sides of (4.144) when we drop the infinitesimal quantity $O(h^2)$.

This substitution shows that the resulting finite-difference scheme corresponds exactly to the node Eq. (4.143) associated with any function φ_i characterising a node x_i strictly in the interior of the interval $[0, 1]$. Furthermore, the finite-difference scheme (4.143) is second order, since the approximation consists in neglecting the term $O(h^2)$ in (4.144).

▶ **Basis Function φ_{N+1} Characterising a Node at $x_{N+1} = 1$**

10) The node equation associated with the basis function φ_{N+1} characterising the node at $x_{N+1} = 1$ can be written

$$(\widetilde{\text{VP}}_{\text{N+1}}) \quad A_{N+1,N}\, \tilde{u}_N + A_{N+1,N+1}\, \tilde{u}_{N+1} = b_{N+1}. \tag{4.145}$$

This node Eq. (4.145) involves the coefficients $A_{N+1,N}$, $A_{N+1,N+1}$, and b_{N+1} to be estimated below.

Approximate Calculation of the Coefficients $A_{N+1,N}$ and $A_{N+1,N+1}$.

1. *Approximation of the coefficient $A_{N+1,N+1}$.*

We have

$$A_{N+1,\,N+1} = \int_0^1 \left(\varphi_{N+1}'^2 + \varphi_{N+1}^2 \right) dx + k\varphi_{N+1}^2(1)$$

$$= \int_{x_N}^{x_{N+1}} \left(\varphi_{N+1}'^2 + \varphi_{N+1}^2 \right) dx + k$$

$$\approx \left(\frac{1}{h^2} \times h \right) + \frac{h}{2}(0+1) + k$$

$$= \frac{1}{h} + \frac{h}{2} + k. \tag{4.146}$$

Here we used the characteristic property of the basis function φ_{N+1} at the point x_{N+1}, namely $\varphi_{N+1}(x_{N+1}) = 1$.

2. *Approximation of the coefficient $A_{N+1,N}$.*

Here we have

$$A_{N+1,\,N} = \int_0^1 \left(\varphi'_N \varphi'_{N+1} + \varphi_N \varphi_{N+1} \right) dx + k\varphi_N(1)\varphi_{N+1}(1)$$

$$= \int_{\text{supp}\,\varphi_N \cap \text{supp}\,\varphi_{N+1}} \left(\varphi'_N \varphi'_{N+1} + \varphi_N \varphi_{N+1} \right) dx$$

$$\approx -\frac{1}{h^2} \times h + \frac{h}{2}\big[(0 \times 1) + (1 \times 0)\big] = -\frac{1}{h}. \tag{4.147}$$

Once again, we used the characterising property of the basis function φ_N, viz., $\varphi_N(x_{N+1}) = 0$.

Estimation of b_{N+1}. Here we have

$$b_{N+1} = \int_0^1 f\varphi_{N+1}\, dx + \varphi_{N+1}(1) = \int_{x_N}^{x_{N+1}} f\varphi_{N+1}\, dx + 1$$

$$\approx \frac{h}{2}(0 + f_{N+1}) + 1 = \frac{h}{2} f_{N+1} + 1. \tag{4.148}$$

11) The node equation associated with the basis function φ_{N+1} is obtained by bringing together the results (4.146–4.148), whence we find

$$-\frac{1}{h}\tilde{u}_N + \left(\frac{1}{h} + \frac{h}{2} + k \right)\tilde{u}_{N+1} = \frac{h}{2}f_{N+1} + 1. \tag{4.149}$$

12) We now discretise the Fourier boundary condition at the point $x_{N+1} = 1$ using the finite-difference method. To do this, we consider a backward Taylor expansion at

x_{N+1}, expressing the solution u of the continuous problem **(CP)** at x_N as a fonction of the values of u and its derivatives at x_{N+1}.

The expansion is

$$u(x_N) = u(x_{N+1}) - hu'(x_{N+1}) + \frac{h^2}{2}u''(x_{N+1}) + O(h^3). \qquad (4.150)$$

We then replace the second derivative of the solution u at x_{N+1}, which appears in (4.150), assuming that we can write the differential equation of the continuous problem **(CP)** on the boundary of the integration domain $(0, 1)$, i.e., in this case, at the point x_{N+1}:

$$u(x_N) = u(x_{N+1}) - hu'(x_{N+1}) + \frac{h^2}{2}\left[u(x_{N+1}) - f(x_{N+1})\right] + O(h^3). \quad (4.151)$$

In addition, we express the first derivative of the solution u at x_{N+1} using the Fourier boundary condition:

$$u'(x_{N+1}) = 1 - ku(x_{N+1}). \qquad (4.152)$$

Equation (4.151) then takes the form

$$u(x_N) = u(x_{N+1}) - h\left[1 - ku(x_{N+1})\right] + \frac{h^2}{2}\left[u(x_{N+1}) - f(x_{N+1})\right] + O(h^3). \quad (4.153)$$

Rearranging, this becomes

$$-\frac{1}{h}u(x_N) + \left(\frac{1}{h} + \frac{h}{2} + k\right)u(x_{N+1}) = \frac{h}{2}f_{N+1} + 1 + O(h^2). \qquad (4.154)$$

We then obtain the node Eq. (4.149) associated with the basis function φ_{N+1} by replacing the values $u(x_i)$ of the solution u to the continuous problem **(CP)** by the corresponding approximations $\tilde{u}_i$ in (4.154).

This yields

$$-\frac{1}{h}\tilde{u}_N + \left(\frac{1}{h} + \frac{h}{2} + k\right)\tilde{u}_{N+1} = \frac{h}{2}f_{N+1} + 1. \qquad (4.155)$$

4.4 Periodic Problem

4.4.1 Statement of the Problem

The aim in this section is to use the finite-element method for a second order differential problem with periodic boundary conditions.

More exactly, we shall be concerned with the solutions to the following continuous problem (**CP**):

Find $u \in H^2(0, 1)$ solution of

$$(\textbf{CP}) \quad \begin{cases} -u''(x) + u(x) = f(x), & 0 \le x \le 1, \\ u(0) = u(1), & u'(0) = u'(1), \end{cases} \tag{4.156}$$

where f is a given function in $L^2(0, 1)$.

▶ **Theoretical Part: Variational Formulation**

1) Let v be a real-valued test function defined on $[0, 1]$ and belonging to a variational space V. Show that the continuous problem (**CP**) can be cast in the following variational form (**VP**):

$$(\textbf{VP}) \quad \begin{cases} \text{Find } u \in V \text{ solution of} \\ a(u, v) = L(v), \quad \forall v \in V. \end{cases} \tag{4.157}$$

Specify the bilinear form $a(., .)$, the linear form $L(.)$, and the function space V.

2) Establish the existence and uniqueness of a weak solution to the variational problem (**VP**) in $H^1_{\text{per}}(0, 1)$ defined by

$$H^1_{\text{per}}(0, 1) = \left\{ v : (0, 1) \to \mathbb{R}, \ v \in L^2(0, 1), \ v' \in L^2(0, 1), \ v(0) = v(1) \right\}.$$

3) Show that the weak solution to the variational problem (**VP**) also belongs to $H^2(0, 1)$.

4) Deduce the equivalence of the strong formulation (**CP**), set in $H^2(0, 1)$ and the weak formulation (**VP**), considered in $H^1_{\text{per}}(0, 1) \cap H^2(0, 1)$.

▶ **Numerical Part: P_1 Lagrange Finite-Element Analysis**

5) The variational problem (**VP**) is approximated by a P_1 Lagrange finite-element method. To this end, we introduce a regular mesh with constant spacing h on the interval $[0, 1]$:

$$\begin{cases} x_0 = 0, & x_{N+1} = 1, \\ x_{i+1} = x_i + h, & i = 0, \ldots, N. \end{cases} \tag{4.158}$$

We now define the approximation space $\widetilde{V}$ by

$$\widetilde{V} = \left\{ \tilde{v} : [0, 1] \longrightarrow \mathbb{R}, \ \tilde{v} \in C^0([0, 1]), \ \tilde{v}|_{[x_i, x_{i+1}]} \in P_1([x_i, x_{i+1}]), \ \tilde{v}(0) = \tilde{v}(1) \right\}, \tag{4.159}$$

where $P_1([x_i, x_{i+1}])$ is the space of polynomials on $[x_i, x_{i+1}]$ of degree less than or equal to one.

– What is the dimension of $\widetilde{V}$?

6) In order to find a numerical solution to the variational problem **(VP)** by finite elements, we temporarily forget the periodic boundary conditions on the values of the functions u and v at $x = 0$ and $x = 1$.

To do this, we introduce the approximation space $\widetilde{W}$ defined by

$$\widetilde{W} = \left\{ \tilde{w} : [0, 1] \longrightarrow \mathbb{R}, \ \tilde{w} \in C^0([0, 1]), \ \tilde{w} \in P_1([x_i, x_{i+1}]) \right\}. \tag{4.160}$$

– What is the dimension of $\widetilde{W}$?

7) Let φ_i, $(i = 0, \ldots, \dim \widetilde{W} - 1)$, be the basis of $\widetilde{W}$ satisfying $\varphi_i(x_j) = \delta_{ij}$. Having written down the approximate variational formulation of the solution $\tilde{u}$, which we seek for the time being in $\widetilde{W}$, associated with the problem **(VP)**, show that by choosing

$$\tilde{w}(x) = \varphi_i(x), \quad i = 0, \ldots, \dim \widetilde{W} - 1, \qquad \tilde{u}(x) = \sum_{j=0,\ldots,\dim \widetilde{W}-1} \tilde{u}_j \varphi_j, \tag{4.161}$$

we obtain the following system:

$$\textbf{(}\widetilde{\textbf{VP}}\textbf{)} \qquad \sum_{j=0,\ldots,\dim \widetilde{W}-1} A_{ij} \tilde{u}_j = b_i, \quad \forall i \in \{0, \ldots, \dim \widetilde{W} - 1\}, \tag{4.162}$$

where

$$A_{ij} = \int_0^1 (\varphi_i' \varphi_j' + \varphi_i \varphi_j) \, \mathrm{d}x, \qquad b_i = \int_0^1 f \varphi_i \, \mathrm{d}x. \tag{4.163}$$

▶ **Function φ_i Characterising a Node Strictly in the Interior of $[0, 1]$**

8) Given the regularity of the mesh, the generic node equation of the system $\widetilde{\textbf{(VP)}}$ associated with any basis function φ_i, $(i = 1, \ldots, \dim \widetilde{W} - 2)$, characterising a node strictly in the interior of $[0, 1]$ can be written

$$(\widetilde{\textbf{VP}_{\text{int}}}) \quad A_{i,i-1}\tilde{u}_{i-1} + A_{i,i}\tilde{u}_i + A_{i,i+1}\tilde{u}_{i+1} = b_i, \quad \forall i = 1, \ldots, \dim \widetilde{W} - 2.$$
(4.164)

Using the trapezoidal rule, calculate the four coefficients (A_{ij}, b_i). Recall that the trapezoidal rule is

$$\int_a^b \xi(s)\mathrm{d}s \approx \frac{b-a}{2}\big[\xi(a) + \xi(b)\big].$$

9) Bring together all the above results to write down the corresponding node equation.

10) Show that we recover the centred finite-difference scheme associated with the differential equation of the continuous problem **(CP)**. What is the order of accuracy?

▶ **Function φ_0 Characterising the Node at $x_0 = 0$**

11) We proceed in the same way for the basis function φ_0 characterising the initial node x_0. The corresponding equation of the system $(\widetilde{\textbf{VP}})$ is then

$$(\widetilde{\textbf{VP}_0}) \quad A_{00}\tilde{u}_0 + A_{01}\tilde{u}_1 = b_0.$$
(4.165)

Using the trapezoidal rule, calculate the coefficients A_{00}, A_{01}, and b_0.

12) Bring together the above results by writing down the corresponding node equation.

▶ **Function φ_{N+1} Characterising the Node at x_{N+1}**

13) We proceed as for the basis function φ_{N+1} characterising the final node x_{N+1}. The corresponding equation of the system $(\widetilde{\textbf{VP}})$ is then

$$(\widetilde{\textbf{VP}_{N+1}}) \quad A_{N+1,N}\tilde{u}_N + A_{N+1,N+1}\tilde{u}_{N+1} = b_{N+1}.$$
(4.166)

Use the trapezoidal rule to calculate the coefficients $A_{N+1,N}$, $A_{N+1,N+1}$, and b_{N+1}.

14) Bring together the above results to write down the corresponding node equation.

15) Taking into account the periodicity properties for the node equations characterising the nodes x_0 and x_{N+1}, write down the algebraic relation **(R)** between the unknowns $(\tilde{u}_0, \tilde{u}_1, \tilde{u}_N)$ and the data $(\tilde{f}_0, \tilde{f}_{N+1})$.

16) Use the finite-difference method to discretise to order two the periodic boundary conditions of the continuous problem **(CP)** and show that we recover precisely the algebraic relation **(R)** found in the last question.

4.4.2 Solution to Theoretical Part

▶ **Theoretical Part: Variational Formulation**

1) Let v be a sufficiently regular real-valued test function defined on $[0, 1]$ and belonging to a function space V. We multiply the differential equation of the continuous problem **(CP)** by v and integrate over the interval $[0, 1]$ to obtain

$$-\int_0^1 u''v\,dx + \int_0^1 uv\,dx = \int_0^1 fv\,dx, \quad \forall v \in V. \qquad (4.167)$$

Integrating by parts,

$$\int_0^1 u'v'\,dx + u'(0)v(0) - u'(1)v(1) + \int_0^1 uv\,dx = \int_0^1 fv\,dx, \quad \forall v \in V. \ (4.168)$$

We now observe that the periodic boundary conditions on the derivative of u, viz., $u'(0) = u'(1)$, can be inserted directly in the integral formulation (4.168). We thus have

$$\int_0^1 (u'v' + uv)\,dx + u'(0)\big[v(0) - v(1)\big] = \int_0^1 fv\,dx, \quad \forall v \in V. \qquad (4.169)$$

Regarding the periodic boundary conditions on u, we are forced to impose the same boundary conditions on the test functions v, viz.,

$$v(0) = v(1). \qquad (4.170)$$

The main motivation to introduce condition (4.170) is that it looks likes to a kind of homogeneous Dirichlet condition if one write it: $v(0) - v(1) = 0$.

It follows that u solves the following *formal* variational formulation:

$$\int_0^1 (u'v' + uv)\,dx = \int_0^1 fv\,dx, \quad \forall v \text{ such that } v(0) = v(1). \qquad (4.171)$$

Finally, we note that, if v and v' are functions in $L^2(0, 1)$, the Cauchy–Schwarz inequality will as usual guarantee the convergence of the various integrals featuring in the variational formulation (4.171).

In other words, we assume in what follows that the test functions v, and hence also the solution u, belong to $H^1(0, 1)$. Given the periodic boundary conditions (4.170) that we have also imposed, it follows that the variational space V is defined by

$$V \equiv H^1_{\mathrm{per}}(0, 1) = \left\{ v : (0, 1) \to \mathbb{R} , \quad v \in L^2(0, 1) , \quad v' \in L^2(0, 1), \ v(0) = v(1) \right\}. \tag{4.172}$$

Finally, the variational formulation **(VP)** becomes:

(VP)
$$\begin{cases} \text{Find } u \text{ in } V \text{ solution of } a(u, v) = L(v) , \quad \forall v \in V, \\[2mm] a(u, v) \equiv \displaystyle\int_0^1 \left[u'(x)v'(x) + u(x)v(x) \right] \mathrm{d}x, \\[2mm] L(v) \equiv \displaystyle\int_0^1 f(x)v(x) \, \mathrm{d}x, \\[2mm] V \equiv H^1_{\mathrm{per}}(0, 1). \end{cases} \tag{4.173}$$

2) The existence and uniqueness of the solution to the variational formulation **(VP)** are established using the Lax–Milgram theorem (Theorem 1.11), in a completely analogous way to what happened for the Dirichlet problem in Sect. 4.1.

For this purpose, we equip the space $H^1_{\mathrm{per}}(0, 1)$ with the $H^1(0, 1)$ norm defined by (4.20) and show that $H^1_{\mathrm{per}}(0, 1)$ is a closed subspace of $H^1(0, 1)$, whence it has a Hilbert structure for the H^1 norm.

We thus consider a sequence v_n in $H^1_{\mathrm{per}}(0, 1)$ which converges to some function v in $H^1(0, 1)$ for the H^1 norm. Showing that $H^1_{\mathrm{per}}(0, 1)$ is closed in $H^1(0, 1)$ amounts to proving that the limit v of this arbitrary sequence is also an element of $H^1_{\mathrm{per}}(0, 1)$.

Now it can be shown that [1], if v is a function in $H^1(0, 1)$, then it is also a continuous function (more precisely, there is a continuous representative of the class of functions equal to v almost everywhere).

Furthermore, since v_n converges to v in $H^1(0, 1)$, we deduce that there is a sub-sequence v_{n_k} of v_n such that v_{n_k} converges pointwise to v for almost every x in $[0, 1]$—in fact, it can be shown that the convergence of v_n to v is uniform [1].

But since v_n and v are 'continuous' functions, we deduce that the pointwise convergence occurs at all points x of the interval $[0, 1]$.

We thus have pointwise convergence at $x = 0$ and $x = 1$:

$$\lim_{n \to +\infty} v_{n_k}(0) = v(0), \qquad \lim_{n \to +\infty} v_{n_k}(1) = v(1). \tag{4.174}$$

It remains only to evaluate the difference $v(0) - v(1)$, and in particular, to show that it is zero, in order to ensure that the limit v of the sequence v_n does indeed belong to $H^1_{\mathrm{per}}(0, 1)$:

$$\begin{aligned} |v(0) - v(1)| &= \left| v(0) - v_{n_k}(0) + v_{n_k}(1) - v(1) \right| \\[2mm] &\leq \left| v(0) - v_{n_k}(0) \right| + \left| v_{n_k}(1) - v(1) \right|. \end{aligned} \tag{4.175}$$

Here we used the periodicity of the sequence v_n, viz., $v_n(0) = v_n(1)$, applied to the subsequence v_{n_k}.

We conclude here by taking the limit in the inequality (4.175), whereupon we obtain finally

$$v(0) = v(1). \qquad (4.176)$$

This ends the proof and we may conclude that $H^1_{\text{per}}(0, 1)$ is a Hilbert space for the H^1 norm defined in (4.20), being a closed vector subspace of $H^1(0, 1)$.

We can now apply the Lax–Milgram theorem to the space $H^1_{\text{per}}(0, 1)$ equipped with the H^1 norm, and there is no formal difference with the discussion for the Dirichlet problem.

There is therefore one and only one solution to the variational formulation **(VP)** in $H^1_{\text{per}}(0, 1)$.

3) The regularity result for the solution to the variational formulation **(VP)** is obtained in the same way as was done for the Dirichlet problem in Sect. 4.1. We only need to adapt to the functional context $H^1_{\text{per}}(0, 1)$.

Indeed, if u solves the problem **(VP)**, we have

$$\int_0^1 u'(x)v'(x)\,dx = \int_0^1 [f(x) - u(x)]v(x)\,dx, \quad \forall v \in H^1_{\text{per}}(0, 1). \qquad (4.177)$$

We can then choose v in $C^1_0(]0, 1[)$, which is indeed contained in $H^1_{\text{per}}(0, 1)$, and we return once again to the family of variational Eq. (4.31).

The rest of the demonstration remains the same and we deduce that the solution to the variational problem **(VP)** belongs to $H^1_{\text{per}}(0, 1) \cap H^2(0, 1)$.

4) As for the last question, the equivalence of the strong and weak solutions is treated in the same way as for the Dirichlet problem, but adapting to the relevant functional framework, viz., $H^1_{\text{per}}(0, 1)$.

So, if u is the solution to the variational problem **(VP)**, integration by parts yields

$$\int_0^1 (-u'' + u - f)v(x)\,dx + u'(1)v(1) - u'(0)v(0) = 0, \quad \forall v \in H^1_{\text{per}}(0, 1), \quad (4.178)$$

and since $v(0) = v(1)$,

$$\int_0^1 (-u'' + u - f)v(x)\,dx + [u'(1) - u'(0)]v(0) = 0, \quad \forall v \in H^1_{\text{per}}(0, 1). \quad (4.179)$$

Once again, we can legitimately choose v in $\mathscr{D}(0, 1)$ since $\mathscr{D}(0, 1) \subset H^1_{\text{per}}(0, 1)$. The rest of the demonstration remains unchanged. We use a density argument and deduce that the solution u of the variational formulation **(VP)** satisfies the differential equation of the continuous problem **(CP)** as a functional equation in $L^2(0, 1)$.

If in addition f belongs to $L^2(0, 1) \cap C^0(]0, 1[)$, then the differential equation is satisfied for any x in $(0, 1)$ and the solution u is the classic solution of the continuous problem **(CP)** belonging to $C^2(]0, 1[)$.

4.4.3 Solution to Numerical Part

► Numerical Part: P_1 Lagrange Finite-Element Analysis

5) A function $\tilde{v}$ in $\tilde{V}$ is a continuous and piecewise affine function on the interval $[0, 1]$. For this reason, without the periodic boundary conditions, $\tilde{V}$ would be isomorphic to $\mathbb{R}^{N+2}$.

To see this, note that any function $\tilde{v} \in \tilde{V}$ is fully determined by specifying its values at the $N+2$ points x_i of the mesh. Indeed, the difference between two functions in $\tilde{V}$ corresponds inevitably to a change in one of the values of these functions at one of the nodes x_i, $(i = 0, \ldots, N + 1)$, of the mesh.

The periodicity constraint on the functions $\tilde{v}$ in $\tilde{V}$ then leads to the loss of one degree of freedom. In other words, we obtain finally

$$\dim \tilde{V} = N + 1. \qquad (4.180)$$

6) The last question implies immediately that the dimension of $\widetilde{W}$ is $N + 2$, due to the removal of the periodicity constraint.

7) The approximate variational formulation $\widetilde{\textbf{(VP)}}$ is obtained by replacing the functions u and v in the variational formulation **(VP)**, by the corresponding approximation functions $\tilde{u}$ and $\tilde{w}$.

We then obtain

$$\int_0^1 \tilde{u}'\tilde{w}' \, dx + \int_0^1 \tilde{u}\tilde{w} \, dx = \int_0^1 f\tilde{w} \, dx \,, \quad \forall \tilde{w} \in \widetilde{W}, \qquad (4.181)$$

or using the specific values given by (4.161),

$$\sum_{j=0,N+1} \left[\int_0^1 \left(\varphi_i'\varphi_j' + \varphi_i\varphi_j \right) dx \right] \tilde{u}_j = \int_0^1 f\varphi_i \, dx \,, \quad \forall i = 0, \ldots, N+1.$$
$$\qquad (4.182)$$

This is precisely what we had to show, when we introduce the quantities A_{ij} and b_i defined by (4.163).

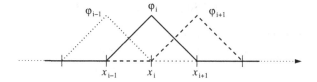

Fig. 4.7 Supports of the basis functions φ_{i-1}, φ_i, and φ_i

▶ **Function φ_i Characterising a Node Strictly in the Interior of [0, 1]**

8) Due to the regularity of the mesh, the node equations associated with functions φ_i characterising nodes x_i in the interior of the interval [0, 1] can all be treated on an equal footing.

Indeed, given the properties of the supports of the basis functions φ_i, $(i = 1, \ldots, N)$, and for fixed i, only the values $j = i - 1$, $j = i$, and $j = i + 1$ in the sum of Eq. (4.182) can produce non-zero contributions (see Fig. 4.7).

For this reason, the approximate variational equation $\widetilde{(\mathbf{VP})}$ can be expressed in the form of $\widetilde{(\mathbf{VP_{int}})}$, for all i between 1 and N inclusive.

We then recover strictly the same formalism as we considered for the Dirichlet, Neumann, and Fourier–Dirichlet problems.

The results proven in our earlier discussions can thus be applied directly to obtain the coefficients of the matrix A_{ij} and the right-hand side b_i.

In other words, we have

$$A_{i,i} \approx \frac{2}{h} + h, \qquad A_{i,i-1} = A_{i,i+1} \approx -\frac{1}{h}, \qquad b_i \approx h f_i. \tag{4.183}$$

9) We may likewise quote exactly the same node equation corresponding to the coefficients listed above:

$$-\frac{\tilde{u}_{i-1} - 2\tilde{u}_i + \tilde{u}_{i+1}}{h^2} + \tilde{u}_i = f_i, \quad i = 1, \ldots, N. \tag{4.184}$$

10) The finite-difference discretisation can also be taken from the discussion of the Dirichlet problem, whence

$$-\frac{u(x_{i+1}) - 2u(x_i) + u(x_{i-1})}{h^2} + u(x_i) = f(x_i) + O(h^2), \ i = 1, \ldots, N. \tag{4.185}$$

We then replace the values u_i of u at the nodes x_i by the approximations $\tilde{u}_i \approx u_i$, in order to preserve the equality in (4.185) when the infinitesimal $O(h^2)$ is dropped. This substitution leads immediately to the node Eq. (4.184).

▶ **Basis Function φ_0 Characterising the Node at $x_0 = 0$**

11) Now consider the first equation of the linear system (4.162), i.e., corresponding to $i = 0$. Given the properties of the supports of the basis functions φ_i, only the functions φ_0 and φ_1 can yield non-zero contributions when integrated with the function φ_0.

For this reason, the generic equation of the system (4.162) can be expressed in the present case by the formula (4.165), viz.,

$$A_{00}\, \tilde{u}_0 + A_{01}\, \tilde{u}_1 = b_0. \tag{4.186}$$

Once again, the calculation of the coefficients A_{00}, A_{01}, and b_0 is the same as for the Neumann problem in Sect. 4.2.

We thus obtain

$$A_{00} \approx \frac{1}{h} + \frac{h}{2}, \qquad A_{01} \approx -\frac{1}{h}, \qquad b_0 \approx \frac{h}{2} f_0. \tag{4.187}$$

12) The resulting node equation is then

$$\left(\frac{1}{h} + \frac{h}{2} \right) \tilde{u}_0 - \frac{1}{h} \tilde{u}_1 = \frac{h}{2} f_0. \tag{4.188}$$

▶ **Basis Function φ_{N+1} Characterising the Node at $x_{N+1} = 1$**

13) For similar reasons to those mentioned in connection with the node equation associated with the basis function φ_0, the results can be taken directly from the discussion of the Neumann problem in Sect. 4.2.

The coefficients $A_{N,N+1}$, $A_{N+1,N+1}$, and b_{N+1} are thus given by

$$A_{N+1,\,N+1} \approx \frac{1}{h} + \frac{h}{2}, \qquad A_{N+1,N} \approx -\frac{1}{h}, \qquad b_{N+1} \approx \frac{h}{2} f_{N+1}. \tag{4.189}$$

14) The resulting node equation is then

$$-\frac{1}{h} \tilde{u}_N + \left(\frac{1}{h} + \frac{h}{2} \right) \tilde{u}_{N+1} = \frac{h}{2} f_{N+1}. \tag{4.190}$$

15) We now consider the periodicity properties of the solution u of the continuous problem **(CP)**, which we impose on the approximation $\tilde{u}$, namely, $\tilde{u}_0 = \tilde{u}_{N+1}$.

Adding together the two node Eqs. (4.188) and (4.190), we obtain the algebraic relation **(R)**:

$$\left(\frac{2}{h} + h \right) \tilde{u}_0 - \frac{1}{h} \left(\tilde{u}_1 + \tilde{u}_N \right) = \frac{h}{2} \left(f_0 + f_{N+1} \right). \tag{4.191}$$

16) The finite-difference method is applied to the periodic boundary conditions of the continuous problem **(CP)** by considering the forward and backward Taylor expansions at the points x_0 and x_{N+1}, respectively, to obtain

$$u(x_1) = u(x_0) + hu'(x_0) + \frac{h^2}{2}u''(x_0) + O(h^3), \tag{4.192}$$

$$u(x_N) = u(x_{N+1}) - hu'(x_{N+1}) + \frac{h^2}{2}u''(x_{N+1}) + O(h^3). \tag{4.193}$$

We then assume that the solution u of the continuous problem **(CP)** is regular enough to justify writing the differential Eq. (4.156) at the points x_0 and x_{N+1}:

$$u''(x_0) = u(x_0) - f(x_0), \tag{4.194}$$

$$u''(x_{N+1}) = u(x_{N+1}) - f(x_{N+1}). \tag{4.195}$$

We then substitute the expressions (4.194) and (4.195) for the second derivative at the points x_0 and x_{N+1} into the Eqs. (4.192) and (4.193):

$$u(x_1) = u(x_0) + hu'(x_0) + \frac{h^2}{2}\left[u(x_0) - f(x_0)\right] + O(h^3), \tag{4.196}$$

$$u(x_N) = u(x_{N+1}) - hu'(x_{N+1}) + \frac{h^2}{2}\left[u(x_{N+1}) - f(x_{N+1})\right] + O(h^3). \tag{4.197}$$

We then introduce the approximations $\tilde{u}_i$, whereupon we may neglect the infinitesimals while maintaining equality.

Finally, we use the periodicity conditions for the $\tilde{u}_i$, and this leads to the algebraic relation **(R)**, identifying (4.196) and (4.197):

$$\frac{h}{2}(f_0 + f_{N+1}) = \left(\frac{2}{h} + h\right)\tilde{u}_0 - \frac{1}{h}(\tilde{u}_1 + \tilde{u}_N). \tag{4.198}$$

Reference

1. H. Brézis, *Analyse fonctionnelle, théorie et applications* (Masson, Paris, 1983)

Chapter 5
Mechanics of Deformable Solids

5.1 Mixed Stress–Strain Problem

5.1.1 Statement of the Problem

A homogeneous and isotropic elastic solid with given Lamé parameters $\lambda > 0$ and $\mu > 0$ occupies a region Ω represented by a bounded open subset of three-dimensional space with coordinates (x_1, x_2, x_3) (see Fig. 5.1). We consider the following linear elastostatic problem:

Find $(\sigma, \mathbf{u})$, the stress and displacement fields in Ω, that solve the continuous problem (**CP**) specified by

(**CP**)
$$
\begin{cases}
\dfrac{\partial \sigma_{ij}}{\partial x_j} + f_i = 0, & \text{in } \Omega, & (5.1) \\[2mm]
\mathbf{u} = \mathbf{0}, & \text{on } \Sigma_1, & (5.2) \\[2mm]
\sigma_{ij} n_j = F_i, & \text{on } \Sigma_2, & (5.3) \\[2mm]
\sigma_{ij} = \lambda \varepsilon_{ll}(\mathbf{u})\delta_{ij} + 2\mu \varepsilon_{ij}(\mathbf{u}), & & (5.4) \\[2mm]
\varepsilon_{ij}(\mathbf{u}) = \dfrac{1}{2}\left[\dfrac{\partial u_i}{\partial x_j} + \dfrac{\partial u_j}{\partial x_i}\right], & & (5.5)
\end{cases}
$$

where we use the Einstein summation (or repeated index) convention.

We thus have

$$
\frac{\partial \sigma_{ij}}{\partial x_j} \equiv \sum_{j=1,2,3} \frac{\partial \sigma_{ij}}{\partial x_j}, \tag{5.6}
$$

$$
\varepsilon_{ll}(\mathbf{u}) \equiv \operatorname{Tr} \varepsilon(\mathbf{u}) \equiv \varepsilon_{11}(\mathbf{u}) + \varepsilon_{22}(\mathbf{u}) + \varepsilon_{33}(\mathbf{u}), \tag{5.7}
$$

J. Chaskalovic, *Mathematical and Numerical Methods for Partial Differential Equations*, Mathematical Engineering, DOI: 10.1007/978-3-319-03563-5_5,
© Springer International Publishing Switzerland 2014

Fig. 5.1 Three-dimensional
elastic medium

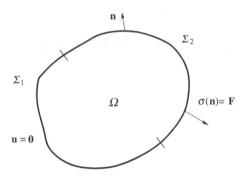

where Tr $\varepsilon(\mathbf{u})$ is the trace of the matrix associated with the strain tensor $\varepsilon(\mathbf{u})$, with generic element $\varepsilon_{ij}(\mathbf{u})$.

Furthermore, the force per unit volume $\mathbf{f}$, with components f_i, is given, and it is such that f_i, ($i = 1, 2, 3$), belong to $L^2(\Omega)$. Likewise, the force per unit area $\mathbf{F}$, with components F_i, is given and belongs to $L^2(\Sigma_2)$.

▶ **Mixed Stress–Strain Variational Formulation**

1) Let $\mathbf{v}$ be a sufficiently regular field of test (or virtual) vectors with components v_i. By the symmetry of the stress tensor σ, solution of the continuous problem **(CP)**, show that

$$\int_{\Omega} \sigma_{ij} \frac{\partial v_i}{\partial x_j} \, d\Omega = \int_{\Omega} \sigma_{ij} \varepsilon_{ij}(\mathbf{v}) \, d\Omega. \tag{5.8}$$

2) Deduce that for every displacement field $\mathbf{v}$ vanishing on the boundary Σ_1, a variational formulation **(VP)** can be associated with the continuous problem **(CP)** in the following way:

$$\int_{\Omega} \sigma_{ij} \varepsilon_{ij}(\mathbf{v}) \, d\Omega = \int_{\Omega} f_i v_i \, d\Omega + \int_{\Sigma_2} F_i v_i \, d\Gamma, \quad \forall \mathbf{v} \in V. \tag{5.9}$$

3) Eliminate the stress tensor from the variational formulation (5.9) and write the displacement variational formulation **(VP)$_\mathbf{u}$** in the following form:

$$\text{Find } \mathbf{u} \in V \text{ solution of } a(\mathbf{u}, \mathbf{v}) = L(\mathbf{v}), \quad \forall \mathbf{v} \in V. \tag{5.10}$$

Specify the forms $a(., .)$ and $L(.)$ and also the function space V.

4) Show that there is a minimisation problem **(MP)** equivalent to the variational problem **(VP)$_\mathbf{u}$**.

Fig. 5.2 Elastic square
$]0, 1[\times]0, 1[$

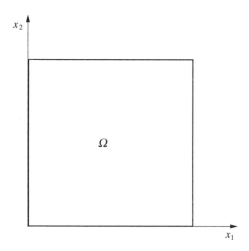

► **Variational Formulation Associated with the Navier Equation**

Throughout the remainder of this problem, we consider only the two-dimensional case (x_1, x_2).

We discuss here the mechanical system specified by the square $\Omega \equiv]0, 1[\times]0, 1[$. The edge Σ_2 of $\partial\Omega$ is just the segment $x_2 = 0$, and the edge Σ_1 corresponds to the complement of Σ_2 in $\partial\Omega$, i.e., Σ_1 consists of the three other sides of the square Ω (see Fig. 5.2).

5) Show that the continuous problem **(CP)** can be solved in displacements using the vectorial form of the Navier equation:

$$(\lambda + \mu)\,\mathbf{grad}(\mathrm{div}\,\mathbf{u}) + \mu\Delta\mathbf{u} + \mathbf{f} = 0, \quad \text{in } \Omega. \tag{5.11}$$

6) Deduce that the displacement continuous problem **(DCP)**, with unknown displacements $\mathbf{u}$, has the following structure:

(DCP) $\begin{cases} (\lambda + \mu)\,\mathbf{grad}(\mathrm{div}\,\mathbf{u}) + \mu\Delta\mathbf{u} + \mathbf{f} = \mathbf{0}, \quad \text{in } \Omega, & (5.12) \\[2mm] \mathbf{u} = \mathbf{0}, \quad \text{on } \Sigma_1, & (5.13) \\[2mm] \dfrac{\partial u_1}{\partial x_2}(x_1, 0) + \dfrac{\partial u_2}{\partial x_1}(x_1, 0) = -\dfrac{F_1}{\mu}, & (5.14) \\[2mm] \lambda\dfrac{\partial u_1}{\partial x_1}(x_1, 0) + (\lambda + 2\mu)\dfrac{\partial u_2}{\partial x_2}(x_1, 0) = -F_2, & (5.15) \end{cases}$

where u_1 and u_2 are the components of the displacement field $\mathbf{u}$ in the coordinate plane (x_1, x_2).

Fig. 5.3 Mesh for P_1
finite-element analysis

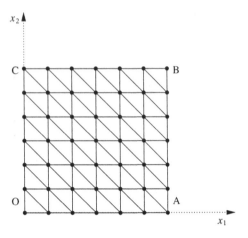

7) Let **v** be an arbitrary vector field in V_d. Show that a variational formulation (**DVP**) can be associated with the displacement continuous problem (**DCP**) in the following way:

$$
\textbf{(DVP)} \quad \begin{cases} \text{Find } \mathbf{u} \in V_d \text{ solution of} \\ a(\mathbf{u}, \mathbf{v}) = L(\mathbf{v}), \quad \forall \mathbf{v} \in V_d, \end{cases} \tag{5.16}
$$

with

$$
a(\mathbf{u}, \mathbf{v}) = \mu \int_{\Omega} (\nabla u_1 \cdot \nabla v_1 + \nabla u_2 \cdot \nabla v_2) \, d\Omega + (\lambda + \mu) \int_{\Omega} \operatorname{div} \mathbf{u} \, \operatorname{div} \mathbf{v} \, d\Omega \tag{5.17}
$$

$$
+ \mu \int_0^L \left(\frac{\partial u_1}{\partial x_1} v_2 - \frac{\partial u_2}{\partial x_1} v_1 \right) dx_1,
$$

$$
L(\mathbf{v}) = \int_{\Omega} f_1 v_1 \, d\Omega + \int_{\Omega} f_2 v_2 \, d\Omega + \int_0^L F_1 v_1 \, dx_1 + \int_0^L F_2 v_2 \, dx_1. \tag{5.18}
$$

Specify the space V_d, i.e., the regularity of and boundary conditions on the vector fields **v** in V_d.

▶ P_1 Lagrange Finite-Element Analysis

8) The variational problem (**DVP**) is treated approximately using a P_1 Lagrange finite-element analysis. To do this, we introduce a regular mesh on the square Ω (see Fig. 5.3) in the form of a triangulation $\mathfrak{I}$ made up of equilateral triangles T_k of side h, such that

$$\begin{cases} x_1^{(0)} = x_2^{(0)} = 0, & x_1^{(N+1)} = x_2^{(N+1)} = L, \\ x_1^{(i+1)} = x_1^{(i)} + h, & x_2^{(i+1)} = x_2^{(i)} + h, \quad i = 0, \dots, N. \end{cases} \tag{5.19}$$

Now define the approximation space $\widetilde{V}$ by

$$\widetilde{V} = \left\{ \tilde{\mathbf{v}} : \Omega \longrightarrow \mathbb{R}^2, \ \tilde{\mathbf{v}} \in [C^0(\Omega)]^2, \ \tilde{v}_i |_{T_k} \in P_1(T_k), \ \tilde{\mathbf{v}} = \mathbf{0} \text{ on } \Sigma_1 \right\}, \tag{5.20}$$

where $P_1(T_k)$ is the space of polynomials on the generic triangle T_k, with degree less than or equal to 1 in the pair of space variables (x_1, x_2).

– What is the dimension of $\widetilde{V}$?

► **Approximate Variational Formulation**

9) Set $P = N(N + 1)$ and let φ_i, $(i = 1, \dots, P)$, be the basis functions used classically for the P_1 finite-element method, satisfying $\varphi_i(M_j) = \delta_{ij}$.

Write down the approximate variational formulation $\widetilde{\textbf{(DVP)}}$ with solution $\tilde{\mathbf{u}}$, associated with the variational problem **(DVP)**, and show that by expressing $\tilde{\mathbf{u}}$ in the form

$$\tilde{\mathbf{u}} = \left(\sum_{j=1,\dots,P} \tilde{u}_1^j \varphi_j, \ \sum_{j=1,\dots,P} \tilde{u}_2^j \varphi_j \right), \tag{5.21}$$

and successively choosing $\tilde{\mathbf{v}}$ in the form

$$\tilde{\mathbf{v}} = (\varphi_i, 0), \quad \text{then } \tilde{\mathbf{v}} = (0, \varphi_i), \quad i = 1, \dots, P, \tag{5.22}$$

we obtain the following linear system:

$$\widetilde{\textbf{(DVP)}} \quad \begin{cases} A_{ij}^{(1,1)} \tilde{u}_{1j} + A_{ij}^{(2,1)} \tilde{u}_{2j} = b_i^{(1)}, \\ A_{ij}^{(1,2)} \tilde{u}_{1j} + A_{ij}^{(2,2)} \tilde{u}_{2j} = b_i^{(2)}, \quad \forall i = 1, \dots, P, \end{cases} \tag{5.23}$$

with

$$A_{ij}^{(1,1)} = \mu \int_{\Omega} \nabla \varphi_i \cdot \nabla \varphi_j \, \mathrm{d}\Omega + (\lambda + \mu) \int_{\Omega} \frac{\partial \varphi_i}{\partial x_1} \frac{\partial \varphi_j}{\partial x_1} \, \mathrm{d}\Omega, \tag{5.24}$$

$$A_{ij}^{(2,1)} = (\lambda + \mu) \int_{\Omega} \frac{\partial \varphi_i}{\partial x_1} \frac{\partial \varphi_j}{\partial x_2} \, \mathrm{d}\Omega - \mu \int_{0}^{L} \varphi_i \frac{\partial \varphi_j}{\partial x_1} \, \mathrm{d}x_1, \tag{5.25}$$

$$A_{ij}^{(1,2)} = (\lambda + \mu) \int_{\Omega} \frac{\partial \varphi_i}{\partial x_2} \frac{\partial \varphi_j}{\partial x_1} \, \mathrm{d}\Omega + \mu \int_{0}^{L} \varphi_i \frac{\partial \varphi_j}{\partial x_1} \, \mathrm{d}x_1, \tag{5.26}$$

Fig. 5.4 Local numbering system associated with a node strictly in the interior of Ω

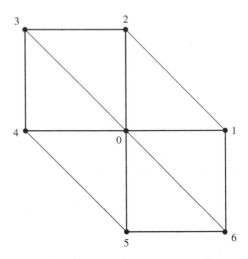

$$A_{ij}^{(2,2)} = \mu \int_{\Omega} \nabla\varphi_i \cdot \nabla\varphi_j \, d\Omega + (\lambda + \mu) \int_{\Omega} \frac{\partial\varphi_i}{\partial x_2} \frac{\partial\varphi_j}{\partial x_2} \, d\Omega, \tag{5.27}$$

$$b_i^{(1)} = \int_{\Omega} f_1 \varphi_i \, d\Omega + \int_0^L F_1 \varphi_i \, dx_1, \quad b_i^{(2)} = \int_{\Omega} f_2 \varphi_i \, d\Omega + \int_0^L F_2 \varphi_i \, dx_1. \tag{5.28}$$

▶ **Basis Function φ_i Characterizing a Node in the Interior of Ω**

10) Due to the regularity of the mesh, the two general node equations of the system $\widetilde{(\mathbf{DVP})}$ can be rewritten in the following local forms, using the local numbering system presented in Fig. 5.4:

$$\widetilde{(\mathbf{DVP})}_{int} \quad \begin{cases} \displaystyle\sum_{j=0,6} \left[A_{0,j}^{(1,1)} \tilde{u}_1^j + A_{0,j}^{(2,1)} \tilde{u}_2^j \right] = b_0^{(1)}, \\[2ex] \displaystyle\sum_{j=0,6} \left[A_{0,j}^{(1,2)} \tilde{u}_1^j + A_{0,j}^{(2,2)} \tilde{u}_2^j \right] = b_0^{(2)}. \end{cases} \tag{5.29}$$

In the rest of the problem, we shall deal explicitly with only the first node equation of the system (5.29), i.e., corresponding to the particular choice of $\tilde{\mathbf{v}}$ equal to $(\varphi_i, 0)$, given that the second node equation can be handled in a completely analogous way.

Fig. 5.5 Local numbering system associated with a node in the interior of the edge Σ_2

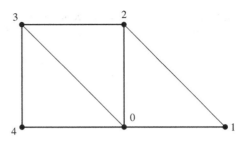

Calculate the 14 coefficients $A_{0,j}^{(1,1)}$, $A_{0,j}^{(2,1)}$, $(j = 0, \ldots, 6)$, and also the associated quantity $b_0^{(1)}$.

Note that the trapezoidal rule applied to a triangle T_{123} with vertices A_1, A_2, and A_3 is

$$\iint_{T_{123}} f(x, y)\, dxdy \approx \frac{\text{area}\,(T_{123})}{3} \sum_{i=1,2,3} f(A_i), \qquad (5.30)$$

where area (T_{123}) is the area of the triangle T_{123}.

11) Combine the above results to write down the node equation for the system $\widetilde{(\text{DVP})}_{int}$.

12) Show that we recover the centered finite-difference scheme associated with the first partial differential equation of the continuous problem **(DCP)**. What is the level of accuracy?

▶ **Basis Function φ_i Characterizing a Node in the Interior of Σ_2**

13) Using the same local numbering system (see Fig. 5.5) and taking into account the geometric features of the edge Σ_2, we see that the node equations corresponding to the basis function characterizing nodes on the edge Σ_2 can be written in the form

$$\widetilde{(\text{DVP})}_{\Sigma_2} \qquad \begin{cases} \displaystyle\sum_{j=0,4} \left[A_{0,j}^{(1,1)} \tilde{u}_1^j + A_{0,j}^{(2,1)} \tilde{u}_2^j \right] = b_0^{(1)}, \\[2mm] \displaystyle\sum_{j=0,4} \left[A_{0,j}^{(1,2)} \tilde{u}_1^j + A_{0,j}^{(2,2)} \tilde{u}_2^j \right] = b_0^{(2)}. \end{cases} \qquad (5.31)$$

Calculate the ten coefficients $A_{0,j}^{(1,1)}$, $A_{0,j}^{(2,1)}$, $(j = 0, \ldots, 4)$, together with $b_0^{(1)}$ on the right-hand side.

14) Combine the above results to write down the corresponding node equation of the system $\widetilde{(\text{DVP})}_{\Sigma_2}$.

15) Show that we recover the second-order centered finite-difference scheme associated with the system of two Eqs. (5.14) and (5.15) of the continuous problem **(DCP)** on Σ_2.

———————————————

5.1.2 Solution

▶ **Mixed Stress–Strain Variational Formulation**

1) Let **v** be a test vector field with all the regularity needed for the integrals arising in the following questions. Equation (5.8) is obtained by bringing in the strain tensor $\varepsilon(\mathbf{v})$ from the left-hand integral of (5.8).

To this end, we evaluate the integral of the stress tensor σ multiplied by the partial derivative $\partial v_j/\partial x_i$ appearing in the Definition (5.5) of the strain tensor $\varepsilon(\mathbf{v})$:

$$\int_\Omega \sigma_{ij} \frac{\partial v_j}{\partial x_i}\, d\Omega = \int_\Omega \sigma_{ji} \frac{\partial v_i}{\partial x_j}\, d\Omega = \int_\Omega \sigma_{ij} \frac{\partial v_i}{\partial x_j}\, d\Omega, \qquad (5.32)$$

where we have used the symmetry of the stress tensor σ, viz., $\sigma_{ij} = \sigma_{ji}$.

We thus deduce that

$$\int_\Omega \sigma_{ij} \left(\frac{\partial v_j}{\partial x_i} + \frac{\partial v_i}{\partial x_j} \right) d\Omega = 2 \int_\Omega \sigma_{ij} \frac{\partial v_i}{\partial x_j}\, d\Omega. \qquad (5.33)$$

Equation (5.8) follows, provided that we refer to the Definition (5.5) of the strain tensor $\varepsilon(\mathbf{v})$.

2) To construct a variational formulation (**VP**), we multiply the partial differential Eq. (5.1) by the component v_i of the test displacement field **v** and integrate over the domain Ω to obtain

$$\int_\Omega \frac{\partial \sigma_{ij}}{\partial x_j} v_i\, d\Omega + \int_\Omega f_i v_i\, d\Omega = 0. \qquad (5.34)$$

We then use Green's formula to transform the first integral of (5.34), whence

$$-\int_\Omega \sigma_{ij} \frac{\partial v_i}{\partial x_j}\, d\Omega + \int_{\partial\Omega} \sigma_{ij} n_j v_i\, d\Gamma + \int_\Omega f_i v_i\, d\Omega = 0. \qquad (5.35)$$

If we now take into account (5.8), on the one hand, and the stress boundary conditions (5.3), on the other, we find that

$$\int_\Omega \sigma_{ij} \varepsilon_{ij}(\mathbf{v})\, d\Omega = \int_{\Sigma_1} \sigma_{ij} n_j v_i\, d\Gamma + \int_\Omega f_i v_i\, d\Omega + \int_{\Sigma_2} F_i v_i\, d\Gamma. \qquad (5.36)$$

Since it is impossible to take into account the boundary conditions (5.2) on the solution displacement field **u** in the formulation (5.36), we construct the variational space V in such a way that the elements **v** are identically zero on Σ_1.

This means that the solution **u**, being a particular element of V, will also vanish on the edge Σ_1 of Ω.

It then follows that

$$\int_{\Omega} \sigma_{ij}\varepsilon_{ij}(\mathbf{v})\,d\Omega = \int_{\Omega} f_i v_i\,d\Omega + \int_{\Sigma_2} F_i v_i\,d\Gamma, \quad \forall \mathbf{v} \in V, \qquad (5.37)$$

where the space V is defined by $V \equiv \{\mathbf{v} \text{ such that } \mathbf{v} = \mathbf{0} \text{ on } \Sigma_1\}$.

3) We now use the constitutive equation of the material, viz., (5.35), to replace the stress tensor σ by the strain tensor $\varepsilon(\mathbf{u})$, and then the solution displacement field $\mathbf{u}$.

Equation (5.37) thereby delivers the variational formulation $(\mathbf{VP})_{\mathbf{u}}$,

Find $\mathbf{u} \in V$ solution of

$$\begin{cases} \forall \mathbf{v} \in V = \{\mathbf{v} \text{ such that } \mathbf{v} = \mathbf{0} \text{ on } \Sigma_1\} \\ \int_{\Omega} \left[\lambda\varepsilon_{ll}(\mathbf{u})\delta_{ij} + 2\mu\varepsilon_{ij}(\mathbf{u})\right]\varepsilon_{ij}(\mathbf{v})\,d\Omega = \int_{\Omega} f_i v_i\,d\Omega + \int_{\Sigma_2} F_i v_i\,d\Gamma, \end{cases} \qquad (5.38)$$

or again, using the properties of the Kronecker symbol,

$$a(\mathbf{u}, \mathbf{v}) = L(\mathbf{v}), \quad \forall \mathbf{v} \in V, \qquad (5.39)$$

with

$$a(\mathbf{u}, \mathbf{v}) \equiv \int_{\Omega} \left[\lambda\varepsilon_{ll}(\mathbf{u})\varepsilon_{mm}(\mathbf{v}) + 2\mu\varepsilon_{ij}(\mathbf{u})\varepsilon_{ij}(\mathbf{v})\right]d\Omega, \qquad (5.40)$$

$$L(\mathbf{v}) \equiv \int_{\Omega} f_i v_i\,d\Omega + \int_{\Sigma_2} F_i v_i\,d\Gamma. \qquad (5.41)$$

Properties of Functions in the Space V. The time has come to ensure convergence of the various integrals arising in the formulation of (5.40) and (5.41). We therefore describe sufficient conditions on the regularity of the vector fields $\mathbf{v}$ in V.

On several occasions (see the Dirichlet problem in Sect. 4.1 and the Neumann problem in Sect. 4.2, etc.), we have shown how the Cauchy–Schwarz inequality can be used to provide bounds on integrals with structure similar to those appearing in the expression for $L(\mathbf{v})$, provided that the components f_i and F_i of the force per unit volume $\mathbf{f}$ and per unit area $\mathbf{f}$ belong to $L^2(\Omega)$ and $L^2(\Sigma_2)$, respectively.

But this is precisely what was assumed at the outset.

Concerning the two terms appearing in the integral of the quantity $a(\mathbf{u}, \mathbf{v})$, we note that only linear combinations of first-order partial derivatives turn up in the expression for $a(\mathbf{u}, \mathbf{v})$. Hence we may use the Cauchy–Schwarz inequality to write down the general formula

$$\left| \int_{\Omega} \frac{\partial u_i}{\partial x_j} \frac{\partial v_k}{\partial x_l} \, d\Omega \right| \leq \left(\int_{\Omega} \left| \frac{\partial u_i}{\partial x_j} \right|^2 d\Omega \right)^{1/2} \left(\int_{\Omega} \left| \frac{\partial v_k}{\partial x_l} \right|^2 d\Omega \right)^{1/2}. \tag{5.42}$$

We thus observe that it suffices to consider the vector fields $\mathbf{v}$ in a function space for which the first partial derivatives of the components of $\mathbf{v}$ belong to $L^2(\Omega)$.

So recalling from above that the components v_i of the field $\mathbf{v}$ must also belong to $L^2(\Omega)$, we set finally

$$V = \left\{ \mathbf{v} : \Omega \to \mathbb{R}^2, \text{ such that } v_i \in L^2(\Omega), \frac{\partial v_i}{\partial x_j} \in L^2(\Omega), \ v_i = 0 \text{ on } \Sigma_1 \right\}. \tag{5.43}$$

This is the Sobolev space $H^1(\Omega) \times H^1(\Omega)$ for vector fields in $\mathbb{R}^2$.

4) The existence of a minimisation problem **(MP)** equivalent to and associated with the variational formulation **(VP)$_\mathbf{u}$** is a consequence of the following properties of the forms $a(.,.)$ and $L(.)$ defined in (5.40) and (5.41), respectively:

1. $a(.,.)$ and $L(.)$ are bilinear and linear, respectively.

2. $a(.,.)$ is symmetric.

3. $a(.,.)$ is positive.

In general, we require a little more of the form $a(.,.)$ whenever we wish to implement the Lax–Milgram theorem (Theorem 1.11). Indeed, in this case, we must show that the form $a(.,.)$ is V-elliptic. We then note that a V-elliptic form is automatically positive.

Furthermore, the equivalence of the problems **(MP)** and **(VP)$_\mathbf{u}$** under the conditions listed above is guaranteed by Theorem 1.13.

▶ **Variational Formulation Associated with the Navier Equation**

5) The Navier equation is obtained by replacing the stress tensor σ in (5.1) by the formula provided by Hooke's law (5.4), which expresses it as a function of the strain tensor $\varepsilon(\mathbf{u})$, and hence as a function of the solution displacement field $\mathbf{u}$.

We thus have

$$\frac{\partial}{\partial x_j} \left[\lambda \varepsilon_{ll}(\mathbf{u}) \delta_{ij} + 2\mu \varepsilon_{ij}(\mathbf{u}) \right] + f_i = 0, \quad \forall i = 1, 2, 3, \tag{5.44}$$

or again,

$$\lambda \frac{\partial \varepsilon_{ll}(\mathbf{u})}{\partial x_j} \delta_{ij} + 2\mu \frac{\partial \varepsilon_{ij}(\mathbf{u})}{\partial x_j} + f_i = 0, \quad \forall i = 1, 2, 3. \tag{5.45}$$

The properties of the Kronecker symbol, on the one hand, and the possibility of permuting the second partial derivative in the case of a sufficiently regular solution, on the other, mean that (5.45) can be rewritten in the form

$$\lambda \frac{\partial (\operatorname{div} \mathbf{u})}{\partial x_i} + \mu \frac{\partial^2 u_i}{\partial x_j \partial x_j} + \mu \frac{\partial}{\partial x_i} \left(\frac{\partial u_j}{\partial x_j} \right) + f_i = 0, \quad \forall i = 1, 2, 3, \tag{5.46}$$

using the definition (5.5) of the linear strain tensor.

Finally,

$$(\lambda + \mu) \frac{\partial (\operatorname{div} \mathbf{u})}{\partial x_i} + \mu \Delta u_i + f_i = 0, \quad \forall i = 1, 2, 3. \tag{5.47}$$

6) The formulation of the displacement continuous problem **(DCP)** is immediate. We simply replace the equilibrium Eq. (5.1) of the continuous problem **(CP)** with the Navier Equation (5.47) and express the stress conditions (5.3) on Σ_2 in terms of the solution displacement field **u**.

Then, since the edge Σ_2 corresponds to the segment $x_2 = 0$, the outward normal vector is $-\mathbf{x_2}$.

The boundary condition (5.3) can therefore be written

$$\begin{pmatrix} \sigma_{11}(x_1, 0) & \sigma_{12}(x_1, 0) \\ \sigma_{21}(x_1, 0) & \sigma_{22}(x_1, 0) \end{pmatrix} \begin{pmatrix} 0 \\ -1 \end{pmatrix} = \begin{pmatrix} F_1(x_1) \\ F_2(x_1) \end{pmatrix}, \tag{5.48}$$

or again,

$$\begin{cases} -\sigma_{12}(x_1, 0) = F_1(x_1), \\ -\sigma_{22}(x_1, 0) = F_2(x_1). \end{cases} \tag{5.49}$$

So expressing the stresses σ_{12} and σ_{22} in terms of the components of the solution displacement field **u**, we obtain the Eqs. (5.14) and (5.15) in the formulation of the continuous problem **(DCP)**.

7) We now establish a variational formulation **(DVP)** corresponding to the continuous problem **(DCP)**. To do this, we consider an arbitrary vector field **v** belonging to a function space V_d to be specified later.

We then multiply the Navier Eq. (5.47) by the component v_i to obtain

$$(\lambda + \mu) \int_\Omega \frac{\partial (\operatorname{div} \mathbf{u})}{\partial x_i} v_i \, d\Omega + \mu \int_\Omega \Delta u_i v_i \, d\Omega + \int_\Omega f_i v_i \, d\Omega = 0. \tag{5.50}$$

Green's formula now yields

$$- (\lambda + \mu) \int_{\Omega} \text{div } \mathbf{u} \, \frac{\partial v_i}{\partial x_i} \, d\Omega + (\lambda + \mu) \int_{\partial\Omega} (\text{div } \mathbf{u}) v_i n_i \, d\Gamma$$

$$- \mu \int_{\Omega} \nabla u_i \nabla v_i \, d\Omega + \mu \int_{\partial\Omega} \frac{\partial u_i}{\partial n} v_i \, d\Gamma + \int_{\Omega} f_i v_i \, d\Omega = 0. \qquad (5.51)$$

We consider that the displacement field $\mathbf{v}$ satisfies the homogeneous Dirichlet condition on Σ_1, i.e., $\mathbf{v} = \mathbf{0}$ on Σ_1. The two boundary integrals on $\partial\Omega$ arising in the variational Eq. (5.51) are evaluated as follows:

$$\int_{\partial\Omega} (\text{div } \mathbf{u}) v_i n_i \, d\Gamma = \int_0^L \text{div } \mathbf{u}(-v_2) \, dx_1 = -\int_0^L \left(\frac{\partial u_1}{\partial x_1} v_2 + \frac{\partial u_2}{\partial x_2} v_2 \right) dx_1,$$

$$(5.52)$$

$$\int_{\partial\Omega} \frac{\partial u_i}{\partial n} v_i \, d\Omega = \int_0^L -\frac{\partial u_i}{\partial x_2} v_i \, dx_1 = -\int_0^L \left(\frac{\partial u_1}{\partial x_2} v_1 + \frac{\partial u_2}{\partial x_2} v_2 \right) dx_1. \qquad (5.53)$$

This result is inserted in (5.52) and (5.53) of the formulation (5.51) to yield

$$(\lambda + \mu) \int_{\Omega} \text{div} \mathbf{u} \, \text{div } \mathbf{v} \, d\Omega + (\lambda + \mu) \int_0^L (\text{div } \mathbf{u}) v_2 \, dx_1$$

$$+ \mu \int_{\Omega} \nabla u_i \cdot \nabla v_i \, d\Omega + \mu \int_0^L \left(\frac{\partial u_1}{\partial x_2} v_1 + \frac{\partial u_2}{\partial x_2} v_2 \right) dx_1 = \int_{\Omega} f_i v_i \, d\Omega. \qquad (5.54)$$

The boundary conditions on Σ_2, expressed in terms of the components (u_1, u_2) of the solution vector field $\mathbf{u}$ and the components of the force density $\mathbf{F}$ on the edge Σ_2, can now be used to rewrite the variational formulation (5.54), and after some elementary rearrangement, we obtain the following:

(DVP)

$$\begin{cases} \text{Find } \mathbf{u} = (u_1, u_2) \in V \text{ solution of } a(\mathbf{u}, \mathbf{v}) = L(\mathbf{v}), \quad \forall \mathbf{v} \in V \\[2mm] a(\mathbf{u}, \mathbf{v}) = \mu \int_{\Omega} (\nabla u_1 \cdot \nabla v_1 + \nabla u_2 \cdot \nabla v_2) \, d\Omega \qquad\qquad (5.55) \\[2mm] \qquad\quad + (\lambda + \mu) \int_{\Omega} \text{div } \mathbf{u} \cdot \text{div } \mathbf{v} \, d\Omega + \mu \int_0^L \left(\frac{\partial u_1}{\partial x_1} v_2 - \frac{\partial u_2}{\partial x_1} v_1 \right) dx_1, \\[2mm] L(\mathbf{v}) = \int_{\Omega} (f_1 \cdot v_1 + f_2 \cdot v_2) d\Omega + \int_0^L (F_1 \cdot v_1 + F_2 \cdot v_2) dx_1. \qquad (5.56) \end{cases}$$

▶ **P_1 Lagrange Finite-Element Analysis**

8) The approximation space $\tilde{V}$ is simply the Cartesian product with itself of the classical P_1 finite-element approximation space for scalar approximation in a square.

In other words, the dimension of $\tilde{V}$ is just $N(N+1)+N(N+1)=2N(N+1)$, since each component of the solution vector field $\mathbf{u}$ is approximated by a scalar approximation function $\tilde{u}_i$ that describes the classical P_1 finite-element approximation space just mentioned.

▶ **Approximate Variational Formulation**

9) The approximate variational formulation $\widetilde{(\mathbf{DVP})}$ is obtained immediately when we replace $(\mathbf{u}, \mathbf{v})$ in the variational formulation $(\mathbf{DVP})$ by the respective approximation fields $(\tilde{\mathbf{u}}, \tilde{\mathbf{v}})$.

Furthermore, to obtain the two generic equations corresponding to each of the components of the approximate displacement field $\tilde{\mathbf{u}}$, we first choose the test vector field $\tilde{\mathbf{v}}$ in the form $\tilde{\mathbf{v}} = (\varphi_i, 0)$, and subsequently $\tilde{\mathbf{v}} = (0, \varphi_i)$.

So when $\tilde{\mathbf{v}} = (\varphi_i, 0)$, we have

$$a\big[(u_1, u_2), (\varphi_i, 0)\big] = \sum_j \left[\mu \int_\Omega \nabla\varphi_i \cdot \nabla\varphi_j \, d\Omega + (\lambda + \mu) \int_\Omega \frac{\partial\varphi_i}{\partial x_1} \frac{\partial\varphi_j}{\partial x_1} \, d\Omega \right] \tilde{u}_1^j$$

$$+ \sum_j \left[(\lambda + \mu) \int_\Omega \frac{\partial\varphi_i}{\partial x_1} \frac{\partial\varphi_j}{\partial x_2} \, d\Omega - \mu \int_0^L \varphi_i \frac{\partial\varphi_j}{\partial x_1} \, dx_1 \right] \tilde{u}_2^j,$$

$$\tag{5.57}$$

$$L\big[(\varphi_i, 0)\big] = \int_\Omega f_1 \varphi_i \, d\Omega + \int_0^L F_1 \varphi_i \, dx_1. \tag{5.58}$$

In the same way, choosing $\tilde{v} = (0, \varphi_i)$, we obtain

$$a\big[(u_1, u_2), (0, \varphi_i)\big] = \sum_j \left[(\lambda + \mu) \int_\Omega \frac{\partial\varphi_i}{\partial x_2} \frac{\partial\varphi_j}{\partial x_1} \, d\Omega + \mu \int_0^L \varphi_i \frac{\partial\varphi_j}{\partial x_1} \, dx_1 \right] \tilde{u}_1^j$$

$$+ \sum_j \left[\mu \int_\Omega \nabla\varphi_i \cdot \nabla\varphi_j \, d\Omega + (\lambda + \mu) \int_\Omega \frac{\partial\varphi_i}{\partial x_2} \frac{\partial\varphi_j}{\partial x_2} \, d\Omega \right] \tilde{u}_2^j,$$

$$\tag{5.59}$$

$$L\big[(0, \varphi_i)\big] = \int_\Omega f_2 \varphi_i \, d\Omega + \int_0^L F_2 \varphi_i \, d\Omega. \tag{5.60}$$

Hence, the system of Eq. (5.29) is obtained by simple identification.

▶ **Basis Function φ_i Characterizing a Node in the Interior of Ω**

10) The following calculations are taken from the account of the P_1 finite-element approximation for the Laplace–Dirichlet problem in [1], evaluating integrals using the triangular mesh.

For this purpose, we shall systematically exploit the local numbering system in the statement of the problem (see Fig. 5.4).

Calculation of the Coefficients $A_{0j}^{(1,1)}$.

We have

$$A_{0j}^{(1,1)} = \mu \int_{\Omega} \nabla\varphi_0 \cdot \nabla\varphi_j \, d\Omega + (\lambda + \mu) \int_{\Omega} \frac{\partial\varphi_0}{\partial x_1} \frac{\partial\varphi_j}{\partial x_1} \, d\Omega. \tag{5.61}$$

1. *Calculation of the coefficient $A_{00}^{(1,1)}$.*

We have

$$A_{00}^{(1,1)} = \mu \int_{\text{supp } \varphi_0} |\nabla\varphi_0|^2 \, d\Omega + (\lambda + \mu) \int_{\Omega} \left(\frac{\partial\varphi_0}{\partial x_1}\right)^2 d\Omega. \tag{5.62}$$

But,

$$\int_{\text{supp } \varphi_0} |\nabla\varphi_0|^2 \, d\Omega = \int_{012} |\nabla\varphi_0|^2 + \int_{023} |\nabla\varphi_0|^2 + \int_{034} |\nabla\varphi_0|^2$$

$$+ \int_{045} |\nabla\varphi_0|^2 + \int_{056} |\nabla\varphi_0|^2 + \int_{061} |\nabla\varphi_0|^2$$

$$= \frac{h^2}{2} \left(\frac{2}{h^2} + \frac{1}{h^2} + \frac{1}{h^2} + \frac{2}{h^2} + \frac{1}{h^2} + \frac{1}{h^2}\right) = 4. \tag{5.63}$$

In addition,

$$\int_{\Omega} \left(\frac{\partial\varphi_0}{\partial x_1}\right)^2 d\Omega = \left(2 \times \frac{1}{h^2} \times \frac{h^2}{2}\right) + 0 + \left(\frac{1}{h^2} \times \frac{h^2}{2} \times 2\right) + 0. \tag{5.64}$$

Putting together the results from (5.63) and (5.64), we obtain the following expression for the coefficient $A_{00}^{(1,1)}$, with the help of the definition (5.61):

$$A_{00}^{(1,1)} = 4\mu + 2(\lambda + \mu). \tag{5.65}$$

Proceeding in a similar way for the other coefficients $A_{0j}^{(1,1)}$, but adapting the calculations to take into account the intersections of the supports of the basis functions φ_j, we obtain the following.

2. *Calculation of the coefficients $A_{0j}^{(1,1)}$, ($j = 1, \ldots, 6$).*

To begin with, we have

$$A_{01}^{(1,1)} = -\mu - (\lambda + \mu), \tag{5.66}$$

$$A_{02}^{(1,1)} = -\mu, \tag{5.67}$$

$$A_{03}^{(1,1)} = 0. \tag{5.68}$$

The other coefficients $A_{0j}^{(1,1)}$, $(j = 4, 5, 6)$, are obtained using symmetry and translational invariance of the mesh in the two generating directions of the plane (x_1, x_2).

We obtain

$$A_{04}^{(1,1)} = A_{01}^{(1,1)}, \tag{5.69}$$

$$A_{05}^{(1,1)} = A_{02}^{(1,1)}, \tag{5.70}$$

$$A_{06}^{(1,1)} = A_{03}^{(1,1)}. \tag{5.71}$$

3. *Calculation of the coefficients $A_{0j}^{(2,1)}$.*

   Here we obtain

$$A_{0j}^{(2,1)} = (\lambda + \mu) \int_{\Omega} \frac{\partial \varphi_0}{\partial x_1} \frac{\partial \varphi_j}{\partial x_2} \, d\Omega - \mu \int_0^L \varphi_0 \frac{\partial \varphi_j}{\partial x_1} dx_1. \tag{5.72}$$

Since we are considering basis functions characterizing nodes strictly in the interior of Ω, the boundary integral between $x_1 = 0$ and $x_1 = L$ in the expression (5.72) is identically zero.

In this case, the coefficients $A_{0j}^{(2,1)}$ are given by

$$A_{0j}^{(2,1)} = (\lambda + \mu) \int_{\Omega} \frac{\partial \varphi_0}{\partial x_1} \frac{\partial \varphi_j}{\partial x_2} \, d\Omega. \tag{5.73}$$

4. *Calculation of the coefficient $A_{0j}^{(2,1)}$, $(j = 0, \ldots, 6)$.*

Proceeding in the same way as for the coefficients $A_{0j}^{(1,1)}$, we obtain

$$A_{00}^{(2,1)} = (\lambda + \mu), \tag{5.74}$$

$$A_{01}^{(2,1)} = A_{02}^{(2,1)} = A_{04}^{(2,1)} = A_{05}^{(2,1)} = -\frac{1}{2}(\lambda + \mu), \tag{5.75}$$

$$A_{03}^{(2,1)} = A_{06}^{(2,1)} = \frac{1}{2}(\lambda + \mu). \tag{5.76}$$

Estimation of $b_0^{(1)}$. For a function characterizing a node in the interior of Ω, the right-hand side of (5.60) is

$$b_0^{(1)} = \int_\Omega f_1 \varphi_0 \, d\Omega. \tag{5.77}$$

Using the trapezoidal rule, we have

$$b_0^{(1)} \approx 6 \times \frac{h^2}{2} \times \frac{1}{3} \times f_1^{(0)} = h^2 f_1^{(0)}. \tag{5.78}$$

11) The node equation corresponding to a basis function characterizing a node strictly in the interior of Ω is obtained by combining the results of all the above calculations. This yields

$$h^2 f_1^{(0)} = 4\mu \tilde{u}_1^{(0)} - \mu \left[\tilde{u}_1^{(1)} + \tilde{u}_1^{(2)} + \tilde{u}_1^{(4)} + \tilde{u}_1^{(5)} \right] + 2(\lambda + \mu)\tilde{u}_1^{(0)} - (\lambda + \mu) \left[\tilde{u}_1^{(1)} + \tilde{u}_1^{(4)} \right]$$

$$+ (\lambda + \mu)\tilde{u}_2^{(0)} - \frac{\lambda + \mu}{2} \left[\tilde{u}_2^{(1)} + \tilde{u}_2^{(2)} - \tilde{u}_2^{(3)} + \tilde{u}_2^{(4)} + \tilde{u}_2^{(5)} - \tilde{u}_2^{(6)} \right]. \tag{5.79}$$

12) The projection of the Navier equation on the $(O; \mathbf{x}_1)$ axis is

$$(\lambda + 2\mu)\frac{\partial^2 u_1}{\partial x_1^2} + (\lambda + \mu)\frac{\partial^2 u_2}{\partial x_1 \partial x_2} + \mu \frac{\partial^2 u_1}{\partial x_2^2} + f_1 = 0. \tag{5.80}$$

In order to obtain a second-order approximation of a mixed second partial derivative, we carry out Taylor expansions of a sufficiently regular solution u to (5.80) between the points $(x_1 - h, x_2 + h)$ and (x_1, x_2) on the one hand, and between the points $(x_1 + h, x_2 - h)$ and (x_1, x_2) on the other. There are two reasons for this choice.

Given that we hope to make a second-order approximation, we must expand to order 4, so the choice of points $M_{-h,+h} \equiv (x_1 - h, x_2 + h)$, $M_{h,-h} \equiv (x_1 + h, x_2 - h)$, and $M_{0,0} \equiv (x_1, x_2)$ will allow us to exploit the symmetry of odd-order Taylor expansions, and in particular of order 3, which we need to eliminate.

In addition, to simplify the notation, we shall use the convention $u_2 \equiv u_2(M_{0,0})$. Now,

$$u_2(M_{-h,+h}) = u_2 - h\frac{\partial u_2}{\partial x_1} + h\frac{\partial u_2}{\partial x_2} \tag{5.81}$$

$$+ \frac{h^2}{2}\frac{\partial^2 u_2}{\partial x_1^2} + \frac{h^2}{2}\frac{\partial^2 u_2}{\partial x_2^2} - h^2\frac{\partial^2 u_2}{\partial x_1 \partial x_2}$$

$$+ \frac{1}{3!}\left(-h^3\frac{\partial^3 u_2}{\partial x_1^3} + 3h^3\frac{\partial^3 u_2}{\partial x_1^2 \partial x_2} - 3h^3\frac{\partial^3 u_2}{\partial x_1 x_2^2} + h^3\frac{\partial^3 u_2}{\partial x_2^3} \right) + O(h^4).$$

Likewise,

$$u_2(M_{h,-h}) = u_2 + h\frac{\partial u_2}{\partial x_1} - h\frac{\partial u_2}{\partial x_2} \tag{5.82}$$

$$+ \frac{h^2}{2}\frac{\partial^2 u_2}{\partial x_1^2} + \frac{h^2}{2}\frac{\partial^2 u_2}{\partial x_2^2} - h^2\frac{\partial^2 u_2}{\partial x_1 \partial x_2}$$

$$+ \frac{1}{3!}\left(h^3\frac{\partial^3 u_2}{\partial x_1^3} - 3h^3\frac{\partial^3 u_2}{\partial x_1^2 \partial x_2} + 3h^3\frac{\partial^3 u_2}{\partial x_1 x_2^2} - h^3\frac{\partial^3 u_2}{\partial x_2^3}\right) + O(h^4).$$

Adding together (5.81) and (5.82), we obtain

$$u_2(M_{-h,+h}) + u_2(M_{h,-h}) = 2u_2 + h^2\Delta u_2 - 2h^2\frac{\partial^2 u_2}{\partial x_1 \partial x_2} + O(h^4). \tag{5.83}$$

We replace the Laplacian Δu_2 at the point $M_{0,0}$ by its evaluation involving the second-order centered finite differences, and this yields the following expression for the mixed second partial derivative of u_2 at the point $M_{0,0}$:

$$\frac{\partial^2 u_2}{\partial x_1 \partial x_2}(M_{0,0}) = + \frac{u_2(M_{h,0}) + u_2(M_{-h,0}) + u_2(M_{0,h}) + u_2(M_{0,-h})}{2h^2} \tag{5.84}$$

$$- \frac{2u_2(M_{0,0}) + u_2(M_{-h,+h}) + u_2(M_{h,-h})}{2h^2} + O(h^2).$$

A second-order approximation for the mixed partial derivative of u_2 is obtained by neglecting the $O(h^2)$ in (5.84) and substituting the sequence of approximations $\tilde{u}_2^{(i)}$ for the true values of the solution $u_2^{(i)}$ at the corresponding points, with the notation introduced in the previous questions:

$$\frac{\partial^2 u_2}{\partial x_1 \partial x_2}(M_{0,0}) \approx \frac{\tilde{u}_2^{(1)} + \tilde{u}_2^{(4)} + \tilde{u}_2^{(2)} + \tilde{u}_2^{(5)}}{2h^2} - \frac{2\tilde{u}_2^{(0)} + \tilde{u}_2^{(3)} + \tilde{u}_2^{(6)}}{2h^2}. \tag{5.85}$$

A finite-difference approximation of (5.80) is then obtained by replacing the mixed second partial derivative by the approximation (5.85), and the other second partial derivatives by their classical approximations.

This yields

$$0 = (\lambda + 2\mu)\frac{\tilde{u}_1^{(1)} - 2\tilde{u}_1^{(0)} + \tilde{u}_1^{(4)}}{h^2} + \mu\frac{\tilde{u}_1^{(2)} - 2\tilde{u}_1^{(0)} + \tilde{u}_1^{(5)}}{h^2} \tag{5.86}$$

$$+ (\lambda + \mu)\frac{\tilde{u}_2^{(1)} + \tilde{u}_2^{(4)} + \tilde{u}_2^{(2)} + \tilde{u}_2^{(5)} - 2\tilde{u}_2^{(0)} - \tilde{u}_2^{(3)} - \tilde{u}_2^{(6)}}{2h^2} + f_1^{(0)}.$$

With one final rearrangement of (5.86), we obtain the node Eq. (5.79) corresponding to a basis function characterizing a node strictly in the interior of Ω.

Fig. 5.6 Local numbering
system associated with a node
in the interior of Ω

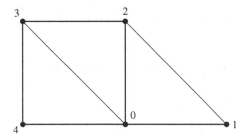

Given that we have neglected the terms of order $O(h^2)$ in all the approximations
of the partial derivatives, the finite-difference scheme is globally of order 2.

▶ **Basis Function φ_i Characterizing a Node in the Interior of Σ_2**

13) Since the edge Σ_2 is the segment parameterized by the equation $x_2 = 0$, the
support of a basis function characterizing a node of the mesh lying in this segment
comprises the triangles T_{012}, T_{023}, and T_{034}, with the local numbering system shown
in Fig. 5.6.

The coefficients $A_{0j}^{(1,1)}$ can then be calculated by adapting the results of Question
10 to the support of the basis functions φ_i specified above.

Calculation of the Coefficients $A_{0j}^{(1,1)}$.

We obtain

$$A_{00}^{(1,1)} = 2\mu + (\lambda + \mu), \tag{5.87}$$

$$A_{01}^{(1,1)} = -\frac{\mu}{2} - \frac{1}{2}(\lambda + \mu), \tag{5.88}$$

$$A_{02}^{(1,1)} = -\mu, \tag{5.89}$$

$$A_{03}^{(1,1)} = 0, \tag{5.90}$$

$$A_{04}^{(1,1)} = A_{01}^{(1,1)}. \tag{5.91}$$

Calculation of the Coefficients $A_{0j}^{(2,1)}$.

The expression for the coefficients $A_{0j}^{(2,1)}$ differs from that used in Question 10
by the boundary integral over Σ_2. Indeed, we have

$$A_{0j}^{(2,1)} = (\lambda + \mu) \int_{\Omega} \frac{\partial \varphi_0}{\partial x_1} \frac{\partial \varphi_j}{\partial x_2} \, d\Omega - \mu \int_0^L \varphi_0 \frac{\partial \varphi_j}{\partial x_1} \, dx_1, \tag{5.92}$$

or again,

$$A_{0j}^{(2,1)} = (\lambda + \mu) \int_\Omega \frac{\partial \varphi_0}{\partial x_1} \frac{\partial \varphi_j}{\partial x_2} \, d\Omega - \mu \int_{40} \varphi_0 \frac{\partial \varphi_j}{\partial x_1} \, dx_1 - \mu \int_{01} \varphi_0 \frac{\partial \varphi_j}{\partial x_1} \, dx_1, \quad (5.93)$$

where $\int_{40}$ denotes the integral over the segment between nodes 4 and 0, and similarly for $\int_{01}$.

Using the trapezoidal rule for these two one-dimensional integrals, it follows that

$$A_{00}^{(2,1)} = \frac{\lambda + \mu}{2} - \left(\frac{\mu}{h} \times \frac{h}{2} \right) + \left(\frac{\mu}{h} \times \frac{h}{2} \right) = \frac{\lambda + \mu}{2}, \quad (5.94)$$

$$A_{01}^{(2,1)} = 0 - 0 - \left(\frac{\mu}{h} \times \frac{h}{2} \right) = -\frac{\mu}{2}, \quad (5.95)$$

$$A_{02}^{(2,1)} = -\frac{\lambda + \mu}{2}, \quad (5.96)$$

$$A_{03}^{(2,1)} = \frac{\lambda + \mu}{2}, \quad (5.97)$$

$$A_{04}^{(2,1)} = -\frac{\lambda + \mu}{2} - \mu \int_{40} \varphi_0 \frac{\partial \varphi_4}{\partial x_1} \, dx_1 = -\frac{\lambda + \mu}{2} + \frac{\mu}{2}. \quad (5.98)$$

Estimation of $b_0^{(1)}$. Once again, the calculation of $b_0^{(1)}$ must take into account the boundary integral on Σ_2, i.e., on the interval $[0, L]$:

$$b_0^{(1)} = \int_\Omega f_1 \varphi_0 \, d\Omega + \int_0^L F_0 \varphi_0 \, dx_1. \quad (5.99)$$

A numerical integration applying the trapezoidal rule to the triangles T_{ijk} and segments $[x_i, x_{i+1}]$ leads to the following approximation:

$$b_0^{(1)} = \frac{h^2}{2} f_1^{(0)} + h F_1^{(0)}. \quad (5.100)$$

14) The node equation corresponding to a basis function characterizing a node of the mesh on the edge Σ_2 is now obtained by combining the estimates (5.86)–(5.91), (5.95)–(5.98), and (5.100):

$$\frac{h^2}{2} f_1^{(0)} + h F_1^{(0)} = (\lambda + 3\mu) \tilde{u}_1^{(0)} - \left(\frac{\lambda}{2} + \mu \right) \left[\tilde{u}_1^{(1)} + \tilde{u}_1^{(4)} \right] - \mu \tilde{u}_1^{(2)} + \frac{\lambda + \mu}{2} \tilde{u}_2^{(0)}$$

$$- \frac{\mu}{2} \tilde{u}_2^{(1)} + \frac{\lambda + \mu}{2} \left[\tilde{u}_2^{(3)} - \tilde{u}_2^{(2)} \right] - \frac{\lambda}{2} \tilde{u}_2^{(4)}. \quad (5.101)$$

15) In order to obtain the node Eq. (5.101) by the finite-difference method, we consider the projection $\sigma(-\mathbf{x_2}) \cdot \mathbf{x_1}$ of the boundary condition (5.14) on Σ_2:

$$\frac{\partial u_1}{\partial x_2}(x_1, 0) + \frac{\partial u_2}{\partial x_1}(x_1, 0) = -\frac{F_1}{\mu}. \tag{5.102}$$

We then use Taylor expansions to replace the combination of partial derivatives in (5.102) with finite differences. To do this, we evaluate the component u_1 at the point $(x_1, x_2 + h)$ and the component u_2 at the point $(x_1 + h, x_2)$, as functions of the central point (x_1, x_2).

As for the expansions presented in the answer to Question 12, we adopt the notational simplifications

$$u_1 \equiv u_1(M_{0,0}), \quad u_2 \equiv u_2(M_{0,0}), \quad M_{0,+h} \equiv (x_1, x_2 + h), \quad M_{+h,0} \equiv (x_1 + h, x_2).$$

Then

$$u_1(M_{0,+h}) = u_1 + h\frac{\partial u_1}{\partial x_2} + \frac{h^2}{2}\frac{\partial^2 u_1}{\partial x_2^2} + O(h^3), \tag{5.103}$$

$$u_2(M_{+h,0}) = u_2 + h\frac{\partial u_2}{\partial x_1} + \frac{h^2}{2}\frac{\partial^2 u_2}{\partial x_1^2} + O(h^3), \tag{5.104}$$

where upon we have

$$\frac{u_1(M_{0,+h}) + u_2(M_{+h,0}) - u_1(M_{0,0}) - u_2(M_{0,0})}{h} = \tag{5.105}$$

$$\left(\frac{\partial u_1}{\partial x_2} + \frac{\partial u_2}{\partial x_1}\right)(M_{0,0}) + \frac{h}{2}\frac{\partial^2 u_1}{\partial x_2^2}(M_{0,0}) + \frac{h}{2}\frac{\partial^2 u_2}{\partial x_1^2}(M_{0,0}) + O(h^2).$$

The first substitution we make now is to replace the sum of the first partial derivatives at the point $(M_{0,0})$ with the right-hand side of the boundary condition (5.102):

$$\frac{u_1(M_{0,+h}) + u_2(M_{+h,0}) - u_1(M_{0,0}) - u_2(M_{0,0})}{h} = \tag{5.106}$$

$$-\frac{F_1(M_{0,0})}{\mu} + \frac{h}{2}\frac{\partial^2 u_1}{\partial x_2^2}(M_{0,0}) + \frac{h}{2}\frac{\partial^2 u_2}{\partial x_1^2}(M_{0,0}) + O(h^2).$$

The second substitution provides an evaluation of the second partial derivative of

$$\frac{\partial^2 u_1}{\partial x_2^2}(M_{0,0})$$

as a function of the mixed second partial derivative

$$\frac{\partial^2 u_2}{\partial x_1 \partial x_2}(M_{0,0}),$$

using the partial differential Eq. (5.80), for which we have a second-order finite-difference approximation in (5.85).

We thus obtain

$$\frac{u_1(M_{0,+h}) + u_2(M_{+h,0}) - u_1(M_{0,0}) - u_2(M_{0,0})}{h} = \tag{5.107}$$

$$-\frac{F_1(M_{0,0})}{\mu} - \frac{h}{2\mu} f_1(M_{0,0}) - \frac{\lambda + 2\mu}{2\mu} h \frac{\partial^2 u_1}{\partial x_1^1}(M_{0,0})$$

$$-\frac{\lambda + 2\mu}{2\mu} h \frac{\partial^2 u_2}{\partial x_1 \partial x_2}(M_{0,0}) + \frac{h}{2} \frac{\partial^2 u_2}{\partial x_1^2}(M_{0,0}) + O(h^2).$$

With the notation arising from our local numbering system, the finite-difference scheme corresponding to the discretization of the boundary condition (5.102) becomes finally

$$\mu \left[\tilde{u}_1^{(2)} + \tilde{u}_2^{(1)} - \tilde{u}_1^{(0)} - \tilde{u}_2^{(0)} \right] = -h F_1^{(0)} - \frac{h^2}{2} f_1^{(0)} - \frac{\lambda + 2\mu}{2} \left[\tilde{u}_1^{(1)} - 2\tilde{u}_1^{(0)} + \tilde{u}_1^{(4)} \right]$$

$$- \frac{\lambda + \mu}{2} \left[\tilde{u}_2^{(2)} - \tilde{u}_2^{(3)} - \tilde{u}_2^{(0)} + \tilde{u}_2^{(4)} \right]$$

$$+ \frac{\mu}{2} \left[\tilde{u}_2^{(1)} - 2\tilde{u}_2^{(0)} + \tilde{u}_2^{(4)} \right]. \tag{5.108}$$

Rearranging (5.108), we then recover precisely the node Eq. (5.101) associated with a basis function characterizing a node in the interior of the segment [0, 1] of the boundary Σ_2.

Fig. 5.7 Clamped plate

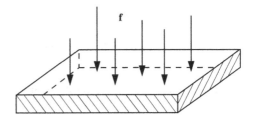

5.2 Clamped Plate

5.2.1 Statement of the Problem

The aim here is to study two types of variational formulation for the equation governing the vertical displacement u, along the $(O; \mathbf{z})$ axis, of a square elastic plate Ω that is perfectly clamped around the edges and subjected to a density of forces, perpendicular to the plate, of the form $\mathbf{f} = f(x, y)\mathbf{z}$ (see Fig. 5.7).

We shall be interested in the scalar function u of the variables (x, y) that solves the fourth-order partial differential equation in the following continuous problem:

Find $u \in H^4(\Omega)$ solution of

$$(\mathbf{CP_1}) \quad \begin{cases} \Delta^2 u = f \text{ in } \Omega, \\ u = \dfrac{\partial u}{\partial n} = 0 \text{ on } \partial\Omega, \end{cases} \tag{5.109}$$

where the biharmonic (or bi-Laplacian) operator Δ^2 is defined by

$$\Delta^2 = \Delta(\Delta) = \frac{\partial^4}{\partial x^4} + 2\frac{\partial^4}{\partial x^2 \partial y^2} + \frac{\partial^4}{\partial y^4}. \tag{5.110}$$

Here Ω is the square $]0, 1[\times]0, 1[$ with outward normal $\mathbf{n}$, f is a given function in $L^2(\Omega)$, and $H^4(\Omega)$ is the function space defined by

$$H^4(\Omega) = \left\{ v : \Omega \subset \mathbb{R}^n \to \mathbb{R}, \quad \frac{\partial^k v}{\partial x_{i_1} \ldots \partial x_{i_k}} \in L^2(\Omega), \quad k = 0, \ldots, 4 \right\}, \tag{5.111}$$

where the zeroth derivative is just the function itself.

▶ First Variational Formulation

1) If v is a sufficiently regular test function of the variables (x, y), show that a variational formulation **(VP)** of the following form can be associated with the continuous problem **(CP)**:

Find $u \in V$ solution of

$$\textbf{(VP)} \quad \int_\Omega \Delta u \, \Delta v \, d\Omega = \int_\Omega f v \, d\Omega, \quad \forall v \in V. \tag{5.112}$$

Specify the space V, i.e., regularity of and boundary conditions on the functions v in V.

2) Show that there is one and only one solution u in V of the variational formulation **(VP)**.

Hint. To establish the coercivity of a suitable bilinear form $a(., .)$ introduced here, use the fact that for every function v in

$$H^2(\Omega) \cap \left\{ v \text{ such that } v = 0 = \frac{\partial v}{\partial n} \text{ on } \partial \Omega \right\},$$

we have the identity

$$\int_\Omega (\Delta v)^2 \, d\Omega = \sum_{i,j=1}^{2} \int_\Omega \left(\frac{\partial^2 v}{\partial x_i \partial x_j} \right)^2 d\Omega. \tag{5.113}$$

3) Can one use a P_1 finite-element analysis to obtain an approximate solution to the variational formulation **(VP)**?

▶ **Second Variational Formulation**

4) We now introduce a function φ defined by

$$-\Delta u = \varphi. \tag{5.114}$$

Show that if u solves the problem **(CP$_1$)**, then the pair (φ, u) solves the problem **(CP$_2$)** given as follows:
Find $(\varphi, u) \in H^2(\Omega) \times H^2(\Omega)$ solution of

$$\textbf{(CP}_2\textbf{)} \quad \begin{cases} -\Delta \varphi = f \text{ in } \Omega, \\ -\Delta u = \varphi \text{ in } \Omega, \\ u = \dfrac{\partial u}{\partial n} = 0 \text{ on } \partial \Omega, \end{cases} \tag{5.115}$$

where

$$H^2(\Omega) \equiv \left\{ v : \Omega \longrightarrow \mathbb{R} \text{ such that } v \in L^2(\Omega), \frac{\partial v}{\partial x_i} \in L^2(\Omega), \frac{\partial^2 v}{\partial x_i x_j} \in L^2(\Omega) \right\}.$$

Fig. 5.8 P_1 finite-element
mesh

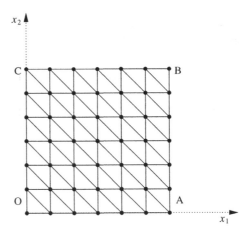

5) We now move to a double variational formulation by introducing a pair of test
functions (ψ, v) in $H_0^1(\Omega) \times H^1(\Omega)$, where $H_0^1(\Omega)$ is the Sobolev space defined by

$$H_0^1(\Omega) = H^1(\Omega) \cap \left\{ v \text{ such that } v = 0 \text{ on } \partial\Omega \right\}. \qquad (5.116)$$

Show that if (φ, u) solves the continuous problem **(CP₂)**, then (φ, u) solves the
following variational problem **(VP₂)**:

Find $(\varphi, u) \in H^1(\Omega) \times H_0^1(\Omega)$ solution of

$$\textbf{(VP}_2\textbf{)} \quad \begin{cases} a(\varphi, \psi) = L_f(\psi), & \forall \psi \in H_0^1(\Omega), \\ a(u, v) = L_\varphi(v), & \forall v \in H^1(\Omega), \end{cases} \qquad (5.117)$$

where the bilinear form $a(., .)$ and the linear form $L_f(.)$ are to be specified. Note
that the linear form $L_\varphi(.)$ is identical to $L_f(.)$ whenever we replace f by φ.

▶ **P_1 Finite-Element Approximation**

We now approximate the variational problem **(VP₂)** by a P_1 finite-element analysis.
To do this, we introduce a constant discretization interval h and cover the square Ω
with a uniform mesh of isosceles right triangles T_k, $(k = 1, \ldots, N_{\text{triangles}})$, of side h
(see Fig. 5.8).

This generates the sequence of points $M_{i,j}$ with coordinates (x_i, y_j) defined by

$$\begin{cases} x_0 = y_0 = 0, \quad x_{N+1} = y_{N+1} = 1, \\ x_{i+1} = x_i + h, \quad i = 1, \ldots, N+1, \\ y_{j+1} = y_j + h, \quad j = 1, \ldots, N+1. \end{cases} \qquad (5.118)$$

Fig. 5.9 Local numbering system associated with a node strictly in the interior of Ω

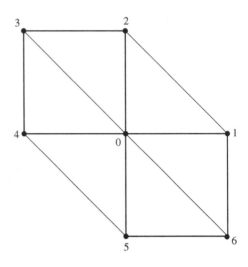

6) Define the space $\tilde{V}$ by

$$\tilde{V} = \left\{\tilde{\psi} : \Omega \longrightarrow \mathbb{R}, \quad \tilde{\psi} \in C^0(\Omega), \quad \tilde{\psi}|_{T_k} \in P_1(T_k), \quad k = 1, \ldots, N_{\text{triangles}}\right\},$$
(5.119)

where $P_1(T_k)$ is the set of polynomials of degree less than or equal to 1 in the pair of variables (x, y).

In addition, let ξ_i, $(i = 1, \ldots, (N + 2)^2)$, be the canonical basis of the space $\tilde{V}$, i.e., satisfying the property $\xi_l(M_m) = \delta_{lm}$, and let $\tilde{V}_0$ be the space of functions in $\tilde{V}$ that vanish on the boundary of Ω.

Setting

$$\tilde{\varphi} = \sum_{j=1,\ldots,(N+2)^2} \tilde{\varphi}_j \xi_j, \qquad \tilde{u} = \sum_{j=1,\ldots,N^2} \tilde{u}_j \xi_j, \qquad (5.120)$$

show that the approximate variational formulation of the problem **(VP$_2$)** can be expressed in the following form:

$$
\widetilde{(\mathbf{VP_2})} \quad
\begin{cases}
\displaystyle\sum_{j=1,\ldots,(N+2)^2} A_{ij}\tilde{\varphi}_j = b_i^{(1)}, \quad i = 1, \ldots, N^2, \\[2mm]
\displaystyle\sum_{j=1,\ldots,N^2} A_{ij}\tilde{u}_j = b_i^{(2)}, \quad i = 1, \ldots, (N+2)^2, \\[2mm]
\text{where } A_{ij} = \displaystyle\int_\Omega \nabla \xi_i \cdot \nabla \xi_j \, d\Omega, \quad b_i^{(1)} = \int_\Omega f \xi_i \, d\Omega, \quad b_i^{(2)} = \int_\Omega \tilde{\varphi} \xi_i \, d\Omega.
\end{cases}
$$
(5.121)

Fig. 5.10 Local numbering
system associated with a node
on the boundary $\partial\Omega$

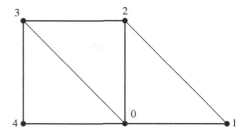

▶ **System of Equations for $\tilde{\varphi}$**

7) Given the regularity of the mesh, i.e., horizontal and vertical translational invariance, we adopt the local numbering system shown in Fig. 5.9. Express the part of the system (5.121) dealing with $\tilde{\varphi}$ in its local form:

$$\sum_{j=0,\ldots,6} A_{0j}\tilde{\varphi}_j = b_0^{(1)}. \tag{5.122}$$

Now calculate the coefficients A_{0j} exactly, and evaluate the right-hand side $b_0^{(1)}$ approximately using the trapezoidal rule.

8) What is the key feature of the system (5.122) that makes it impossible to implement an inversion algorithm for linear systems?

▶ **System of Equations for $\tilde{u}$: Node Equation Associated with a Basis Function Characterizing a Node in the Interior of Ω**

9) Using the experience gained from the last question, write down directly the equations of the system (5.121) corresponding to the basis functions characterizing a node situated in the interior of the mesh.

▶ **System of Equations for $\tilde{u}$: Node Equation Associated with a Basis Function Characterizing a Node in the Interior of the Segment OA**

10) Still using the local numbering system of Fig. 5.10, express the system of equations (5.121) in its local form:

$$\sum_{j=2,3} A_{0j}\tilde{u}_j = b_0^{(2)}. \tag{5.123}$$

11) Recover directly all the discrete equations for the two approximation functions $(\tilde{\varphi}, \tilde{u})$ using finite-difference discretization.

5.2.2 Solution

▶ **First Variational Formulation**

1) We multiply the partial differential equation of the continuous problem (**CP₁**) by a real-valued test function v on Ω and integrate over Ω to obtain

$$\int_\Omega (\Delta^2 u) v \, d\Omega = \int_\Omega f v \, d\Omega. \tag{5.124}$$

We then apply Green's formula twice, whence the left-hand side of (5.124) becomes

$$\int_\Omega (\Delta^2 u) v \, d\Omega = \int_\Omega \Delta(\Delta u) v \, d\Omega = -\int_\Omega \nabla(\Delta u) \cdot \nabla v \, d\Omega + \int_{\partial\Omega} \frac{\partial(\Delta u)}{\partial n} v \, d\Gamma \tag{5.125}$$

$$= \int_\Omega \Delta u \, \Delta v \, d\Omega - \int_{\partial\Omega} \Delta u \frac{\partial v}{\partial n} \, d\Gamma + \int_{\partial\Omega} \frac{\partial(\Delta u)}{\partial n} v \, d\Omega.$$

As we have done systematically up to now, we must construct the function space V that characterizes the functions v in such a way as to preserve all the information contained in the formulation of the continuous problem (**CP₁**).

Now it turns out that none of the boundary conditions expressing the clamping of the plate along its boundary $\partial\Omega$ appear in the integral formulation (5.125).

For this reason, we impose the following boundary conditions on the functions v of V:

$$v = 0 = \frac{\partial v}{\partial n}, \quad \text{on } \partial\Omega. \tag{5.126}$$

Hence, since we seek the solution to the variational formulation in the space V, as soon as we have proven the existence of such a solution, we can be sure that it will satisfy the right boundary conditions.

So taking into account the conditions (5.126), the last equation of (5.125) becomes

$$\int_\Omega (\Delta^2 u) v \, d\Omega = \int_\Omega \Delta u \, \Delta v \, d\Omega. \tag{5.127}$$

We then use (5.127) in (5.124) to obtain the following variational formulation:

$$\text{Find } u \in V \text{ solution of } \int_\Omega \Delta u \, \Delta v \, d\Omega = \int_\Omega f v \, d\Omega. \tag{5.128}$$

The final step in constructing the variational formulation is to define the function space V. To do this, we have only to add regularity properties to the boundary conditions (5.126) in order to ensure convergence of the integrals in (5.128).

The left-hand integral in (5.128) can be bounded as follows:

$$\left| \int_\Omega \Delta u \, \Delta v \, d\Omega \right| \le \int_\Omega |\Delta u \, \Delta v| \, d\Omega \le \left[\int_\Omega (\Delta u)^2 \, d\Omega \right]^{1/2} \left[\int_\Omega (\Delta v)^2 \, d\Omega \right]^{1/2},$$

$$(5.129)$$

where we have used the Cauchy–Schwarz inequality.

We thus consider the natural norm on the Sobolev space $H^2(\Omega)$:

$$\forall u \in H^2(\Omega), \ \|u\|_{H^2(\Omega)}^2 = \left[\|u\|_{L^2(\Omega)}^2 + \sum_{i=1}^2 \left\| \frac{\partial u}{\partial x_i} \right\|_{L^2(\Omega)}^2 + \sum_{i,j=1}^2 \left\| \frac{\partial^2 u}{\partial x_i \partial x_j} \right\|_{L^2(\Omega)}^2 \right],$$

$$(5.130)$$

with

$$\forall u \in L^2(\Omega), \quad \|u\|_{L^2(\Omega)}^2 = \int_\Omega |u|^2 \, d\Omega. \tag{5.131}$$

We thus have

$$\forall u \in H^2(\Omega), \quad \left[\int_\Omega (\Delta u)^2 \, d\Omega \right] \le \|u\|_{H^2(\Omega)}^2. \tag{5.132}$$

Returning to the integral on the left in (5.128) and introducing the inequality (5.129), we now have

$$\left| \int_\Omega \Delta u \, \Delta v \, d\Omega \right| \le \|u\|_{H^2(\Omega)} \|v\|_{H^2(\Omega)}. \tag{5.133}$$

In other words, we need to consider only u and v in $H^2(\Omega)$ in order to ensure the convergence of the integral on the left in the variational Eq. (5.128).

Likewise, applying the Cauchy–Schwarz inequality once again, we see that the right-hand side of (5.128) is bounded by

$$\left| \int_\Omega f v \, d\Omega \right| \le \int_\Omega |f v| \, d\Omega \le \|f\|_{L^2(\Omega)} \|u\|_{L^2(\Omega)}. \tag{5.134}$$

We then introduce the Sobolev space $H_0^2(\Omega)$ defined by

$$H_0^2(\Omega) = H^2(\Omega) \cap \left\{ v \text{ such that } v = 0 = \frac{\partial v}{\partial n} \text{ on } \partial \Omega \right\}. \tag{5.135}$$

Finally, the variational problem (**VP**) becomes the following:

$$\textbf{(VP)} \quad \begin{cases} \text{Find } u \in H_0^2(\Omega) \text{ solution of} \\ \displaystyle \int_\Omega \Delta u \, \Delta v \, d\Omega = \int_\Omega f v \, d\Omega, \quad \forall v \in H_0^2(\Omega). \end{cases} \tag{5.136}$$

2) To establish the existence and uniqueness of the variational formulation **(VP)**, we introduce the forms $a(.,.)$ and $L(.)$ defined by

$$a : H_0^2(\Omega) \times H_0^2(\Omega) \longrightarrow \mathbb{R},$$
$$(u, v) \longmapsto a(u, v) = \int_\Omega \Delta u \, \Delta v \, d\Omega, \qquad (5.137)$$

$$L : H_0^2(\Omega) \longrightarrow \mathbb{R},$$
$$v \longmapsto L(v) \equiv \int_\Omega f v \, d\Omega. \qquad (5.138)$$

We then check the properties required of the forms $a(.,.)$ and $L(.)$, and also those of the space V, to see whether they fulfill the premises of the Lax–Milgram theorem (Theorem 1.11):

1. In the present case, V is the space $H_0^2(\Omega)$ defined by (5.135). We shall show that it is a closed vector subspace of $H^2(\Omega)$. It will follow that $H_0^2(\Omega)$ is itself a Hilbert space for the norm H^2.

   We thus suppose that v_n is a sequence of functions in $H_0^2(\Omega)$ that converges to v in $H^2(\Omega)$. The aim will be to show that v is also an element of $H_0^2(\Omega)$. The proof follows the same lines as the proof that the space $H_0^1(\Omega)$ is a Hilbert space, as a closed vector subspace of $H^1(\Omega)$ (see Lemma 1.20).

   The limit v of the sequence v_n will thus belong to $H_0^2(\Omega)$ if

$$v = 0 = \frac{\partial v}{\partial n},$$

   or, using a more precise formulation in terms of the trace operators γ_0 and γ_1 (see Theorems 1.3 and 1.6),

$$\gamma_0 v = \gamma_1 v = 0.$$

Now,

$$\gamma_0 v = \gamma_0 (v - v_n + v_n) = \gamma_0 (v - v_n), \qquad (5.139)$$

since $v_n \in H_0^2(\Omega)$, whence $\gamma_0 v_n = 0$.

Moreover, since the trace map γ_0 is continuous from $H^1(\Omega)$ to $L^2(\partial\Omega)$ (see Theorem 1.3), we have

$$\exists C_0 > 0 \text{ such that } \forall w \in H^1(\Omega), \ \|\gamma_0 w\|_{L^2(\partial\Omega)} \leq C_0 \|w\|_{H^1(\Omega)} \leq C_0 \|w\|_{H^2(\Omega)}.$$

Using this property in (5.139), it follows that

$$\|\gamma_0 v\|_{L^2(\partial\Omega)} = \|\gamma_0 (v - v_n)\|_{L^2(\partial\Omega)} \leq C_0 \|v - v_n\|_{H^2(\Omega)}. \qquad (5.140)$$

Taking the limit as n tends to $+\infty$ in the last inequality, we have finally

$$\gamma_0 v \equiv v|_{\partial\Omega} = 0. \tag{5.141}$$

In the same way, we have

$$\gamma_1 v = \gamma_1(v - v_n + v_n) = \gamma_1(v - v_n), \tag{5.142}$$

since $v_n \in H_0^2(\Omega)$, whence $\gamma_1 v_n = 0$.

Moreover, since the trace map γ_1 is continuous from $H^2(\Omega)$ to $L^2(\partial\Omega)$ (see Theorem 1.6), it follows that

$$\exists\, C_1 > 0 \text{ such that } \forall w \in H^2(\Omega), \quad \|\gamma_1 w\|_{L^2(\partial\Omega)} \le C_1 \|w\|_{H^2(\Omega)}.$$

Then using this property in (5.142), we have

$$\|\gamma_1 v\|_{L^2(\partial\Omega)} = \|\gamma_1(v - v_n)\|_{L^2(\partial\Omega)} \le C_1 \|v - v_n\|_{H^2(\Omega)}. \tag{5.143}$$

Taking the limit as n tends to $+\infty$ in this last inequality, we obtain finally

$$\gamma_1 v \equiv \frac{\partial v}{\partial \mathbf{n}}\bigg|_{\partial\Omega} \equiv \nabla v|_{\partial\Omega} \cdot \mathbf{n} = 0. \tag{5.144}$$

It follows that the limit v of the sequence v_n, which is an element of $H_0^2(\Omega)$, also belongs to $H_0^2(\Omega)$.

As a consequence, $H_0^2(\Omega)$ is closed in $H^2(\Omega)$, and it is therefore a Hilbert space for the norm induced from $H^2(\Omega)$.

2. The form $L(.)$ defined by (5.138) is trivially linear, and its continuity for the H^2 norm follows from the inequalities

$$\left|\int_\Omega fv\, d\Omega\right| \le \int_\Omega |fv|\, d\Omega \le \|f\|_{L^2(\Omega)} \|v\|_{L^2(\Omega)} \le \|f\|_{L^2(\Omega)} \|v\|_{H^2(\Omega)}. \tag{5.145}$$

There is thus a constant $C_L \equiv \|f\|_{L^2(\Omega)}$, independent of v, such that

$$\forall v \in H_0^2(\Omega), \quad |L(v)| \le C_L \|v\|_{H^2(\Omega)}. \tag{5.146}$$

3. The form $a(.,.)$ defined by (5.137) is clearly bilinear, and its continuity for the H^2 norm follows from

$$\left|\int_\Omega \Delta u \Delta v\, d\Omega\right| \le \int_\Omega |\Delta u \Delta v|\, d\Omega \le \|\Delta u\|_{L^2(\Omega)} \|\Delta v\|_{L^2(\Omega)} \tag{5.147}$$

$$\le \|u\|_{H^2(\Omega)} \|v\|_{H^2(\Omega)}.$$

Put another way, there is a constant $C_a \equiv 1$ such that

$$\forall (u, v) \in H_0^2(\Omega) \times H_0^2(\Omega), \quad |a(u, v)| \leq C_a \|u\|_{H^2(\Omega)} \|v\|_{H^2(\Omega)}. \quad (5.148)$$

4. It remains to demonstrate the coercivity of the bilinear form $a(.,.)$. We do this by two applications of the Poincaré inequality (1.130), which is valid in $H_0^1(\Omega)$.

Indeed, we begin with the following observation. If v belongs to $H_0^2(\Omega)$, then v belongs to $H_0^1(\Omega)$, but so also does each partial derivative $\partial v/\partial x_i$, $(i = 1, 2)$. The Poincaré inequality then provides the upper bounds

$$\forall v \in H_0^2(\Omega), \quad \|v\|_{L^2(\Omega)}^2 \leq \sum_{i=1}^2 \left\| \frac{\partial v}{\partial x_i} \right\|_{L^2(\Omega)}^2, \quad (5.149)$$

$$\forall v \in H_0^2(\Omega), \quad \sum_{i=1}^2 \left\| \frac{\partial v}{\partial x_i} \right\|_{L^2(\Omega)}^2 \leq \alpha \sum_{i,j=1}^2 \left\| \frac{\partial^2 v}{\partial x_i \partial x_j} \right\|_{L^2(\Omega)}^2, \quad (5.150)$$

where α is a positive constant depending only on Ω. The two inequalities (5.149) and (5.150) then imply

$$\|v\|_{L^2(\Omega)}^2 \leq \alpha^2 \sum_{i,j=1}^2 \left\| \frac{\partial^2 v}{\partial x_i \partial x_j} \right\|_{L^2(\Omega)}^2. \quad (5.151)$$

Furthermore, the inequality (5.150) implies that

$$\sum_{i=1}^2 \left\| \frac{\partial v}{\partial x_i} \right\|_{L^2(\Omega)}^2 + \sum_{i,j=1}^2 \left\| \frac{\partial^2 v}{\partial x_i \partial x_j} \right\|_{L^2(\Omega)}^2 \leq (1+\alpha) \sum_{i,j=1}^2 \left\| \frac{\partial^2 v}{\partial x_i \partial x_j} \right\|_{L^2(\Omega)}^2. \quad (5.152)$$

If we now add the square of the L^2 norm of v to each side of the inequality (5.152), we obtain the H^2 norm of v defined in (5.130):

$$\|v\|_{H^2(\Omega)}^2 \leq (1 + \alpha + \alpha^2) \sum_{i,j=1}^2 \left\| \frac{\partial^2 v}{\partial x_i \partial x_j} \right\|_{L^2(\Omega)}^2, \quad (5.153)$$

where we have also used the inequality (5.151). The identity (5.113) leads finally to

$$a(v, v) = \sum_{i,j=1}^2 \int_\Omega \left(\frac{\partial^2 v}{\partial x_i \partial x_j} \right)^2 \, d\Omega \geq \frac{1}{(1 + \alpha + \alpha^2)} \|v\|_{H^2(\Omega)}^2. \quad (5.154)$$

It thus follows that the bilinear form $a(.,.)$ is indeed coercive.

To conclude, then, all the premises of the Lax–Milgram theorem (Theorem 1.11) are satisfied, and there thus exists one and only one function u in $H_0^2(\Omega)$ that solves the variational formulation (**VP**) specified by (5.136).

3) Using a P_1 finite-element method to obtain an approximate solution to the variational formulation (**VP**) would lead to the left-hand side of (5.136) being identically zero.

Indeed, since the second partial derivative of a function of degree less than or equal to 1 with respect to a pair of variables (x, y) is necessarily zero, it follows that the Laplacian of such a function is also identically zero!

It is not therefore recommended to use such a finite-element analysis for numerical solution by variational approximation of the problem (**VP**).

One solution here would be to use a P_2 method. The details are left to the reader. Note, however, that the trapezoidal rule is poorly suited to such finite-element analysis and should be replaced by a more precise numerical quadrature, corresponding to the accuracy of the P_2 analysis (see the Bramble–Hilbert lemma on p. 88).

▶ **Second Variational Formulation**

4) The formulation of the problem (**CP₂**) is immediate. Indeed, we just insert the change of function (5.114) in the partial differential equation for the plate in the problem (**CP₁**) to obtain the first partial differential equation of the problem (**CP₂**), i.e., the Laplace equation

$$-\Delta\varphi = f.$$

The other equations constituting the rest of the problem (**CP₂**) are trivial.

5) Let (ψ, v) be a pair of test functions in $H_0^1(\Omega) \times H^1(\Omega)$. The function ψ will be the test function associated with the function φ, while the function v will be associated with the second unknown u.

To obtain a variational formulation (**VP₂**) associated with the continuous problem (**CP₂**), we multiply each partial differential equation of the continuous problem (**CP₂**) by its corresponding test function and integrate:

$$(\textbf{VP}_2) \quad \begin{cases} -\displaystyle\int_\Omega \Delta\varphi.\psi \, d\Omega = \int_\Omega \nabla\varphi\cdot\nabla\psi \, d\Omega + \int_{\partial\Omega} \frac{\partial\varphi}{\partial n}.\psi \equiv \int_\Omega f.\psi \, d\Omega, \\[3mm] -\displaystyle\int_\Omega \Delta u.v \, d\Omega = \int_\Omega \nabla u\cdot\nabla.v \, d\Omega + \int_{\partial\Omega} \frac{\partial u}{\partial n}.v \equiv \int_\Omega \varphi.v \, d\Omega. \end{cases}$$

$$(5.155)$$

We then use the boundary conditions on u and its normal derivative, both identically zero on the boundary $\partial\Omega$ of Ω. In fact, regarding the normal derivative of u, given that the values of u appear nowhere in the integrals of the double formulation (5.155), we introduce the relevant properties in the function space in which the function ψ will

be the generic element in order to preserve the memory of the variational formulation
(**VP$_2$**).

We refer here to the variational space associated with ψ because it is indeed this
function that multiplies the plate equation, even though it is rewritten with a change
of function φ.

So if we replace the boundary condition that u vanishes on the boundary $\partial\Omega$ in
the system (5.155), and if we choose the function spaces $H^1(\Omega) \times H_0^1(\Omega)$ for the
unknowns (φ, u) on the one hand, and $H_0^1(\Omega) \times H^1(\Omega)$ for the pair of test functions
(ψ, v) on the other, the double variational formulation becomes the following:

Find $(\varphi, u) \in H^1(\Omega) \times H_0^1(\Omega)$ solution of

(VP$_2$)
$$\begin{cases} a(\varphi, \psi) \equiv \displaystyle\int_\Omega \nabla\varphi\cdot\nabla\psi \, d\Omega = \int_\Omega f.\psi \, d\Omega \equiv L_f(\psi), \quad \forall \psi \in H_0^1(\Omega) \\[2mm] a(u, v) \equiv \displaystyle\int_\Omega \nabla u\cdot\nabla v \, d\Omega = \int_\Omega \varphi.v \, d\Omega \equiv L_\varphi(v), \quad \forall v \in H^1(\Omega). \end{cases}$$
(5.156)

6) The transition from the variational formulation (**VP$_2$**) to the approximate variational formulation $\widetilde{(\mathbf{VP_2})}$ is made by replacing the unknowns (φ, u) with the corresponding approximation functions $(\tilde\varphi, \tilde u)$:

Find $(\tilde\varphi, \tilde u) \in \tilde V \times \tilde V_0$ solution of

$$\widetilde{(\mathbf{VP_2})} \qquad \begin{cases} a(\tilde\varphi, \tilde\psi) = L_f(\tilde\psi), \quad \forall \tilde\psi \in \tilde V_0, \\[2mm] a(\tilde u, \tilde v) = L_{\tilde\varphi}(\tilde v), \quad \forall \tilde v \in \tilde V. \end{cases}$$
(5.157)

We then use the expansions of $\tilde\varphi$ and $\tilde u$ in terms of the relevant bases, according to
(5.120). In addition, given the bilinearity of the form $a(.,.)$ and the linearity of the
form $L_f(.)$ [the form $L_\varphi(.)$ clearly has the same property], the expressions in (5.121)
follow without difficulty.

We note here that the approximation $\tilde u$ is expressed in terms of N^2 basis functions,
while the approximation $\tilde\varphi$ is expressed in terms of $(N+2)^2$ basis functions. Indeed,
this is due to the fact that the unknown $\tilde u$ must be identically zero on the boundary
$\partial\Omega$.

As a consequence, the basis functions characterizing nodes located on the boundary of the square Ω must not be counted in the decomposition of the function $\tilde u$.

As a result, only the N^2 points strictly in the interior of the mesh and the basis
functions characterizing these nodes should be considered in the decomposition of
the approximation $\tilde u$.

The same arguments lead to the N^2 equations with $b_i^{(1)}$ on the right-hand side in
the formulation of the variational problem and the $(N+2)^2$ equations of the second
system with $b_i^{(2)}$ on the right-hand side.

7) We now turn to the constitution of the linear system associated with the unknown $\tilde{\varphi}$ in the problem $\widetilde{(\mathbf{VP_2})}$, expressed in its local form. We begin by noting that since we are considering a regular mesh (see Fig. 5.8), and since we adopt the local numbering system shown in Fig. 5.9, if 0 becomes the local number of the function numbered i in the global numbering system, then only the numbers of the triangle vertices where the node 0 is one of the vertices will need to be considered, given the properties of the supports of basis functions that are affine on each triangle.

Hence,

$$\sum_{j=1,\dots,(N+2)^2} A_{ij}\varphi_j = b_i^{(1)}, \quad i = 1,\dots,N^2 \quad \Longrightarrow \quad \sum_{j=0,\dots,6} A_{0j}\tilde{\varphi}_j = b_0^{(1)}.$$

$$(5.158)$$

Calculation of the Coefficients A_{0j}.

All the coefficients will be correctly calculated given that the gradients of the basis functions ξ_i are constant for each triangle. These are standard calculations, and the reader is referred to [1] or the previous problem for further details.

1. *Calculation of the coefficient A_{00}.*

We obtain

$$A_{00} = \int_{\text{supp } \xi_0} \nabla\xi_0^2 \, d\Omega$$

$$= \int_{012} \nabla\xi_0^2 \, d\Omega + \int_{023} \nabla\xi_0^2 \, d\Omega + \int_{034} \nabla\xi_0^2 \, d\Omega$$

$$+ \int_{045} \nabla\xi_0^2 \, d\Omega + \int_{056} \nabla\xi_0^2 \, d\Omega + \int_{061} \nabla\xi_0^2 \, d\Omega, \quad (5.159)$$

whence

$$A_{00} = \frac{h^2}{2}\left(\frac{2}{h^2} + \frac{1}{h^2} + \frac{1}{h^2} + \frac{2}{h^2} + \frac{1}{h^2} + \frac{1}{h^2}\right) = 4. \quad (5.160)$$

2. *Calculation of the coefficients A_{01} and A_{02}.*

Here we obtain

$$A_{01} = \int_{\text{supp } \xi_0 \cap \text{supp } \xi_1} \nabla\xi_0\cdot\nabla\xi_1 \, d\Omega = \int_{012} \nabla\xi_0\cdot\nabla\xi_1 \, d\Omega + \int_{061} \nabla\xi_0\cdot\nabla\xi_1 \, d\Omega,$$

and

$$A_{01} = \frac{h^2}{2}\left(-\frac{1}{h^2} - \frac{1}{h^2}\right) = -1. \quad (5.161)$$

By similar reasoning, we find that

$$A_{02} = \int_{\text{supp } \xi_0 \cap \text{supp } \xi_2} \nabla \xi_0 \cdot \nabla \xi_2 \, d\Omega = \int_{012} \nabla \xi_0 \cdot \nabla \xi_2 \, d\Omega + \int_{023} \nabla \xi_0 \cdot \nabla \xi_2 \, d\Omega$$

$$= \frac{h^2}{2} \left(-\frac{1}{h^2} - \frac{1}{h^2} \right) = -1. \tag{5.162}$$

3. *Calculation of the coefficient A_{03}.*

   Here we have

$$A_{03} = \int_{\text{supp } \xi_0 \cap \text{supp } \xi_3} \nabla \xi_0 \cdot \nabla \xi_3 \, d\Omega = \int_{023} \nabla \xi_0 \cdot \nabla \xi_3 \, d\Omega + \int_{34} \nabla \xi_0 \cdot \nabla \xi_3 \, d\Omega$$

$$= \frac{h^2}{2} (0 + 0) = 0.$$

4. *Calculation of the coefficients A_{04}, A_{05}, and A_{06}.*

   For symmetry reasons relating to the translational invariance of the mesh in the two directions of the plane on the one hand, and the symmetry of the bilinear form $a(., .)$, and hence of the matrix A_{0j}, on the other, we have

$$\begin{cases} A_{04} = A_{40} = A_{01} = -1, \\ A_{05} = A_{50} = A_{02} = -1, \\ A_{06} = A_{60} = A_{03} = 0. \end{cases} \tag{5.164}$$

Estimation of $b_0^{(1)}$. The right-hand side b_0^1 is estimated using the trapezoidal rule:

$$\iint_T f(\chi) \, d\chi \approx \frac{\text{area } T}{3} \left[f(A_i) + f(A_j) + f(A_k) \right], \tag{5.165}$$

where T is an arbitrary triangle of the mesh with vertices A_i, A_j, and A_k. Hence,

$$b_0^{(1)} = \int_{\text{supp } \xi_0} f \xi_0 \, d\Omega = \int_{012} f \xi_0 \, d\Omega + \int_{023} f \xi_0 \, d\Omega + \int_{034} f \xi_0 \, d\Omega$$

$$+ \int_{045} f \xi_0 \, d\Omega + \int_{056} f \xi_0 \, d\Omega + \int_{061} f \xi_0 \, d\Omega,$$

$$\approx \frac{h^2}{2} \times 6 \times \frac{1}{3} \times (f_0 \times 1), \tag{5.166}$$

where f_0 is the value of f at the node 0.

The generic Eq. (5.158) associated with a basis function ξ_0 characterizing an interior node, i.e., equal to 1 at this node and 0 at the other nodes of the mesh, denoted by 0 in the local numbering system, can be written in the form

$$4\tilde{\varphi}_0 - \tilde{\varphi}_1 - \tilde{\varphi}_4 - \tilde{\varphi}_2 - \tilde{\varphi}_5 = h^2 f_0. \tag{5.167}$$

As usual, and for any regular mesh, we recover the finite-difference scheme associated with the Laplacian:

$$-\frac{\tilde{\varphi}_1 - 2\tilde{\varphi}_0 + \tilde{\varphi}_4}{h^2} - \frac{\tilde{\varphi}_2 - 2\tilde{\varphi}_0 + \tilde{\varphi}_5}{h^2} = f_0. \tag{5.168}$$

8) The linear system described by the generic Eq. (5.167) is a system comprising $(N+2)^2$ equations in N^2 unknowns. In other words, it is a rectangular linear system. It cannot therefore be solved numerically, since the number of equations differs from the number of unknowns.

Only by supplementing this system by the one resulting from Questions 9 and 10 can we proceed to a complete solution by simultaneously determining the pair of unknowns $(\tilde{\varphi}, \tilde{u})$. However, everything in its own time.

9) Given that the linear system governing the approximation function $\tilde{u}$ has the same matrix germ, we can directly write down the node equation for basis functions characterizing a node strictly in the interior of Ω, with the result

$$4\tilde{u}_0 - \tilde{u}_1 - \tilde{u}_4 - \tilde{u}_2 - \tilde{u}_5 = h^2 \tilde{\varphi}_0. \tag{5.169}$$

Note that we have replaced the right-hand side f_0 of the system (5.167) with the appropriate value of $\tilde{\varphi}_0$.

10) We now consider a function characterizing a node in the segment OA one of the segments making up the boundary $\partial\Omega$ (see Fig. 5.8). In this case, only the functions corresponding to nodes numbered locally 0, 1, 2, 3, and 4 have the potential to contribute to the node equation associated with the characteristic function of node 0. In other words, the node equation is in this case

$$\sum_{j=0,\ldots,4} A_{0j}\tilde{u}_j = \sum_{j=2,3} A_{0j}\tilde{u}_j = b_0^{(2)}. \tag{5.170}$$

Note that the contributions of nodes 0, 1, and 4 have already been eliminated, since they are situated on the boundary of the domain Ω, where the unknown u must be identically zero. The same goes for the approximation function $\tilde{u}$.

The two coefficients A_{02} and A_{03} are then obtained trivially using the calculations presented in detail in Question 6.

We thus have immediately

$$A_{02} = \int_{\text{supp } \xi_0 \cap \text{supp } \xi_2} \nabla \xi_0 \cdot \nabla \xi_2 \, d\Omega \tag{5.171}$$

$$= \int_{012} \nabla \xi_0 \cdot \nabla \xi_2 \, d\Omega + \int_{023} \nabla \xi_0 \cdot \nabla \xi_2 \, d\Omega = -1.$$

Likewise,

$$A_{03} = \int_{\text{supp } \xi_0 \cap \text{supp } \xi_3} \nabla \xi_0 \cdot \nabla \xi_3 \, d\Omega \tag{5.172}$$

$$= \int_{023} \nabla \xi_0 \cdot \nabla \xi_3 \, d\Omega + \int_{034} \nabla \xi_0 \cdot \nabla \xi_3 \, d\Omega = 0.$$

Since the same causes produce the same effects, the right-hand side $b_0^{(2)}$ is estimated in the same way as $b_0^{(1)}$, except for taking into account the fact that there are in the present case only half as many triangles and the function $\tilde{\varphi}$ replaces the function f in the integral of $b_0^{(1)}$.

Finally, the node equation is

$$-\tilde{u}_2 = \frac{h^2}{2} \tilde{\varphi}_0. \tag{5.173}$$

11) The finite-difference discretization of the problem (CP$_2$) is a standard exercise. It involves successive second-order discretizations of two Laplacians and one Neumann condition. The reader is referred to the solution of the Dirichlet and Neumann problems presented in Sects. 4.1 and 4.2 for further discussion of finite-difference discretizations.

Reference

1. D. Euvrard, *Résolution des équations aux dérivées partielles de la physique, de la mécanique et des sciences de l'ingénieur* (Masson, Paris, 1994)

Chapter 6
Strength of Materials

In this chapter, we apply the finite-element method to problems in the strength of materials. The main aim is to counter as far as possible a certain state of confusion that often arises in the minds of MSc and engineering students.

Indeed, experience shows that the finite-element method often looks radically different when presented as part of a standard numerical analysis course or when introduced in a course on the mechanics of materials, frequently causing unjustified confusion in the minds of otherwise diligent students.

So, the motivation behind the following account is to eliminate once and for all any doubt that might remain among newcomers to this area, by showing that all applications of the finite-element method have the same foundation and the same form, especially since the method itself takes its roots in the mechanics of deformable media.

As is often the case, it is only indifference on the part of one's teachers that can lead to this kind of confusion, and it the responsibility of all those entrusted to the task to see that this particular fate does not befall their students.

So, to remove all final ambiguity, even when applied to solid mechanics, the finite-element method must be presented in the unified way it deserves, not only in the interest of students, but also to facilitate and improve applications, which can only benefit from this indisputably effective and flexible method.

In order to achieve this unification, the aim in the present chapter is to present a twofold implementation of the finite-element method to the theory of beams.

We thus begin with a 'numerical analysis' version, based on the approximation of a variational formulation, and follow this up with a 'mechanics of deformable solids' version in which we solve the associated minimisation problem, itself equivalent to the variational formulation of the first account.

In addition, regarding the 'solid mechanics' version of the finite-element method, we shall take advantage of this opportunity to discuss the principles underlying the assembly procedure for the linear system obtained when approximating the minimisation problem.

J. Chaskalovic, *Mathematical and Numerical Methods for Partial Differential Equations*, Mathematical Engineering, DOI: 10.1007/978-3-319-03563-5_6, © Springer International Publishing Switzerland 2014

Fig. 6.1 Beam in traction

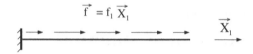

6.1 Beam in Traction

6.1.1 Statement of the Problem

▶ **Clamped Beam: Variational Formulation of the Displacement Field and the Principle of Virtual Work**

We consider a uniform beam Ω of length L, cross-section S, and density ρ. We shall assume isotropic elastic behaviour under small perturbations. Regarding the material from which the beam is made, we denote its Young's modulus by E and its Poisson coefficient by ν.

The beam axis is $(O; \mathbf{X}_1)$, coordinatised by x. The beam is clamped at $x = 0$ and free from stresses at $x = L$. Furthermore, longitudinal forces $\mathbf{f} = f_1 \mathbf{X}_1$ per unit volume are applied along the beam, as shown in Fig. 6.1.

The force density f_1 is given and assumed sufficiently regular to be able to carry out the necessary integrations in the theoretical part of the problem.

1) Choosing a virtual displacement field $\mathbf{U}^*$ of the form

$$\mathbf{U}^* = u^*(x)\mathbf{X}_1 \text{ such that } u^*(x) = 0, \tag{6.1}$$

determine the virtual work T_{int}^* done by the internal forces and also the virtual work T_{ext}^* done by the external forces, as defined by

$$T_{\text{int}}^* = -\int_\Omega \sigma_{ij}\varepsilon_{ij}(\mathbf{U}^*)\,\mathrm{d}\Omega, \qquad T_{\text{ext}}^* = \int_\Omega \mathbf{f} \cdot \mathbf{U}^*\,\mathrm{d}\Omega, \tag{6.2}$$

where σ is the stress tensor and $\varepsilon(\mathbf{U}^*)$ the linear strain tensor associated with the virtual displacement field $\mathbf{U}^*$ defined by

$$\varepsilon_{ij}(\mathbf{U}^*) = \frac{1}{2}\left(\frac{\partial U_i^*}{\partial x_j} + \frac{\partial U_j^*}{\partial x_i}\right). \tag{6.3}$$

We shall require the normal force $N(x)$ and the surface force density f, these being defined by

$$N(x) = \iint_{S(x)} \sigma_{11}\,\mathrm{d}S(x), \qquad f(x) = \iint_{S(x)} f_1\,\mathrm{d}S(x), \tag{6.4}$$

where $S(x)$ is the cross-section as a function of x.

2) Assuming that the various integrated quantities are sufficiently regular, show that the principle of virtual work leads to the following formal variational formulation **(EVP)**:

Find $N : [0, L] \rightarrow \mathbb{R}$ solution of

$$\int_0^L \dot{N}(x)u^*(x)\mathrm{d}x - N(L)u^*(L) + \int_0^L f(x)u^*(x)\mathrm{d}x = 0, \quad \forall u^* \text{ s.t. } u^*(0) = 0,$$

(6.5)

where $\dot{N} \equiv \mathrm{d}N/\mathrm{d}x$.

3) Using the constitutive equation of the material $N = E S \dot{u}(x)$ and assuming that the force density f belongs to $L^2(0, L)$, show that the continuous problem **(CP)** consisting of the equilibrium equations of the beam can be expressed as follows:

Find $u \in H^2(0, L)$ solution of

$$\textbf{(CP)} \quad \begin{cases} -E S \ddot{u}(x) = f(x), & \forall x \in [0, L], \\ u(0) = 0, & \dot{u}(L) = 0. \end{cases}$$

(6.6)

4) Let $V \equiv H_*^1(0, L)$ be the Sobolev space defined by

$$H_*^1(0, L) = \left\{ v \in L^2(0, L) \text{ such that } \dot{v} \in L^2(0, L), \ v(0) = 0 \right\}.$$

(6.7)

Show that a variational formulation **(VP)** for the displacements is given by the following:

$$\textbf{(VP)} \quad \begin{cases} \text{Find } u \in V \text{ solution of} \\ E S \int_0^L \dot{u}(x)\dot{v}(x)\,\mathrm{d}x = \int_0^L f(x)v(x)\,\mathrm{d}x, & \forall v \in V. \end{cases}$$

(6.8)

5) Show likewise that there is a minimisation problem **(MP)** equivalent to the variational formulation **(VP)**, specified as follows:

$$\textbf{(MP)} \quad \begin{cases} \text{Find } u \in V \text{ solution of } J(u) = \min_{v \in V} J(v), \\ \text{where } J(v) = \dfrac{E S}{2} \int_0^L \dot{v}^2(x)\,\mathrm{d}x - \int_0^L f(x)v(x)\,\mathrm{d}x. \end{cases}$$

(6.9)

▶ **Numerical Part**

This part deals with the finite-element approximation of elastic beams. In order to implement the method in a way that extends the discussion in previous chapters, we shall focus on the assembly procedure, and we shall apply it in the specific context provided by approximation of the minimisation problem (**MP**).

The approximation framework is a P_1 finite-element analysis, and we consider a regular mesh on the interval $[0, L]$, with constant spacing h, such that

$$\begin{cases} x_0 = 0, & x_{N+1} = L, \\ x_{i+1} = x_i + h, & i = 0, \ldots, N. \end{cases} \tag{6.10}$$

We now define the approximation space $\widetilde{V}$ by

$$\widetilde{V} = \left\{ \tilde{v} : [0, 1] \longrightarrow \mathbb{R}, \ \tilde{v} \in C^0([0, L]), \ \tilde{v}|_{[x_i, x_{i+1}]} \in P_1([x_i, x_{i+1}]), \ \tilde{v}(0) = 0 \right\}, \tag{6.11}$$

where $P_1([x_i, x_{i+1}])$ is the space of polynomials on $[x_i, x_{i+1}]$ of degree less than or equal to 1.

6) If φ_i, $(i = 1, \ldots, N + 1)$, is the canonical basis of $\widetilde{V}$ satisfying

$$\varphi_i(x_j) = \delta_{ij}, \quad \text{where } \delta_{ii} = 1, \quad \delta_{ij} = 0 \text{ if } i \neq j,$$

then $\tilde{v}$ in $\widetilde{V}$ is defined by

$$\tilde{v} = \sum_{j=1}^{N+1} \tilde{v}_j \varphi_j. \tag{6.12}$$

– Give a mechanical interpretation of the coefficients $\tilde{v}_j$.

7) Let $\widetilde{(\mathbf{MP})}$ be the approximate minimisation problem associated with the problem (**MP**), specified as follows:

$$\widetilde{(\mathbf{MP})} \quad \text{Find } \tilde{u} \in \widetilde{V} \text{ solution of } J(\tilde{u}) = \min_{\tilde{v} \in \widetilde{V}} J(\tilde{v}). \tag{6.13}$$

Show that a necessary condition for $\tilde{u}$ to be a solution of $\widetilde{(\mathbf{MP})}$ can be expressed as follows:

Determine the numerical sequence $(\tilde{u}_i)_{1, \ldots, N+1}$ defining the approximation $\tilde{u}$ in $\widetilde{V}$, solution of the global linear system

$$\sum_{j=1}^{N+1} A_{ij} \tilde{u}_j = b_i, \quad \forall i = 1, \ldots, N + 1, \tag{6.14}$$

where

$$A_{ij} \equiv E S \int_{\text{supp}\,\varphi_i \,\cap\, \text{supp}\,\varphi_j} \dot{\varphi}_i \dot{\varphi}_j \, dx, \qquad b_i \equiv \int_{\text{supp}\,\varphi_i} f \varphi_i \, dx. \tag{6.15}$$

8) We now introduce the elements of the elementary matrix $a^{(i+1)}$, expressing the contribution of the segment $[x_i, x_{i+1}]$, i.e., the $(i+1)$th element of the mesh on the interval $[0, L]$, in the global matrix A of the linear system (6.14), with generic element A_{ij}.

We thus set

$$a^{(i+1)} = \begin{bmatrix} a_{1,1}^{(i+1)} & a_{1,2}^{(i+1)} \\ a_{2,1}^{(i+1)} & a_{2,2}^{(i+1)} \end{bmatrix} \equiv E S \begin{bmatrix} \int_{x_i}^{x_{i+1}} (\dot{\varphi}_i)^2 \, dx & \int_{x_i}^{x_{i+1}} \dot{\varphi}_i \dot{\varphi}_{i+1} \, dx \\ \int_{x_i}^{x_{i+1}} \dot{\varphi}_{i+1} \dot{\varphi}_i \, dx & \int_{x_i}^{x_{i+1}} (\dot{\varphi}_{i+1})^2 \, dx \end{bmatrix}. \tag{6.16}$$

In the same way, we introduce the elementary vector $b^{(i+1)}$ defined by

$$b^{(i+1)} = \begin{bmatrix} b_1^{(i+1)} \\ b_2^{(i+1)} \end{bmatrix} \equiv \begin{bmatrix} \int_{x_i}^{x_{i+1}} f \varphi_i \, dx \\ \int_{x_i}^{x_{i+1}} f \varphi_{i+1} \, dx \end{bmatrix}. \tag{6.17}$$

Determine the relations between the coefficients A_{ij} and $a_{ij}^{(i+1)}$, then between b_i and $b^{(i)}$. Give a qualitative explanation of the resulting assembly process.

9) Calculate exactly the 4 coefficients of the elementary matrix $a^{(i+1)}$, then suggest an approximation of the vector $b^{(i+1)}$ using the trapezoidal rule.

10) Assemble the global matrix A with generic element A_{ij} and also the right-hand side b_i, defined by the problem $\widetilde{(\textbf{MP})}$.

11) Do we recover the node equations obtained by the standard approximation using a P_1 finite-element analysis applied to the variational formulation **(VP)**?

12) What about the finite-difference method applied to the continuous problem **(CP)**?

▶ **Elastic Support**

13) The clamping condition at $x = 0$ is replaced by an elastic support of given stiffness k (see Fig. 6.2). When the spring in the support is at rest, the displacement u_1 of the other end of the beam, at $x_1 = 0$, is zero.

Fig. 6.2 Beam with elastic
support

Given that the beam is subected to a force density $\mathbf{f} = f_1 \mathbf{X}_1$, show that the minimisation of the potential energy of the beam-spring system can be written

(**MP**)
$$
\begin{cases}
\text{Find } u \in V' \text{ solution of } J(u) = \min_{v \in V'} J(v), \\[2mm]
J(v) = \dfrac{ES}{2} \displaystyle\int_0^L \dot{v}^2(x)\,dx - \int_0^L f(x)v(x)\,dx + \dfrac{1}{2}ku_1^2.
\end{cases}
\tag{6.18}
$$

Note that the displacement fields in V' are no longer subject to the homogeneous Dirichlet condition at $x_1 = 0$.

14) Consider the special case where the mesh on the beam Ω consists of a single element $[x_1 = 0, x_2 = L]$. Write down the system of equations satisfied by the approximate displacements $\tilde{u}_1$ and $\tilde{u}_2$ of nodes x_1 and x_2, respectively.

Solve this system of equations and determine the approximate displacements $\tilde{u}_1$ and $\tilde{u}_2$. Then establish an approximation for the displacement field $\tilde{u}$ at all points of the beam.

15) When the stiffness of the elastic support is infinite, show that we recover the results for a beam that is clamped at $x_1 = 0$.

6.1.2 Solution

▶ **Theoretical Part: Variational Formulation of the Displacement Field and Principle of Virtual Work**

1) The choice of virtual field in (6.1) is motivated by the physical nature of the forces acting on the beam Ω. Indeed, it is subject solely to a longitudinal force density $\mathbf{f} = f_1 \mathbf{X}_1$, suggesting that we look for a real displacement field $\mathbf{U}$, and hence virtual fields $\mathbf{U}^*$ of the form (6.1).

Given the form of the virtual fields in (6.1), it is a straightforward matter to deduce the expression for the strain tensor $\varepsilon(\mathbf{U}^*)$ associated with the virtual field $\mathbf{U}^*$, using the definition (6.3):

$$\varepsilon(\mathbf{U}^*) = \varepsilon_{11}^* \mathbf{X}_1 \otimes \mathbf{X}_1 = \dot{u}^*(x) \mathbf{X}_1 \otimes \mathbf{X}_1, \tag{6.19}$$

where the tensor product $\mathbf{X}_1 \otimes \mathbf{X}_1$ indicates that the strain tensor field $\varepsilon(\mathbf{U}^*)$ has only one nonzero component, viz., ε_{11}^*.

We can thus proceed to evaluate the virtual work done by the internal forces:

$$T_{\text{int}}^* = -\int_\Omega \sigma_{ij} \varepsilon_{ij}(\mathbf{U}^*) \, d\Omega = -\int_\Omega \sigma_{11} \varepsilon_{11}^* \, d\Omega \tag{6.20}$$

$$= -\int_0^L \left[\iint_{S(x)} \sigma_{11} \, dS(x) \right] \dot{u}^*(x) \, dx = -\int_0^L N(x) \dot{u}^*(x) \, dx, \tag{6.21}$$

where we use the expression for the normal forces N given in (6.4). Likewise, we can evaluate the virtual work done by the external forces:

$$T_{\text{ext}}^* = \int_\Omega \mathbf{f} \cdot \mathbf{U}^* \, d\Omega = \int_\Omega f_1 u^*(x) \, d\Omega \tag{6.22}$$

$$= \int_0^L \left[\iint_{S(x)} f_1 \, dS(x) \right] u^*(x) \, dx = \int_0^L f(x) u^*(x) \, dx. \tag{6.23}$$

2) Applied to static phenomena, the principle of virtual work states that

$$T_{\text{int}}^* + T_{\text{ext}}^* = 0, \quad \forall u^* \text{ satisfying the conditions (6.1).} \tag{6.24}$$

In other words, with (6.21) and (6.23), we obtain

$$-\int_0^L N(x) \dot{u}^*(x) \, dx + \int_0^L f(x) u^*(x) \, dx = 0, \quad \forall u^* \text{ such that } u^*(0) = 0. \tag{6.25}$$

We then carry out an integration by parts on the first integral of (6.25) and use the homogeneous Dirichlet condition defined in (6.1).

The result is the variational formulation **(EVP)** specified as follows:

$$\int_0^L \dot{N}(x)u^*(x)\,dx - N(L)u^*(L) + \int_0^L f(x)u^*(x)\,dx = 0, \quad \forall u^* \text{ such that } u^*(0) = 0.$$
(6.26)

3) In order to formulate the continuous problem **(CP)**, we return to the Eq. (6.26) of the variational formulation **(EVP)** and replace the normal force N by its expression in terms of the associated displacement field u, using the constitutive law given in the statement of the problem.

This leads to

$$\int_0^L \Big[ES\ddot{u}(x) + f(x)\Big]u^*(x)\,dx - ES\dot{u}(L)u^*(L) = 0, \quad \forall u^* \text{ such that } u^*(0) = 0.$$
(6.27)

Starting from the variational formulation (6.27), which as it stands remains purely formal, we shall suggest a suitable functional framework in which to look for the solution u, that is, a plausible space for u to inhabit, along with the virtual displacement fields u^* arising in (6.27).

For this purpose, given that the force density f is a given function in $L^2(0, L)$, we can ensure its existence by an application of the Cauchy–Schwarz inequality to the integral of Eq. (6.27), provided that $\ddot{u}$ and u^* belong to $L^2(0, L)$. We thus consider the solution u and the virtual fields u^* to inhabit the Sobolev space $H^2(0, L)$.

Note further that, since the functions in $H^2(0, L)$ are C^1 on $[0, L]$ [1], the values of $\dot{u}$ and u^* at the point $x = L$, which arise in the formulation (6.27), are perfectly acceptable.

In addition, the homogeneous Dirichlet condition $u(0) = u^*(0) = 0$ leads us to consider the displacement field u solving (6.27) and also the virtual displacement fields u^* in the space $H_*^2(0, L)$ defined by

$$H_*^2(0, L) \equiv \Big\{ v : [0, L] \to \mathbb{R}, \ v \in H^2(0, L), \ v(0) = 0 \Big\}.$$
(6.28)

The variational formulation **(EVP)** then becomes:
Find $u \in H_*^2(0, L)$, solution of

$$\int_0^L \Big[ES\ddot{u}(x) + f(x)\Big]u^*(x)\,dx - ES\dot{u}(L)u^*(L) = 0, \quad \forall u^* \in H_*^2(0, L).$$
(6.29)

The statement of the continuous problem **(CP)** is obtained in two stages. We first consider (6.29) in the special case where the functions u^* belong to $H_*^2(0, L)$ and vanish at $x = L$.

For such functions, (6.29) has the form

$$\int_0^L \left[E S \ddot{u}(x) + f(x) \right] u^*(x) \, dx = 0, \quad \forall u^* \in H_*^2(0, L) \text{ such that } u^*(L) = 0.$$
(6.30)

We now use a density argument, noting that (6.30) is also satisfied for any function u^* in $\mathcal{D}(0, L) \subset H_*^2(0, L)$, the space of C^∞ functions with compact support.

We then use the fact that $\mathcal{D}(0, L)$ is dense in $L^2(0, L)$:

$$\forall g \in L^2(0, L), \quad \exists g_n \in \mathcal{D}(0, L) \text{ such that } \lim_{n \to +\infty} \| g_n - g \|_{L^2} = 0. \quad (6.31)$$

Note that the limit is taken in the sense of the norm L^2. So, when g is fixed (but still arbitrary) in $L^2(0, L)$, the sequence g_n in $\mathcal{D}(0, L)$ defined by (6.31) satisfies the variational Eq. (6.30) like any function in $\mathcal{D}(0, L)$.

In fact, we would like (6.30) to be satisfied for our function g in $L^2(0, L)$, so that we could then choose among all the functions g in $L^2(0, L)$ the one that would be equal to the particular function G defined by $G \equiv E S \ddot{u}(x) + f(x)$.

The differential equation of the continuous problem **(CP)** would then follow without difficulty. To this end, we bound the left-hand side of (6.30), if u^* is an arbitrary function g in $L^2(0, L)$:

$$\left| \int_0^L G(x) g(x) \, dx \right| \leq \int_0^L |G(x)| |g(x) - g_n(x)| \, dx$$

$$\leq \left[\int_0^L G^2(x) \, dx \right]^{1/2} \left[\int_0^L |g(x) - g_n(x)|^2 \, dx \right]^{1/2}. \quad (6.32)$$

We now take the limit in the inequality (6.32) to obtain the variational Eq. (6.30) for any function g belonging to $L^2(0, L)$.

As already mentioned, the differential equation of the continuous problem **(CP)** is then immediate. To see this, among all the functions g belonging to $L^2(0, L)$ which satisfy (6.30), we simply choose the one equal precisely to G, i.e., $g \equiv G$.

We now return to the variational equation (6.29) in which we have just established that the integral vanishes, since its integrand $E S \ddot{u}(x) + f(x)$ is necessarily zero.

This implies that

$$\dot{u}(L) u^*(L) = 0, \quad \forall u^* \in H_*^2(0, L). \quad (6.33)$$

We conclude straightforwardly that $\dot{u}(L) = 0$.

Summary. We have considered the solution u of the variational equation (6.29) in the space $H_*^2(0, L)$ defined by (6.28). Since we have just shown that such a solution satisfies the differential equation of the continuous problem **(CP)** and the Neumann condition $\dot{u}(L) = 0$, but also the homogeneous Dirichlet condition at $x = 0$ which it inherits as a property of the space $H_*^2(0, L)$, the function u is therefore a solution of the continuous problem **(CP)** defined in (6.6).

4) Given the manner in which we have built up the different formulations of the present problem from mechanical considerations, the variational formulation **(VP)**, which can be obtained by the standard technique presented in the previous chapters, is taken up again here in the following way. In fact, we only need to return to the formulation (6.29), given that we adopt the functional context for this formulation as specified in the previous question, viz., $H_*^2(0, L)$.

Now if u is a solution of the continuous problem **(CP)**, u satisfies the Neumann condition $\dot{u}(L) = 0$, so (6.29) can be reexpressed as follows:

Find $u \in H_*^2(0, L)$ solution of

$$\int_0^L \left[E S \ddot{u}(x) + f(x) \right] u^*(x) \, \mathrm{d}x = 0, \quad \forall u^* \in H_*^2(0, L). \tag{6.34}$$

Integrating by parts in the variational equation (6.34) and using the boundary conditions satisfied by u^* at $x = 0$ and by $\dot{u}$ at $x = L$, we establish the variational formulation **(VP)** defined in (6.8).

Regarding the functional framework of this formulation, following the integration by parts just mentioned, a degree of differentiation is 'lost' in this transformation.

This is why the problem **(VP)** is posed in $H_*^1(0, L)$, to ensure convergence of the integrals featuring in its formulation.

5) To establish the equivalence of the variational formulation **(VP)** and the minimisation problem **(MP)** specified in (6.9), it suffices to note that the variational formulation **(VP)** has the form

$$a(u, v) = L(v), \quad \forall v \in V, \tag{6.35}$$

where the bilinear form $a(., .)$ and the linear form $L(.)$ satisfy the properties ensuring the equivalence of **(VP)** and **(MP)**, in particular, the symmetry and positivity of the form $a(., .)$ (see Theorem 1.13).

Indeed, considering the variational formulation (6.8), we simply set

$$V \equiv H_*^1(0, L), \tag{6.36}$$

$$a(u, v) \equiv E S \int_0^L \dot{u}(x) \dot{v}(x) \, \mathrm{d}x, \tag{6.37}$$

$$L(v) \equiv \int_0^L f(x)v(x)\,\mathrm{d}x. \tag{6.38}$$

Note. The minimisation problem **(MP)** specified in (6.9) presents a mechanical interpretation. Indeed, the functional J to be minimised over the set of displacement fields $\mathbf{v} = v(x)\mathbf{X_1}$ is simply the potential energy E_p of the beam Ω defined by

$$E_p(\mathbf{v}) \equiv E_{\text{strain}}(\mathbf{v}) - W_{\text{ext}}(\mathbf{v}), \tag{6.39}$$

where $E_{\text{strain}}(\mathbf{v})$ is the strain energy corresponding to

$$E_{\text{strain}}(\mathbf{v}) \equiv \frac{1}{2}\int_\Omega \sigma_{ij}\varepsilon_{ij}\,\mathrm{d}\Omega = \frac{1}{2}\int_\Omega \sigma_{11}\varepsilon_{11}\,\mathrm{d}\Omega = \frac{E}{2}\int_\Omega \varepsilon_{11}^2\,\mathrm{d}\Omega \tag{6.40}$$

$$= \frac{E}{2}\int_0^L \left[\iint_{S(x)} \mathrm{d}S(x)\right]\dot{v}^2\,\mathrm{d}x = \frac{ES}{2}\int_0^L \dot{v}^2\,\mathrm{d}x. \tag{6.41}$$

In addition, $W_{\text{ext}}(\mathbf{v})$ represents the work done by known forces in the unknown displacement $\mathbf{v}$, viz.,

$$W_{\text{ext}}(\mathbf{v}) \equiv \int_\Omega \mathbf{f}\cdot\mathbf{v}\,\mathrm{d}\Omega = \int_0^L \left[\iint_{S(x)} f_1\,\mathrm{d}S(x)\right]v(x)\,\mathrm{d}x = \int_0^L f(x)v(x)\,\mathrm{d}x. \tag{6.42}$$

If we then substitute the expressions for (6.41) and (6.42) in the definition of the potential energy $E_p(\mathbf{v})$ given in (6.39), we recover precisely the definition of the functional J in the problem **(MP)** [see (6.9)].

▶ **Numerical Part**

6) From this question on, we consider the P_1 finite-element approximation of the minimisation problem **(MP)**, or equivalently, that of the variational formulation **(VP)**.

With this in mind, we expand any function $\tilde{v}$ in the approximation space $\widetilde{V}$ in terms of the canonical basis containing the functions $(\varphi_i)_{i=1,\,...,\,N+1}$ satisfying

$$\varphi_i(x_j) = \delta_{ij}. \tag{6.43}$$

Any element $\tilde{v}$ of $\widetilde{V}$ is then expanded as follows:

$$\tilde{v} = \sum_{j=1}^{N+1} \tilde{v}_j \varphi_j. \tag{6.44}$$

Given the property (6.43), if we evaluate $\tilde{v}$ in (6.44) at x_i, we obtain

$$\tilde{v}(x_i) = \sum_{j=1}^{N+1} \tilde{v}_j \varphi_j(x_i) = \sum_{j=1}^{N+1} \tilde{v}_j \delta_{ij} = \tilde{v}_i, \tag{6.45}$$

where we have used the properties of the Kronecker symbol

$$\delta_{ij} = \begin{cases} 1, & \text{if } i = j, \\ 0, & \text{if } i \neq j. \end{cases} \tag{6.46}$$

Hence $\tilde{v}_i$ can be interpreted precisely as the value of the approximate displacement field $\tilde{v}$ at the discretisation point x_i.

Therefore, and this is the main consequence of the choice of basis function $(\varphi_i)_{i=1,\ldots,N+1}$ satisfying the conditions (6.43), the unknown coefficients $(\tilde{u}_1, \tilde{u}_2, \ldots, \tilde{u}_{N+1})$ of the linear combination defining the approximate solution $\tilde{u}$ correspond to the $N+1$ approximate displacements $\tilde{u}(x_1), \tilde{u}(x_2), \ldots, \tilde{u}(x_{N+1})$ of the displacement field $\tilde{u}$ at the nodes $x_1, x_2, \ldots, x_{N+1}$ of the mesh.

7) We now consider the approximation of the minimisation problem **(MP)**. For this purpose, we replace any function v of the function space V by its approximation $\tilde{v}$ in the space $\widetilde{V}$ defined by (6.11). The approximate minimisation problem $\widetilde{\textbf{(MP)}}$ is thus the following:

$$\widetilde{\textbf{(MP)}} \begin{cases} \text{Find } \tilde{u} \in \widetilde{V} \text{ solution of } J(\tilde{u}) = \min_{\tilde{v} \in \widetilde{V}} J(\tilde{v}), \\[2mm] \text{with } J(\tilde{v}) = \dfrac{ES}{2} \displaystyle\int_0^L \left(\sum_{j=1}^{N+1} \tilde{v}_j \dot{\varphi}_j \right)^2 dx - \int_0^L f \left(\sum_{j=1}^{N+1} \tilde{v}_j \varphi_j \right) dx. \end{cases} \tag{6.47}$$

In this form, the functional J is a function of the $N+1$ variables $(\tilde{v}_1, \tilde{v}_2, \ldots, \tilde{v}_{N+1})$. Therefore, a necessary condition for minimisation of J at the point $\tilde{u}$ is

$$\frac{\partial J}{\partial \tilde{v}_j}(\tilde{u}_1, \tilde{u}_2, \ldots, \tilde{u}_{N+1}) = 0, \quad \forall j = 1, \ldots, N+1. \tag{6.48}$$

We return to (6.47) and take the partial derivative with respect to each $\tilde{v}_j$, for $j = 1, \ldots, N+1$.

For any i between 1 and $N+1$, we have

$$\frac{\partial J}{\partial \tilde{v}_j}(\tilde{u}_1, \tilde{u}_2, \ldots, \tilde{u}_{N+1}) = 0 \iff ES \sum_{j=1}^{N+1} \left(\int_0^L \dot{\varphi}_i \dot{\varphi}_j \, dx \right) \tilde{u}_j = \int_0^L f(x) \varphi_i \, dx. \tag{6.49}$$

We thus obtain the linear system (6.14).

8) The relations between the coefficients $a_{i,j}^{(i+1)}$ and $A_{i,j}$ are obtained by identification, having noted that a priori only the coefficients $A_{i,i-1}$, $A_{i,i}$, and $A_{i,i+1}$ are nonzero in the global matrix A. These relations then become

$$A_{i,\,i-1} = ES \int_{\text{supp}\,\varphi_{i-1}\,\cap\,\text{supp}\,\varphi_i} \dot{\varphi}_{i-1}\dot{\varphi}_i \, dx = ES \int_{x_{i-1}}^{x_i} \dot{\varphi}_{i-1}\dot{\varphi}_i \, dx \equiv a_{2,\,1}^{(i)}, \qquad (6.50)$$

$$A_{i,\,i} = ES \int_{\text{supp}\,\varphi_i} \dot{\varphi}_i^2 \, dx = ES \left(\int_{x_{i-1}}^{x_i} \dot{\varphi}_i^2 \, dx + \int_{x_i}^{x_{i+1}} \dot{\varphi}_i^2 \, dx \right) = a_{2,\,2}^{(i)} + a_{1,\,1}^{(i+1)}, \tag{6.51}$$

$$A_{i,\,i+1} = ES \int_{\text{supp}\,\varphi_i\,\cap\,\text{supp}\,\varphi_{i+1}} \dot{\varphi}_i\dot{\varphi}_{i+1} \, dx = ES \int_{x_i}^{x_{i+1}} \dot{\varphi}_i\dot{\varphi}_{i+1} \, dx \equiv a_{1,\,2}^{(i+1)}. \tag{6.52}$$

Having established the three relations (6.50), (6.51) and (6.52) between the coefficients of the global matrix A and those of the local matrix $a^{(i+1)}$, expressing the contribution of each element $[x_i, x_{i+1}]$ in the matrix A, it turns out that each coefficient A_{ij} is built up from either one or two contributions of geometric finite elements.

In addition, writing the i th and $(i + 1)$ th lines of the linear system (6.14) using the coefficients of the local matrix $a^{(i)}$, we obtain

$$a_{2,\,1}^{(i)}\tilde{u}_{i-1} + \left[a_{2,\,2}^{(i)} + a_{1,\,1}^{(i+1)}\right]\tilde{u}_i + a_{1,\,2}^{(i+1)}\tilde{u}_{i+1} = b_i, \tag{6.53}$$

$$a_{2,\,1}^{(i+1)}\tilde{u}_i + \left[a_{2,\,2}^{(i+1)} + a_{1,\,1}^{(i+2)}\right]\tilde{u}_{i+1} + a_{1,\,2}^{(i+2)}\tilde{u}_{i+2} = b_{i+1}, \tag{6.54}$$

where b_i and b_{i+1} represent the components of the right-hand side b of the system (6.14), associated with the i th and $(i + 1)$ th lines of the linear system (6.14). Furthermore, we have

$$b_i \equiv \int_0^L f\varphi_i \, dx = \int_{x_{i-1}}^{x_{i+1}} f\varphi_i \, dx \tag{6.55}$$

$$= \int_{x_{i-1}}^{x_i} f\varphi_i \, dx + \int_{x_i}^{x_{i+1}} f\varphi_i \, dx \equiv b_2^{(i)} + b_1^{(i+1)}, \tag{6.56}$$

where we have used the definition (6.17) of the elementary vector $b^{(i+1)}$.

Hence, the quantities $b^{(i+1)}$ express the contribution of the mesh element $[x_i, x_{i+1}]$ in the constitution of the right-hand side b_i, in the same way as the local matrix $a^{(i+1)}$ for the global matrix A.

We have thus identified the contribution of the mesh element $[x_i, x_{i+1}]$ to the global matrix A and also to the right-hand side b. Furthermore, regarding the elements of the local matrix $a^{(i+1)}$, these weight the approximate displacements $\tilde{u}_i$ and $\tilde{u}_{i+1}$:

$$
\begin{array}{c}
\quad\; \tilde{u}_i \qquad\; \tilde{u}_{i+1} \\
\quad\; \downarrow \qquad\quad\; \downarrow
\end{array}
$$

$$
\begin{array}{c}
\tilde{u}_i \;\rightarrow \\
\tilde{u}_{i+1} \;\rightarrow
\end{array}
\begin{bmatrix}
a_{1,1}^{(i+1)} & a_{1,2}^{(i+1)} \\
a_{2,1}^{(i+1)} & a_{2,2}^{(i+1)}
\end{bmatrix},
\quad
\begin{bmatrix}
b_1^{(i+1)} \\
b_2^{(i+1)}
\end{bmatrix}.
\tag{6.57}
$$

The assembly procedure is then to consider each geometric element $[x_i, x_{i+1}]$ one after the other, while retranscribing the contribution of each mesh element in the global matrix A and the in the right-hand side b of the linear system (6.14).

This assembly process is valid because it is based on the linear structure of the global matrix A and the right-hand side b with regard to the contributions of the mesh elements $[x_i, x_{i+1}]$.

This is simply a qualitative interpretation of (6.53)–(6.56), which bring about the above-mentioned contributions in linear combinations.

So we begin the assembly by the first element $[x_0, x_1]$. Since the node x_0 is clampled, whence $u(0) = 0$, the corresponding degree of freedom $\tilde{u}_0$ is zero. In other words, the elementary matrix $\mathbf{a}^{(1)}$ is degenerate in this case, and we have

$$
\left[a^{(1)} \right] (\tilde{u}_1) \equiv \mathbf{a}_{2,2}^{(1)} \tilde{u}_1.
\tag{6.58}
$$

Likewise, for the right-hand side b, the contribution of the mesh element $[x_0, x_1]$ comes from the node x_1. This contribution is equal to $\mathbf{b}_2^{(1)}$ as defined in (6.17).

Using this first result, we insert the coefficient $\mathbf{a}_{2,2}^{(1)}$ in the global matrix A, being careful to put it in the appropriate place, i.e., the intersection of the first line and first column.

We proceed in the same way to incorporate the contribution of the coefficient $\mathbf{b}_2^{(1)}$ in the right-hand side of the system of equations.

We now have

$$
\begin{bmatrix}
a_{2,2}^{(1)} & 0 & \ldots\ldots & 0 \\
0 & 0 & \ldots\ldots & 0 \\
\cdots & \cdots\cdots\cdots & \\
\cdots & \cdots\cdots\cdots & \\
0 & 0 & \ldots\ldots & 0
\end{bmatrix},
\quad
\begin{bmatrix}
b_2^{(1)} \\
\cdots \\
\cdots \\
\cdots \\
\cdots
\end{bmatrix}.
\tag{6.59}
$$

We now move to the second mesh element $[x_1, x_2]$. Its contribution to the global matrix is determined by the elementary matrix $\mathbf{a}^{(2)}$ containing the corresponding approximate displacements $\tilde{u}_1$ and $\tilde{u}_2$:

$$
\begin{array}{cc}
\tilde{\mathbf{u}}_1 & \tilde{\mathbf{u}}_2 \\
\downarrow & \downarrow
\end{array}
$$

$$
\begin{array}{c}
\tilde{\mathbf{u}}_1 \rightarrow \\
\tilde{\mathbf{u}}_2 \rightarrow
\end{array}
\begin{bmatrix}
a_{1,1}^{(2)} & a_{1,2}^{(2)} \\
a_{2,1}^{(2)} & a_{2,2}^{(2)}
\end{bmatrix},
\quad
\begin{bmatrix}
b_1^{(2)} \\
b_2^{(2)}
\end{bmatrix}.
\tag{6.60}
$$

We then insert the contribution of the second mesh element $[x_1, x_2]$ in the global matrix A, placing the coefficients of the elementary matrix $\mathbf{a}^{(2)}$ of (6.60) at the relevant locations in the matrix A.

Doing likewise for the right-hand side b, we obtain

$$
\begin{bmatrix}
(a_{2,2}^{(1)} + \mathbf{a}_{1,1}^{(2)}) & \mathbf{a}_{1,2}^{(2)} & \ldots\ldots & 0 \\
\mathbf{a}_{2,1}^{(2)} & \mathbf{a}_{2,2}^{(2)} & \ldots\ldots & 0 \\
\cdots & \cdots\cdots\cdots\cdots & & \\
\cdots & \cdots\cdots\cdots\cdots & & \\
0 & 0 & \ldots\ldots & 0
\end{bmatrix},
\quad
\begin{bmatrix}
b_2^{(1)} + b_1^{(2)} \\
b_2^{(2)} \\
\cdots \\
\cdots \\
\cdots
\end{bmatrix}.
\tag{6.61}
$$

We thus fill up the global matrix A and the right-hand side b step by step, treating each mesh element $[x_i, x_{i+1}]$ one by one in order to include its contribution, considering in each case the corresponding elementary matrix $\mathbf{a}^{(i+1)}$ and the local right-hand term $\mathbf{b}^{(i+1)}$.

The final result is the matrix A given by

$$
A =
\begin{bmatrix}
(a_{2,2}^{(1)} + a_{1,1}^{(2)}) & a_{1,2}^{(2)} & \ldots & \ldots & 0 \\
a_{2,1}^{(2)} & (a_{2,2}^{(2)} + a_{1,1}^{(3)}) & a_{1,2}^{(3)} & \ldots & 0 \\
0 & a_{2,1}^{(3)} & (a_{2,2}^{(3)} + a_{1,1}^{(4)}) & \ldots & \ldots \\
\ldots & \ldots & \ldots & \ldots & \ldots \\
0 & 0 & (a_{2,2}^{(N-1)} + a_{1,1}^{(N)}) & a_{1,2}^{(N)} & \ldots \\
\ldots & \ldots & a_{2,1}^{(N)} & (a_{2,2}^{(N)} + a_{1,1}^{(N+1)}) & a_{1,2}^{(N+1)} \\
0 & 0 & \ldots & a_{2,1}^{(N+1)} & (a_{1,1}^{(N)} + a_{2,2}^{(N+1)})
\end{bmatrix},
\tag{6.62}
$$

and the right-hand side b by

$$b = \begin{bmatrix} b_2^{(1)} + b_1^{(2)} \\ b_2^{(2)} + b_1^{(3)} \\ \cdots \\ \cdots \\ \cdots \\ b_2^{(N)} + b_1^{(N+1)} \\ b_2^{(N+1)} \end{bmatrix}. \tag{6.63}$$

9) The coefficients of the matrix $a^{(i+1)}$ are calculated without approximation. This is possible because the integrands are just derivatives of functions that are affine on each mesh element $[x_i, x_{i+1}]$.

In other words, the integrands are constant functions on each element of the mesh. It then follows that

$$a_{1,1}^{(i+1)} = ES \int_{x_i}^{x_{i+1}} (\dot{\varphi}_i)^2 \, dx = \frac{ES}{h}, \tag{6.64}$$

$$a_{1,2}^{(i+1)} = a_{2,1}^{(i+1)} = ES \int_{x_i}^{x_{i+1}} \dot{\varphi}_i \dot{\varphi}_{i+1} \, dx = -\frac{ES}{h}, \tag{6.65}$$

$$a_{2,2}^{(i+1)} = ES \int_{x_i}^{x_{i+1}} (\dot{\varphi}_{i+1})^2 \, dx = \frac{ES}{h}. \tag{6.66}$$

We now approximate the elementary vector $b^{(i+1)}$ using the trapezoidal rule:

$$b_1^{(i+1)} = \int_{x_i}^{x_{i+1}} f \varphi_i \, dx \approx \frac{h}{2} \Big[f(x_i)\varphi_i(x_i) + f(x_{i+1})\varphi_i(x_{i+1}) \Big] \tag{6.67}$$

$$\approx \frac{h}{2}(f_i \times 1 + f_{i+1} \times 0) = \frac{h}{2} f_i.$$

$$b_2^{(i+1)} = \int_{x_i}^{x_{i+1}} f \varphi_{i+1} \, dx \approx \frac{h}{2} \Big[f(x_i)\varphi_{i+1}(x_i) + f(x_{i+1})\varphi_{i+1}(x_{i+1}) \Big] \tag{6.68}$$

$$\approx \frac{h}{2}(f_i \times 0 + f_{i+1} \times 1) = \frac{h}{2} f_{i+1}.$$

10) Returning to the matrix A and the right-hand side b obtained in (6.62) and (6.63) and substituting in the coefficients (6.64–6.66) of the elementary matrix $a^{(i+1)}$ and the approximations (6.67) and (6.68) of the components of the elementary vector $b^{(i+1)}$, we obtain the final result

$$A = ES \begin{bmatrix} \dfrac{2}{h} & -\dfrac{1}{h} & 0 & \cdots & \cdots & \cdots & \cdots & 0 \\[2mm] -\dfrac{1}{h} & \dfrac{2}{h} & -\dfrac{1}{h} & 0 & \cdots & \cdots & \cdots & 0 \\[2mm] 0 & -\dfrac{1}{h} & \dfrac{2}{h} & -\dfrac{1}{h} & 0 & \cdots & \cdots & 0 \\[2mm] \cdots & \cdots & \cdots & \cdots & \cdots & \cdots & \cdots & \cdots \\[1mm] \cdots & \cdots & \cdots & \cdots & \cdots & \cdots & \cdots & \cdots \\[1mm] 0 & \cdots & \cdots & 0 & -\dfrac{1}{h} & \dfrac{2}{h} & -\dfrac{1}{h} & 0 \\[2mm] 0 & \cdots & \cdots & \cdots & 0 & -\dfrac{1}{h} & \dfrac{2}{h} & -\dfrac{1}{h} \\[2mm] 0 & \cdots & \cdots & \cdots & \cdots & 0 & -\dfrac{1}{h} & \dfrac{1}{h} \end{bmatrix}, \quad b \approx \begin{bmatrix} hf_1 \\[2mm] hf_2 \\[2mm] \cdots \\[1mm] \cdots \\[1mm] \cdots \\[1mm] \cdots \\[1mm] hf_N \\[2mm] \dfrac{h}{2} f_{(N+1)} \end{bmatrix}. \tag{6.69}$$

Note the edge effects generated by the assembly process, both in the matrix A and in the vector b. In particular, the last coefficient of the diagonal $A_{N+1,\,N+1}$ is ES/h, whereas the other coefficients on the diagonal are equal to $2ES/h$.

This is due to the fact that the coefficient $A_{N+1,\,N+1}$ only gets a contribution from the coefficient $a_{2,2}^{(N+1)}$ of the elementary matrix $a^{(N+1)}$, while the other coefficients $A_{i,i}$ on the diagonal are of the form $a_{2,2}^{(i)} + a_{1,1}^{(i+1)}$. This goes also for the vector b, and in particular, the last component b_{N+1}.

11) In order to compare the results (6.69) of the last question with those that would be obtained by approximating the variational formulation **(VP)** with a P_1 finite-element analysis, we note that the approximate minimisation problem (6.49) constituted by the linear system (6.14) is strictly the same as what would be obtained by carrying out the approximation of the formulation **(VP)**.

To see this, we need only go back to the expression (6.8) of the variational formulation **(VP)** and carry out the standard approximation substitutions:

$$u(x) \longmapsto \tilde{u}(x) = \sum_{j=1}^{N+1} \tilde{u}_j \varphi_j, \tag{6.70}$$

$$v(x) \longmapsto \tilde{v}(x) = \varphi_i. \tag{6.71}$$

Since we now have the same system of equations defined by the matrix A and the vector b (6.69), the node equation characterising a node strictly in the interior x_i, $(i = 1, \ldots, N)$, corresponds to the first N lines of the linear system (6.14), which have the generic form

$$\frac{ES}{h}\left(-\tilde{u}_{i-1} + 2\tilde{u}_i - \tilde{u}_{i+1}\right) = hf_i. \tag{6.72}$$

The node equation corresponding to the basis function φ_{N+1} can be identified as the last equation of the linear system (6.14), viz.,

$$\frac{ES}{h}\left(-\tilde{u}_N + \tilde{u}_{N+1}\right) = \frac{h}{2}f_{N+1}. \tag{6.73}$$

12) We now compare with the finite-difference method. We can see immediately that the node equation (6.72) is strictly identical to the one that would be obtained by discretising the differential equation of the continuous problem **(CP)** by finite differences.

To compare the node equation (6.73) with the discretisation of the Neumann condition $\dot{u}(L) = 0$, we Taylor expand the solution u of the problem **(CP)**, assuming that it is sufficiently regular in the neighbourhood of the point $x = L$:

$$\begin{aligned}
u(x_N) &= u(x_{N+1}) - h\dot{u}(x_{N+1}) + \frac{h^2}{2}\ddot{u}(x_{N+1}) + O(h^3), \\
u(x_N) &= u(x_{N+1}) - \frac{h^2}{2ES}f(x_{N+1}) + O(h^3),
\end{aligned} \tag{6.74}$$

where we have replaced the second derivative $\ddot{u}$ by the expression we get for it from the differential equation of the problem **(CP)**. From (6.74), we move to the approximations $\tilde{u}_i$, neglecting $O(h^3)$, and eventually finding the node equation (6.73).

▶ Elastic Support

13) When we replace the clamp at $x_1 = 0$ by an elastic support of stiffness k, we introduce an extra degree of kinematic freedom at this end of the beam. We thus let u_1 denote the displacement of this end of the beam relative to the wall.

Furthermore, the strain energy of the beam-spring system is given by

$$E_{\text{strain}}(\text{beam–spring}) = E_{\text{strain}}(\text{beam}) + E_{\text{strain}}(\text{spring}). \tag{6.75}$$

But these strain energies can be written

$$E_{\text{strain}}(\text{beam}) = \frac{ES}{2}\int_0^L \dot{u}^2(x)\mathrm{d}x, \quad E_{\text{strain}}(\text{spring}) = \frac{1}{2}ku_1^2, \tag{6.76}$$

where we have used the expression for the strain energy of a beam in traction or compression established in the last part [see (6.40) and (6.41)].

The minimisation problem (6.18) can be formulated immediately as soon as we note that the homogeneous Dirichlet condition at $x_1 = 0$ no longer has its place in the definition of the search space V'.

14) In the special case where the mesh on the beam comprises a single element $[x_1 = 0, x_2 = L]$, we seek the approximate solution $\tilde{u}$ over the whole beam in the set of displacement fields of the form

$$\tilde{v}(x) = \tilde{v}_1 \varphi_1(x) + \tilde{v}_2 \varphi_2(x). \tag{6.77}$$

▶ **System of Equations Satisfied by the Approximate Displacements $\tilde{u}_1$ and $\tilde{u}_2$**

The potential energy of the minimisation problem (6.18) evaluated for the displacement field $\tilde{v}$ in (6.77) is

$$J(\tilde{v}) = \frac{ES}{2} \int_0^L (\tilde{v}_1 \dot{\varphi}_1 + \tilde{v}_2 \dot{\varphi}_2)^2 \, dx - \int_0^L f(\tilde{v}_1 \varphi_1 + \tilde{v}_2 \varphi_2) \, dx + \frac{1}{2} k \tilde{v}_1^2. \tag{6.78}$$

We write down the two necessary conditions for minimisation of the functional J:

$$\frac{\partial J}{\partial \tilde{v}_1}(\tilde{u}_1, \tilde{u}_2) = 0 = \frac{\partial J}{\partial \tilde{v}_2}(\tilde{u}_1, \tilde{u}_2). \tag{6.79}$$

Given (6.78), these two conditions can be written in the form

$$\frac{\partial J}{\partial \tilde{v}_1}(\tilde{u}_1, \tilde{u}_2) = ES \int_0^L (\tilde{u}_1 \dot{\varphi}_1 + \tilde{u}_2 \dot{\varphi}_2) \dot{\varphi}_1 \, dx - \int_0^L f \varphi_1 \, dx + k \tilde{u}_1 = 0, \tag{6.80}$$

$$\frac{\partial J}{\partial \tilde{v}_2}(\tilde{u}_1, \tilde{u}_2) = ES \int_0^L (\tilde{u}_1 \dot{\varphi}_1 + \tilde{u}_2 \dot{\varphi}_2) \dot{\varphi}_2 \, dx - \int_0^L f \varphi_2 \, dx = 0. \tag{6.81}$$

The linear system in the unknowns $\tilde{u}_1$ and $\tilde{u}_2$ can then be written in the form

$$\begin{bmatrix} k + ES \int_0^L \dot{\varphi}_1^2 & ES \int_0^L \dot{\varphi}_1 \dot{\varphi}_2 \\ ES \int_0^L \dot{\varphi}_1 \dot{\varphi}_2 & ES \int_0^L \dot{\varphi}_2^2 \end{bmatrix} \begin{bmatrix} \tilde{u}_1 \\ \tilde{u}_2 \end{bmatrix} = \begin{bmatrix} b_1^{(2)} \\ b_2^{(2)} \end{bmatrix}. \tag{6.82}$$

▶ **Determination of the Approximate Displacements $\tilde{u}_1$ and $\tilde{u}_2$ and Approximation of the Global Displacement Field $\tilde{u}$ Along the Beam**

The linear system (6.82) has the solution

$$\tilde{u}_1 = \frac{L}{2k}(f_1 + f_2), \qquad \tilde{u}_2 = \frac{L^2}{2ES} f_2 + \frac{L}{2k}(f_1 + f_2). \tag{6.83}$$

Therefore, the approximate displacement $\tilde{u}$ over the whole beam is obtained by inserting the values from (6.83) into (6.77), whence we obtain

$$\tilde{u}(x) = \frac{L}{2k}(f_1 + f_2)\varphi_1(x) + \left[\frac{L^2}{2ES}f_2 + \frac{L}{2k}(f_1 + f_2)\right]\varphi_2(x). \qquad (6.84)$$

Note that the approximate displacement field $\tilde{u}$ defined by (6.84) delivers an approximation to the displacement field u at all points of the beam in $[0, L]$, as always when we use the finite-element method.

This distinguishes the latter from the finite-difference method, which produces a series of approximations at fixed points on a discretisation grid, i.e., at the nodes of a predefined mesh on the beam.

15) The limiting case which occurs when the beam is clamped at $x_1 = 0$, viewed as the case of an elastic support whose stiffness constant k tends to infinity, is obtained by taking the limit in the expressions for $\tilde{u}_1$ and $\tilde{u}_2$ defined by (6.83).

The limiting approximations $\tilde{u}_1^\infty$ and $\tilde{u}_2^\infty$ are

$$\tilde{u}_1^\infty = 0, \qquad \tilde{u}_2^\infty = \frac{L^2}{2ES}f_2. \qquad (6.85)$$

We may compare these two values with those obtained directly from the clamped beam problem, discussed in Sect. 6.1.1. To do this, we go back to the global matrix and vector of (6.69), adapting them to the case of a single mesh element $[x_1, x_2]$, clamped at $x_1 = 0$.

We then obtain the degenerate system of a single equation in one unknown, viz., $\tilde{u}_2$ solution of

$$\frac{ES}{L} \times \tilde{u}_2 = \frac{L}{2}f_2. \qquad (6.86)$$

We thus obtain for $\tilde{u}_2$ the same expression as for $\tilde{u}_2^\infty$ given in (6.85):

$$\tilde{u}_2 = \frac{L^2}{2ES}f_2. \qquad (6.87)$$

Finally, we note that, for the clamped beam, we do of course satisfy the boundary condition at $x_1 = 0$, i.e., we do have $\tilde{u}_1 = \tilde{u}_1^\infty = 0$.

———————

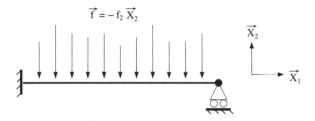

Fig. 6.3 Clamped beam resting on a moving support

6.2 Beam in Bending

6.2.1 Statement of the Problem

▶ **Clamped Beam on a Mobile Support: Variational Formulation of the Displacement Field and Principle of Virtual Work**

We consider a uniform beam Ω of length L, cross-section S, and density ρ, with isotropic elastic behaviour under small perturbations. The beam is made of a material with Young's modulus E and Poisson coefficient v.

The axis of the beam is denoted by $(O; \mathbf{X}_1)$ and coordinatised by x. The cross-section of the beam is coordinatised by the geometric variables (y, z).

We assume the geometric symmetry of the beam and the loads to which it is subjected to be such that the plane $(O; \mathbf{X}_1, \mathbf{X}_2)$ can be treated as a plane of symmetry.

The beam is clamped at the end $x = 0$ and rests on a moving support that prevents any vertical motion at $x = L$.

In addition, a transverse force field $\mathbf{f} = -f_2\mathbf{X}_2$ per unit volume is applied along the beam, as shown in Fig. 6.3. The density f_2 is given and has all the necessary functional regularity to ensure that we may carry out any integrals arising in the first two questions of the theoretical part.

1) We use the natural kinematical theory to describe the displacement field. We thus consider virtual displacement fields $\mathbf{V}^*$ of the form

$$\mathbf{V}^* = -y\Omega^*(x)\mathbf{X}_1 + v^*(x)\mathbf{X}_2, \tag{6.88}$$

satisfying the boundary conditions

$$v^*(0) = \Omega^*(0) = 0, \qquad v^*(L) = 0. \tag{6.89}$$

The boundary conditions (6.89) reflect the presence of the clamp at $x = 0$ and the effect of the movable support preventing any vertical displacement at $x = L$ for any virtual displacement field $\mathbf{V}^*$.

Let $\mathfrak{V}^*$ denote the set of all virtual displacement fields with the properties (6.88) and (6.89).

Determine the virtual work T_{int}^* done by the internal forces and the virtual work T_{ext}^* done by the external forces, these being defined by

$$T_{\text{int}}^* = - \int_{\Omega} \sigma_{ij}\varepsilon_{ij}(\mathbf{V}^*)\, d\Omega, \qquad T_{\text{ext}}^* = \int_{\Omega} \mathbf{f} \cdot \mathbf{V}^*\, d\Omega, \tag{6.90}$$

where σ is the stress tensor and $\varepsilon(\mathbf{V}^*)$ the linear strain tensor associated with the virtual displacement field $\mathbf{V}^*$:

$$\varepsilon_{ij}(\mathbf{V}^*) = \frac{1}{2}\left(\frac{\partial V_i^*}{\partial x_j} + \frac{\partial V_j^*}{\partial x_i} \right). \tag{6.91}$$

We now introduce the bending moment $M(x)$, the shear force $T(x)$, and the surface force density f defined by

$$M(x) = - \int_{S(x)} y\sigma_{11}\, dS(x), \quad T(x) = \int_{S(x)} \sigma_{12}\, dS(x), \quad f(x) = \int_{S(x)} f_2\, dS(x), \tag{6.92}$$

where $S(x)$ is the cross-section at x.

2) Without considering whether the integrals converge, show that the application of the principle of virtual work leads to the formal variational formulation (**EVP**) given by the following:

$$\left\{ \begin{array}{l} \text{Find real-valued } (M, T) \text{ defined on } [0, L] \times [0, L] \text{ solution of} \\[2mm] \displaystyle\int_0^L \left[\dot{M}(x) + T(x)\right]\Omega^*(x)dx + \int_0^L \left[\dot{T}(x) - f(x)\right]v^*(x)dx - M(L)\Omega^*(L) = 0, \\[4mm] \forall\,(v^*, \Omega^*) \text{ such that } v^*(0) = v^*(L) = \Omega^*(0) = 0. \end{array} \right. \tag{6.93}$$

3) We now introduce the constitutive law of the beam material:

$$M(x) = EI\dot{\Omega}(x), \qquad T(x) = \mu S(\dot{v} - \Omega), \tag{6.94}$$

where I is the moment of inertia of the beam about the axis $(O; \mathbf{X}_3)$ and μ is the shear modulus.

Assuming that the force density f belongs to $L^2(0, L)$, show that the continuous problem (**CP**) specified by the equilibrium equations of the beam can be expressed in the following way:

Find (v, Ω) in $H^2(0, L) \times H^2(0, L)$ solution of

$$
\textbf{(CP)} \begin{cases}
EI\ddot{\Omega}(x) + \mu S[\dot{v}(x) - \Omega(x)] = 0, & \forall x \in]0, L[, & (6.95) \\
\mu S[\ddot{v} - \dot{\Omega}] = f(x), & \forall x \in]0, 1[, & (6.96) \\
v(0) = v(L) = 0, & \Omega(0) = \dot{\Omega}(L) = 0. & (6.97)
\end{cases}
$$

4) We introduce the product space $V \equiv V_1 \times V_2 \equiv H_*^1(0, L) \times H_{**}^1(0, L)$, where

$$
H_*^1(0, L) = \left\{ \omega \in L^2(0, L) \text{ such that } \omega' \in L^2(0, L), \ \omega(0) = 0 \right\}, \quad (6.98)
$$

and

$$
H_{**}^1(0, L) = \left\{ h \in L^2(0, L) \text{ such that } h' \in L^2(0, L), \ h(0) = h(L) = 0 \right\}.
$$
$$(6.99)$$

Show that a variational formulation **(VP)** for the displacement field can be written as follows:

$$
\textbf{(VP)} \begin{cases}
\text{Find } (\Omega, v) \in V \text{ solution of} \\
EI \displaystyle\int_0^L \dot{\Omega}\dot{\omega}\, dx + \mu S \int_0^L (\dot{v} - \Omega)(\dot{h} - \omega)\, dx = \int_0^L fh\, dx, \ \forall (\omega, h) \in V.
\end{cases}
$$
$$(6.100)$$

5) Show also that there is a minimisation problem **(MP)**, equivalent to the variational formulation **(VP)**, defined by

$$
\textbf{(MP)} \begin{cases}
\text{Find } (v, \Omega) \in V \text{ solution of } J(v, \Omega) = \displaystyle\min_{(h, \omega) \in V} J(h, \omega), \\
\text{where } J(h, \omega) = \dfrac{EI}{2} \int_0^L \dot{\omega}^2\, dx + \dfrac{\mu S}{2} \int_0^L (\dot{h} - \omega)^2 dx + \int_0^L fh\, dx.
\end{cases}
$$
$$(6.101)$$

▶ **Numerical Part**

We now turn to the P_1 finite-element approximation of elastic beams in bending, as modelled by the natural theory. We thus consider a regular mesh on the interval $[0, L]$, with constant spacing h, so that

$$
\begin{cases}
x_0 = 0, & x_{N+1} = L, \\
x_{i+1} = x_i + h, & i = 0, \ldots, N.
\end{cases}
$$
$$(6.102)$$

We also introduce the approximation spaces $(\widetilde{V}_1, \widetilde{V}_2)$ defined by

$$\widetilde{V}_1 = \left\{ \tilde{h} \in C^0([0, L]) \text{ s.t. } \tilde{h}|_{[x_i, x_{i+1}]} \in P_1([x_i, x_{i+1}]), \ \tilde{h}(0) = \tilde{h}(L) = 0 \right\}, \tag{6.103}$$

$$\widetilde{V}_2 = \left\{ \tilde{\omega} \in C^0([0, L]), \ \tilde{\omega} \in P_1([x_i, x_{i+1}]), \ \tilde{\omega}(0) = 0 \right\}, \tag{6.104}$$

where $P_1([x_i, x_{i+1}])$ is the space of polynomials on $[x_i, x_{i+1}]$ of degree less than or equal to 1.

6) What are the dimensions of the spaces $\widetilde{V}_1$ and $\widetilde{V}_2$?

7) Define $(\tilde{v}, \tilde{\Omega})$ in $\widetilde{V} \equiv \widetilde{V}_1 \times \widetilde{V}_2$ by

$$\tilde{v} = \sum_{j=1}^{N} \tilde{v}_j \varphi_j, \qquad \tilde{\Omega} = \sum_{k=1}^{N+1} \tilde{\Omega}_k \varphi_k, \tag{6.105}$$

where φ_i are functions in the canonical basis associated with the P_1 finite-element analysis and satisfying $\varphi_i(x_j) = \delta_{ij}$.

Let $\widetilde{(\mathbf{MP})}$ be the approximate minimisation problem associated with the problem **(MP)**, defined by as follows:

$$\text{Find } (\tilde{v}, \tilde{\Omega}) \in \widetilde{V} \text{ solution of } J(\tilde{v}, \tilde{\Omega}) = \min_{(\tilde{h}, \tilde{\omega}) \in \widetilde{V}} J(\tilde{h}, \tilde{\omega}), \tag{6.106}$$

where J is the potential energy defined by (6.101).

Show that a necessary condition for $(\tilde{v}, \tilde{\Omega})$ to be a solution of the approximate minimisation problem $\widetilde{(\mathbf{MP})}$ can be expressed as follows:

Determine the two sequences of components $(\tilde{v}_j)_{j=1,\ldots,N}$ and $(\tilde{\Omega}_k)_{k=1,\ldots,N+1}$ specifying the approximation $(\tilde{v}, \tilde{\Omega}) \in \widetilde{V}$ which solves the global linear system:

$$\widetilde{(\mathbf{MP})} \quad \begin{cases} A_{ij}^{(1,1)} \tilde{v}_j + A_{ij}^{(2,1)} \tilde{\Omega}_j = b_i^{(1)}, \quad i = 1, \ldots, N, & (6.107) \\[2mm] A_{ij}^{(1,2)} \tilde{v}_j + A_{ij}^{(2,2)} \tilde{\Omega}_j = b_i^{(2)}, \quad i = 1, \ldots, N+1, & (6.108) \\[2mm] \forall j \text{ such that } \operatorname{supp} \varphi_i \cap \operatorname{supp} \varphi_j \neq \varnothing, \end{cases}$$

where

$$A_{ij}^{(1,1)} = \mu S \int_0^L \dot{\varphi}_i \dot{\varphi}_j \, dx, \qquad A_{ij}^{(1,2)} = -\mu S \int_0^L \dot{\varphi}_i \varphi_j \, dx, \tag{6.109}$$

$$A_{ij}^{(2,\,1)} = -\mu S \int_0^L \varphi_i \dot{\varphi}_j \, dx, \quad A_{ij}^{(2,2)} = EI \int_0^L \ddot{\varphi}_i \ddot{\varphi}_j \, dx + \mu S \int_0^L \varphi_i \varphi_j \, dx,$$

$$\tag{6.110}$$

$$b_i^{(1)} = -\int_0^L f \varphi_i \, dx, \tag{6.111}$$

$$b_i^{(2)} = 0. \tag{6.112}$$

Note that we have used the Einstein summation (or repeated index) convention in this expression of the approximate minimisation problem $\widetilde{(MP)}$.

For example,

$$A_{ij}^{(1,\,1)} \tilde{v}_j \equiv \sum_j A_{ij}^{(1,\,1)} \tilde{v}_j. \tag{6.113}$$

8) For each of the basis functions $(\varphi_i)_{1,\,...,\,N}$ characterising a node strictly in the interior of $[0, L]$, write down the corresponding node equation associated with the discrete Eqs. (6.107) and (6.108).

9) Show that we then recover the finite-difference approximation of the system of differential Eqs. (6.95) and (6.96).

10) Write down the node equation for the basis function φ_{N+1} characterising the node at x_{N+1} associated with the discrete Eq. (6.108).

11) Recover the finite-difference scheme which discretises the Neumann condition (6.97) on the first derivative of Ω at $x = L$.

12) Show that the elementary rigidity matrix expressing the contribution of the local element $[x_i, x_{i+1}]$ and associated with the system (6.107) and (6.108) can be written

$$A^{(i+1)} = \begin{bmatrix} \alpha & \beta & -\alpha & \beta \\ \beta & \delta & -\beta & \gamma \\ -\alpha & -\beta & \alpha & -\beta \\ \beta & \gamma & -\beta & \delta \end{bmatrix}, \tag{6.114}$$

where

$$\alpha = \frac{\mu S}{h}, \quad \beta = \frac{\mu S}{2}, \quad \gamma = -\frac{EI}{h}, \quad \delta = \frac{EI}{h} + \frac{\mu Sh}{2}. \tag{6.115}$$

13) Assemble the discrete approximation system $\widetilde{(MP)}$ specified by (6.107) and (6.108).

6.2.2 Solution

▶ **Theoretical Part: Variational Formulation of the Displacement Field.
Principle of Virtual Work**

1) The form of the virtual displacement field $\mathbf{V}^*$ defined by (6.88) implies that the
strain tensor $\varepsilon(\mathbf{V}^*)$ [see (6.91)] associated with $\mathbf{V}^*$ can be written

$$\varepsilon(\mathbf{V}^*) = \varepsilon_{11}^* \mathbf{X}_1 \otimes \mathbf{X}_1 + \varepsilon_{12}^* [\mathbf{X}_1 \otimes \mathbf{X}_2 + \mathbf{X}_1 \otimes \mathbf{X}_1], \qquad (6.116)$$

where

$$\varepsilon_{11}^* = -y\dot{\Omega}^*(x), \qquad \varepsilon_{12}^* = \frac{1}{2}[\dot{v}^*(x) - \Omega^*(x)]. \qquad (6.117)$$

We have used the tensor product notation $\mathbf{X}_i \otimes \mathbf{X}_j$, reflecting the presence of a
nonzero component ε_{ij}^* in the strain tensor $\varepsilon(\mathbf{U}^*)$.

The virtual work T_{int}^* done by the internal forces is then evaluated as follows:

$$T_{\text{int}}^* = -\int_\Omega \sigma_{ij}\varepsilon_{ij}(\mathbf{V}^*)\,dv = -\left[\int_\Omega \sigma_{11}\varepsilon_{11}(\mathbf{V}^*)\,dv + 2\int_\Omega \sigma_{12}\varepsilon_{12}(\mathbf{V}^*)\,dv\right]$$

$$= -\int_0^L \left[-\int\int_{S(x)} y\sigma_{11}\,dS(x)\right]\dot{\Omega}^*(x)\,dx$$

$$\quad - \int_0^L \left[\int\int_{S(x)} \sigma_{12}\,dS(x)\right]\left[\dot{v}^*(x) - \Omega^*(x)\right]dx,$$

$$= -\int_0^L M(x)\dot{\Omega}^*(x)\,dx - \int_0^L T(x)\left[\dot{v}^*(x) - \Omega^*(x)\right]dx, \qquad (6.118)$$

where we have used the definitions of the bending moment $M(x)$ and the shear force
$T(x)$ in (6.92).

In a similar way, the virtual work done by external forces is

$$T_{\text{ext}}^* = \int_\Omega \mathbf{f}\cdot\mathbf{V}^*\,dv = -\int_\Omega f_2 v^*(x)\,dx \qquad (6.119)$$

$$= -\int_0^L \left[\int\int_{S(x)} f_2\,dS(x)\right]v^*(x)\,dx = -\int_0^L f(x)v^*(x)\,dx.$$

2) In elastostatics, the principle of virtual work takes the form

$$T_{\text{int}}^* + T_{\text{ext}}^* = 0, \quad \forall\,V^* \in \mathfrak{V}^*. \qquad (6.120)$$

Using (6.118) and (6.119), we thus obtain

$$0 = \int_0^L M(x)\dot{\Omega}^*(x)dx + \int_0^L T(x)\left[\dot{v}^*(x) - \Omega^*(x)\right]dx + \int_0^L f(x)v^*(x)\,dx,$$

$\forall\,(v^*, \Omega^*)$ such that $v^*(0) = v^*(L) = \Omega^*(0) = 0$. \hfill (6.121)

Integrating this variational equation by parts, yields

$$0 = \int_0^L \dot{M}\Omega^*dx - \left[M\Omega^*\right]_0^L + \int_0^L \dot{T}v^*dx - \left[Tv^*\right]_0^L + \int_0^L T\Omega^*dx - \int_0^L fv^*dx,$$

$\forall\,(v^*, \Omega^*)$ such that $v^*(0) = v^*(L) = \Omega^*(0) = 0$. \hfill (6.122)

The boundary conditions (6.89) on the virtual fields V^* lead to the variational formulation (**EVP**):

Find real-valued (M, T) on $[0, L] \times [0, L]$ solving:

$$0 = \int_0^L \left[\dot{M}(x) + T(x)\right]\Omega^*(x)dx + \int_0^L \left[\dot{T}(x) - f(x)\right]v^*(x)dx - M(L)\Omega^*(L),$$

$\forall\,(v^*, \Omega^*)$ such that $v^*(0) = v^*(L) = \Omega^*(0) = 0$. \hfill (6.123)

3) The continuous problem (**CP**) is obtained using density arguments of the kind already illustrated on several occasions. To do this in the present case, we first transform the variational Eq. (6.123), replacing the bending moment M and the shear force T by the kinematic quantities (v, Ω) with the help of the constitutive laws (6.94).

The variational problem (**EVP**) then becomes: find real-valued (v, Ω) on $[0, L] \times [0, L]$ solving:

$$0 = \int_0^L \left[EI\ddot{\Omega}(x) + \mu S(\dot{v} - \Omega)(x)\right]\Omega^*(x)dx + \int_0^L \left[\mu S(\ddot{v} - \dot{\Omega})(x) - f(x)\right]v^*(x)dx$$

$$- EI\dot{\Omega}(L)\Omega^*(L) = 0, \ \forall\,(v^*, \Omega^*) \text{ such that } v^*(0) = v^*(L) = \Omega^*(0) = 0.$$

\hfill (6.124)

We now define a functional framework in which the expression of (6.124) will make mathematical sense.

To this end, assume first that v and v^* are elements of $H_{**}^2(0, L)$, defined in a similar way to (6.99), but replacing H^1 by H^2. In fact, it suffices to consider the virtual fields v^* in $L^2(0, L)$. We now seek the rotation Ω of the cross-section in the space $H_*^2(0, L)$, defined along the same lines as (6.98).

Under these conditions, it is easy to see that the integral equation (6.124) is meaningful, essentially by application of the Cauchy–Schwarz inequality.

Indeed, we simply note that the quantities $EI\ddot{\Omega} + \mu S(\dot{v} - \Omega)$ and $\mu S(\ddot{v} - \dot{\Omega}) - f$ are indeed in $L^2(0, L)$, with f itself being a given distribution of forces belonging to $L^2(0, L)$.

We thus consider the particular case where (6.124) is satisfied for identically zero virtual fields Ω^*. The particular formulation of (6.124) is then:

Find $v \in H^2_{**}(0, L)$ solution of

$$\int_0^L \left[\mu S(\ddot{v} - \dot{\Omega})(x) - f(x)\right] v^*(x)\, \mathrm{d}x = 0, \quad \forall v^* \in H^2_{**}(0, L). \tag{6.125}$$

The differential equation (6.96) is subsequently obtained by a density argument. Among all the functions v^* in $H^2_{**}(0, L)$, we consider those in $\mathscr{D}(0, L) \subset H^2_{**}(0, L)$, and we use the fact that $\mathscr{D}(0, L)$ is dense in $L^2(0, L)$. A detailed demonstration of the density method can be found in Sect. 6.1.1 for the beam in traction.

Finally, we obtain

$$\mu S(\ddot{v} - \dot{\Omega})(x) - f(x) = 0, \quad \forall x \in (0, 1). \tag{6.126}$$

By a similar argument, but this time considering a second particular case of (6.124) in which the functions v^* are identically zero, the same density arguments lead to the differential equation (6.95). Once we have the two differential equations of the continuous problem **(CP)**, Eq. (6.124) is degenerate, taking the form

$$-EI\dot{\Omega}(L)\Omega^*(L) = 0, \quad \forall \Omega^* \text{ such that } \Omega^*(0) = 0. \tag{6.127}$$

We deduce immediately that $\dot{\Omega}(L) = 0$.

Summary. We have considered the solution (v, Ω) of the variational equation (6.124) in the product space $H^2_{**}(0, L) \times H^2_*(0, L)$, hence satisfying *ipso facto* the boundary conditions (6.97) at $x = 0$ and $x = L$. In addition, we have shown by density arguments that v and Ω are solutions of the differential system (6.95) and (6.96). Put another way, (v, Ω) solves the continuous problem **(CP)** specified by (6.95)–(6.97).

4) Let (v, Ω) be a solution of the continuous problem **(CP)** defined by (6.95)–(6.97) and let (h, ω) be a pair of test functions. We multiply the differential equation (6.95) of the problem **(CP)** by the test function h and the differential equation (6.96) by the test function ω, then integrate each equation over $[0, L]$ to obtain

$$\textbf{(CP)} \begin{cases} EI \displaystyle\int_0^L \ddot{\Omega}\omega\, \mathrm{d}x = -\mu S \int_0^L (\dot{v} - \Omega)\omega\, \mathrm{d}x, & \forall \omega \in V_1, \tag{6.128} \\[3mm] \mu S \displaystyle\int_0^L (\ddot{v} - \dot{\Omega})h\, \mathrm{d}x = \int_0^L f h\, \mathrm{d}x, & \forall h \in V_2, \tag{6.129} \end{cases}$$

where $V_1 \times V_2$ is a space of test functions (h, ω) to be defined later.

We now integrate by parts in each of the integral equations (6.128) and (6.129) to obtain

(CP)
$$\begin{cases} - EI \int_0^L \dot{\Omega}\dot{\omega}\,dx + EI\left[\dot{\Omega}\omega\right]_0^L = -\mu S \int_0^L (\dot{v} - \Omega)\omega\,dx, \quad \forall\,\omega \in V_1, \quad (6.130) \\[2mm] - \mu S \int_0^L \dot{v}h\,dx + \mu S\left[\dot{v}h\right]_0^L - \mu S \int_0^L \dot{\Omega}h = \int_0^L fv\,dx, \quad \forall\,h \in V_2. \quad (6.131) \end{cases}$$

The formulation (6.130) and (6.131) provides the opportunity to take into account the Neumann boundary condition on Ω, viz., $\dot{\Omega}(L) = 0$. Regarding the other, Dirichlet-type conditions on v and Ω, we have to maintain them in the future variational formulation via the test functions h et ω.

To do this, we require the pair of test functions (ω, h) to belong to the product space $V \equiv V_1 \times V_2 \equiv H^1_*(0, L) \times H^1_{**}(0, L)$ defined by

$$H^1_*(0, L) = \left\{\omega^{(k)} \in L^2(0, L), \quad k = 0, 1, \quad \text{such that } \omega(0) = 0\right\}, \qquad (6.132)$$

$$H^1_{**}(0, L) = \left\{h^{(k)} \in L^2(0, L), \quad k = 0, 1, \quad \text{such that } h(0) = 0 = h(L)\right\}. \tag{6.133}$$

An immediate consequence of this choice of function space is the disappearance of the completely integrated terms in the system (6.130) and (6.131), which thus becomes the following:

Find $(\omega, v) \in V$ solution of

(CP)
$$\begin{cases} - EI \int_0^L \dot{\Omega}\dot{\omega}\,dx = -\mu S \int_0^L (\dot{v} - \Omega)\omega\,dx, \quad \forall\,\omega \in V_1, \qquad (6.134) \\[2mm] - \mu S \int_0^L \dot{v}h\,dx - \mu S \int_0^L \dot{\Omega}h = \int_0^L fv\,dx, \quad \forall\,h \in V_2. \qquad (6.135) \end{cases}$$

Note here that the regularity of the solution pair (Ω, v) has been weakened in going from the strong formulation of the continuous problem **(CP)** to the weak or variational formulation **(VP)**.

Indeed, the regularity of the functions in $V_1 \times V_2$ is enough to ensure convergence of the integrals featuring in the variational formulation **(VP)**. In order to obtain the formulation **(VP)** specified by (6.100), we simply add together the two Eqs. (6.134) and (6.135).

5) The variational formulation **(VP)** specified by (6.100) has the advantage of being expressed in terms of a bilinear form $a[(., .), (., .)]$ and a linear form $L[(., .)]$ defined by

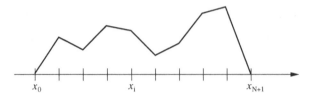

Fig. 6.4 Piecewise affine function in $\widetilde{V}_1$

$$a\big[(\Omega, v), (\omega, h)\big] = EI \int_0^L \dot{\Omega}\dot{\omega}\,dx + \mu S \int_0^L (\dot{v} - \Omega)(\dot{h} - \omega)\,dx, \qquad (6.136)$$

$$L[(\omega, h)] = \int_0^L fh\,dx. \qquad (6.137)$$

We then note that the two forms a and L have all the properties, and in particular symmetry for the bilinear form a (see Theorem 1.13 for a list of these properties), allowing us to conclude as to the equivalence of the minimisation problem **(MP)** specified thus:

$$\textbf{(MP)} \quad \begin{cases} \text{Find } (v, \Omega) \in V \text{ solution of } J(v, \Omega) = \min_{(h,\omega)\,\in V} J(h, \omega), \\[2mm] \text{where } J(h, \omega) = \dfrac{1}{2}a\big[(h, \omega), (h, \omega)\big] - L\big[(h, \omega)\big]. \end{cases} \qquad (6.138)$$

▶ **Numerical Part**

6) The dimensions of the spaces $\widetilde{V}_1$ and $\widetilde{V}_2$ are estimated by the same procedure, except that we must take into account the extra degree of freedom at $x = L$ for the functions in $\widetilde{V}_1$, which leads to a difference of 1 with the dimension of the space $\widetilde{V}_2$.

So the functions belonging to the space $\widetilde{V}_1$ are precisely those defined by the N values at the internal nodes $(x_i)_{i=1,\ldots,N}$ of the mesh (6.102). To see this, let us look more closely at these functions Fig. 6.4.

The functions of the approximation space $\widetilde{V}_1$ are piecewise affine functions, being affine on each mesh element $[x_i, x_{i+1}]$, and such that the internal nodes x_i constitute points of continuity between adjacent pieces.

Moreover, regarding the functions in $\widetilde{V}_1$, we must make the two further requirements that they vanish at $x = 0$ and at $x = L$. This is why only the values at the N nodes strictly in the interior of the interval $[0, L]$ have one degree of freedom for each function $\tilde{v}$ in $\widetilde{V}_1$.

Modifying one of these N values immediately implies changing the relevant element $\tilde{v}$ of $\widetilde{V}_1$ into another function of $\widetilde{V}_1$.

So without formal demonstration, we see that knowing a function $\tilde{v}$ in $\tilde{V}_1$ amounts to specifying a vector in $\mathbb{R}^N$ whose N components are the N values $(\tilde{v}_1, \ldots, \tilde{v}_N)$ of $\tilde{v}$ at the discretisation points $(x_1, \ldots, x_N)$. We conclude therefore that the dimension of $\tilde{V}_1$ is N, while that of $\tilde{V}_2$ is $N + 1$, given the extra degree of freedom at $x = L$.

7) The approximate formulation $\widetilde{(\text{MP})}$ associated with the minimisation problem (MP) is obtained by evaluating the functional J defined by (6.101) at the point $(\tilde{h}, \tilde{\omega})$:

$$J(\tilde{h}, \tilde{\omega}) = \frac{EI}{2} \int_0^L \dot{\tilde{\omega}}^2 \, dx + \frac{\mu S}{2} \int_0^L (\dot{\tilde{h}} - \tilde{\omega})^2 \, dx + \int_0^L f \tilde{h} \, dx, \qquad (6.139)$$

where $(\tilde{h}, \tilde{\omega})$ belong to $\tilde{V}_1 \times \tilde{V}_2$. So given the finite dimension of the spaces $\tilde{V}_1$ and $\tilde{V}_2$, the functions $\tilde{h}$ and $\tilde{v}$ can be expanded in an analogous way to (6.105).

In this case, $J(\tilde{h}, \tilde{v})$ is evaluated as follows:

$$J(\tilde{h}, \tilde{\omega}) = \frac{EI}{2} \int_0^L \left(\sum_{k=1}^{N+1} \tilde{\omega}_k \dot{\varphi}_k \right)^2 dx + \frac{\mu S}{2} \int_0^L \left(\sum_{j=1}^{N} \tilde{h}_j \dot{\varphi}_j - \sum_{k=1}^{N+1} \tilde{\omega}_k \varphi_k \right)^2 dx$$

$$+ \int_0^L f \sum_{j=1}^{N} \tilde{h}_j \varphi_j \, dx. \qquad (6.140)$$

A necessary condition for $(\tilde{v}, \tilde{\Omega})$ to constitute the minimum of the functional J is

$$\frac{\partial J}{\partial \tilde{h}_i}(\tilde{v}, \tilde{\Omega}) = \frac{\partial J}{\partial \tilde{\omega}_l}(\tilde{v}, \tilde{\Omega}) = 0, \quad \forall (i, l) \in \{1, \ldots, N\} \times \{1, \ldots, N+1\}. \quad (6.141)$$

We then obtain

$$\frac{\partial J}{\partial \tilde{h}_i}(\tilde{v}, \tilde{\Omega}) = \mu S \int_0^L \left(\sum_{j=1}^{N} \tilde{v}_j \dot{\varphi}_j - \sum_{k=1}^{N+1} \tilde{\Omega}_k \varphi_k \right) \dot{\varphi}_i \, dx + \int_0^L f \varphi_i = 0, \quad i = 1, \ldots, N,$$

$$(6.142)$$

$$\frac{\partial J}{\partial \tilde{\omega}_l}(\tilde{v}, \tilde{\Omega}) = EI \int_0^L \left(\sum_{k=1}^{N+1} \tilde{\Omega}_k \dot{\varphi}_k \right) \dot{\varphi}_l \, dx - \mu S \int_0^L \left(\sum_{j=1}^{N} \tilde{v}_j \dot{\varphi}_j - \sum_{k=1}^{N+1} \tilde{\Omega}_k \varphi_k \right) \varphi_l \, dx,$$

$$= 0, \quad l = 1, \ldots, N+1. \qquad (6.143)$$

We then note that the support of a basis function φ_i comprises two mesh elements $[x_{i-1}, x_i]$ and $[x_i, x_{i+1}]$, as shown in Fig. 6.5, so only the basis functions φ_{i-1}, φ_i, and φ_{i+1} have supports with non-empty intersection with the support of the function φ_i.

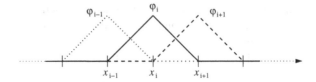

Fig. 6.5 Basis functions φ_{i-1}, φ_i, and φ_i

For this reason, the finite sums arising in (6.142) and (6.143) are degenerate in the following way:

$$\frac{\partial J}{\partial \tilde{h}_i}(\tilde{v}, \tilde{\Omega}) = 0 \iff \mu S \left[\left(\int_0^L \dot{\varphi}_{i-1} \dot{\varphi}_i \right) \tilde{v}_{i-1} - \left(\int_0^L \varphi_{i-1} \dot{\varphi}_i \right) \tilde{\Omega}_{i-1} \right]$$

$$+ \mu S \left[\left(\int_0^L \dot{\varphi}_i^2 \right) \tilde{v}_i - \left(\int_0^L \varphi_i \dot{\varphi}_i \right) \tilde{\Omega}_i \right]$$

$$\mu S \left[\left(\int_0^L \dot{\varphi}_{i+1} \dot{\varphi}_i \right) \tilde{v}_{i+1} - \left(\int_0^L \varphi_{i+1} \dot{\varphi}_i \right) \tilde{\Omega}_{i+1} \right]$$

$$= - \int_0^L f \varphi_i. \tag{6.144}$$

Likewise,

$$\frac{\partial J}{\partial \tilde{\omega}_l}(\tilde{v}, \tilde{\Omega}) = 0 \iff EI \left[\left(\int_0^L \dot{\varphi}_{l-1} \dot{\varphi}_l \right) \tilde{\Omega}_{l-1} + \left(\int_0^L \dot{\varphi}_l^2 \right) \tilde{\Omega}_l \right.$$

$$\left. + \left(\int_0^L \dot{\varphi}_{l+1} \dot{\varphi}_l \right) \tilde{\Omega}_{l+1} \right]$$

$$- \mu S \left[\left(\int_0^L \dot{\varphi}_{l-1} \varphi_l \right) \tilde{v}_{l-1} - \left(\int_0^L \varphi_{l-1} \varphi_l \right) \tilde{\Omega}_{l-1} \right]$$

$$- \mu S \left[\left(\int_0^L \dot{\varphi}_l \varphi_l \right) \tilde{v}_l - \left(\int_0^L \varphi_l^2 \right) \tilde{\Omega}_l \right]$$

$$- \mu S \left[\left(\int_0^L \dot{\varphi}_{l+1} \varphi_l \right) \tilde{v}_{l+1} - \left(\int_0^L \varphi_{l+1} \varphi_l \right) \tilde{\Omega}_{l+1} \right]$$

$$= 0. \tag{6.145}$$

Returning to the notation of (6.109)–(6.112), the necessary conditions (6.144) and (6.145) for minimisation of the functional J correspond exactly to the statement of the approximate minimisation problem $\widetilde{(\mathbf{MP})}$ specified in (6.107) and (6.108).

8) In this question we consider the basis functions characterising each of the internal nodes x_i, $(i = 1, \ldots, N)$. We then proceed to rewrite the system of equations defining the approximate formulation $\widehat{(\mathbf{MP})}$, using the elementary properties of the basis functions φ_i.

Moreover, in order to evaluate, the integrals arising in (6.144) and (6.145), we use the trapezoidal rule. It then follows that $\forall\,(i, l) \in \{1, \ldots, N\}^2$,

$$\left[\frac{\partial J}{\partial \tilde{h}_i}(\tilde{v}, \tilde{\Omega}) = 0\right] \iff \mu S\left[-\frac{\tilde{v}_{i-1} - 2\tilde{v}_i + \tilde{v}_{i+1}}{h^2} + \frac{\tilde{\Omega}_{i+1} - \tilde{\Omega}_{i-1}}{2h}\right] = -f_i,$$
(6.146)

$$\left[\frac{\partial J}{\partial \tilde{\omega}_l}(\tilde{v}, \tilde{\Omega}) = 0\right] \iff EI\left[\frac{\tilde{\Omega}_{l-1} - 2\tilde{\Omega}_l + \tilde{\Omega}_{l+1}}{h^2}\right] + \mu S\left[\frac{\tilde{v}_{l+1} - \tilde{v}_{l-1}}{2h} - \tilde{\Omega}_l\right] = 0.$$
(6.147)

9) In order to obtain the second order finite-difference approximation to the differential equations (6.95) and (6.96) from the node equations (6.146) and (6.147), we note the following identities, which have already been shown on several occasions:

$$\frac{v_{i-1} - 2v_i + v_{i+1}}{h^2} = \ddot{v}(x_i) + O(h^2),$$
(6.148)

$$\frac{\Omega_{i+1} - \Omega_{i-1}}{2h} = \dot{\Omega}(x_i) + O(h^2).$$
(6.149)

The node Eqs. (6.146) and (6.147) can then be written in the form

$$\mu S\left[\ddot{v}(x_i) - \dot{\Omega}(x_i)\right] = f(x_i), \quad i = 1, \ldots, N,$$
(6.150)

$$EI\ddot{\Omega}(x_l) + \dot{v}(x_l) - \Omega(x_l) = 0, \quad l = 1, \ldots, N,$$
(6.151)

to $O(h^2)$.

Equations (6.150) and (6.151) correspond exactly to the differential equations (6.95) and (6.96) of the continuous problem **(CP)** when we replace the approximation functions $(\tilde{v}, \tilde{\Omega})$ by the solutions (v, Ω).

10) In order to write down the node equation corresponding to the basis function φ_{N+1}, we consider the last equation of the system (6.143), which is in this particular case

$$\frac{\partial J}{\partial \tilde{\omega}_{N+1}}(\tilde{v}, \tilde{\Omega}) = 0 \iff EI\left[\left(\int_0^L \dot{\varphi}_N \dot{\varphi}_{N+1}\right)\tilde{\Omega}_N + \left(\int_0^L \dot{\varphi}_{N+1}^2\right)\tilde{\Omega}_{N+1}\right]$$
$$- \mu S\left[\left(\int_0^L \dot{\varphi}_N \varphi_{N+1}\right)\tilde{v}_N - \left(\int_0^L \varphi_N \varphi_{N+1}\right)\tilde{\Omega}_N\right]$$

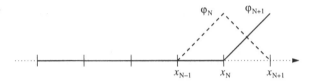

Fig. 6.6 Basis functions φ_N and φ_{N+1}

$$- \mu S \left[\left(\int_0^L \dot{\varphi}_{N+1} \varphi_{N+1} \right) \tilde{v}_{N+1} - \left(\int_0^L \varphi_{N+1}^2 \right) \tilde{\Omega}_{N+1} \right]$$

$$= 0. \tag{6.152}$$

Once again, we use the trapezoidal rule in conjunction with the elementary properties of the basis functions φ_N and φ_{N+1} (see Fig. 6.6), whereupon we obtain

$$EI \frac{\tilde{\Omega}_{N+1} - \tilde{\Omega}_N}{h} + \frac{\mu S}{2} (\tilde{v}_N - \tilde{v}_{N+1}) + \frac{\mu S h}{2} \tilde{\Omega}_{N+1} = 0. \tag{6.153}$$

11) In this question we obtain the finite-difference discretisation of the Neumann condition (6.97), viz., $\dot{\Omega}(x_{N+1}) = 0$. To do this, we write down the backward Taylor expansion at x_{N+1}, i.e.,

$$\Omega(x_N) = \Omega(x_{N+1}) - h\dot{\Omega}(x_{N+1}) + \frac{h^2}{2} \ddot{\Omega}(x_{N+1}) + O(h^3). \tag{6.154}$$

We now evaluate the second derivative $\ddot{\Omega}$ at the point x_{N+1}, assuming that the first differential equation (6.95) of the continuous problem (**CP**) can be extended to this point by continuity.

It then follows that

$$\Omega(x_N) = \Omega(x_{N+1}) + \frac{h^2}{2} \frac{\mu S}{EI} \left[\Omega(x_{N+1}) - \dot{v}(x_{N+1}) \right] + O(h^3). \tag{6.155}$$

In order to maintain an approximation to second order in h, we consider a first order approximation for the first derivative $\dot{v}$ at the point x_{N+1}, this being weighted by a multiple of h^2.

We thus consider the usual forward finite-difference relation:

$$\dot{v}(x_{N+1}) = \frac{v(x_{N+1}) - v(x_N)}{h} + O(h).$$

(6.156)

We then substitute the expression for $\dot{v}(x_{N+1})$ given by (6.156) into the finite expansion (6.155) to obtain

$$EI\left[\Omega(x_N) - \Omega(x_{N+1})\right] - \frac{\mu S h^2}{2}\Omega(x_{N+1}) = \frac{\mu S h}{2}\left[v(x_{N+1}) - v(x_N)\right] + O(h^3),$$

(6.157)

or again,

$$EI\frac{\Omega(x_N) - \Omega(x_{N+1})}{h} - \frac{\mu S h}{2}\Omega(x_{N+1}) - \frac{\mu S}{2}\left[v(x_{N+1}) - v(x_N)\right] = O(h^2).$$

(6.158)

With this last form, we recover precisely the node equation corresponding to the basis function φ_{N+1}, described by the last equation of the system (6.143). To do this, we make our approximations in (6.158), replacing the true values $(v_i, \Omega_i) \equiv (v, \Omega)(x_i)$, $(i = N, N+1)$, by their corresponding approximations $(\tilde{v}_i, \tilde{\Omega}_i)$ and eliminating the Landau remainder $O(h^2)$.

12) To determine the elementary matrix $A^{(i+1)}$ which expresses the contribution of the local element $[x_i, x_{i+1}]$ in the global system of the approximate minimisation problem $\widetilde{(MP)}$ specified in (6.107) and (6.108), we write down the minimisation equations involving this mesh element.

Now only the basis functions φ_i and φ_{i+1} have part of their support which intersects the interval $[x_i, x_{i+1}]$. For this reason, we restrict to the following minimisation equations, the only ones involving the basis functions φ_i and φ_{i+1} on the interval $[x_i, x_{i+1}]$:

$$\frac{\partial J}{\partial \tilde{h}_i}(\tilde{v}, \tilde{\Omega}) = \frac{\partial J}{\partial \tilde{h}_{i+1}}(\tilde{v}, \tilde{\Omega}) = \frac{\partial J}{\partial \tilde{\omega}_i}(\tilde{v}, \tilde{\Omega}) = \frac{\partial J}{\partial \tilde{\omega}_{i+1}}(\tilde{v}, \tilde{\Omega}) = 0.$$

(6.159)

We begin by rewriting the minimisation conditions

$$\frac{\partial J}{\partial \tilde{h}_i}(\tilde{v}, \tilde{\Omega}) = 0 = \frac{\partial J}{\partial \tilde{h}_{i+1}}(\tilde{v}, \tilde{\Omega}).$$

(6.160)

We then have

$$\frac{\partial J}{\partial \tilde{h}_i}(\tilde{v}, \tilde{\Omega}) = 0 \iff \tilde{v}_{i-1}\left(\mu S \int_{x_{i-1}}^{x_i} \dot{\varphi}_{i-1}\dot{\varphi}_i\right) + \tilde{v}_i\left(\mu S \int_{x_{i-1}}^{x_i} \dot{\varphi}_i^2 + \mu S \int_{x_i}^{x_{i+1}} \dot{\varphi}_i^2\right)$$

$$+ \tilde{v}_{i+1}\left(\mu S \int_{x_i}^{x_{i+1}} \dot{\varphi}_{i+1}\dot{\varphi}_i\right)$$

$$+ \tilde{\Omega}_{i-1}\left(-\mu S \int_{x_{i-1}}^{x_i} \varphi_{i-1}\dot{\varphi}_i\right)$$

$$+ \tilde{\Omega}_i\left(-\mu S \int_{x_{i-1}}^{x_i} \varphi_i\dot{\varphi}_i - \mu S \int_{x_i}^{x_{i+1}} \varphi_i\dot{\varphi}_i\right)$$

$$+ \tilde{\Omega}_{i+1}\left(-\mu S \int_{x_i}^{x_{i+1}} \varphi_{i+1}\dot{\varphi}_i\right)$$

$$= -\left(\int_{x_{i-1}}^{x_i} f\varphi_i + \int_{x_i}^{x_{i+1}} f\varphi_i\right) \tag{6.161}$$

and

$$\frac{\partial J}{\partial \tilde{h}_{i+1}}(\tilde{v}, \tilde{\Omega}) = 0 \iff \tilde{v}_i\left(\mu S \int_{x_i}^{x_{i+1}} \dot{\varphi}_i\dot{\varphi}_{i+1}\right)$$

$$+ \tilde{v}_{i+1}\left(\mu S \int_{x_i}^{x_{i+1}} \dot{\varphi}_{i+1}^2 + \mu S \int_{x_{i+1}}^{x_{i+2}} \dot{\varphi}_{i+1}^2\right)$$

$$+ \tilde{v}_{i+2}\left(\mu S \int_{x_{i+1}}^{x_{i+2}} \dot{\varphi}_{i+2}\dot{\varphi}_{i+1}\right)$$

$$+ \tilde{\Omega}_i\left(-\mu S \int_{x_i}^{x_{i+1}} \varphi_i\dot{\varphi}_{i+1}\right)$$

$$+ \tilde{\Omega}_{i+1}\left(-\mu S \int_{x_i}^{x_{i+1}} \varphi_{i+1}\dot{\varphi}_{i+1} - \mu S \int_{x_{i+1}}^{x_{i+2}} \varphi_{i+1}\dot{\varphi}_{i+1}\right)$$

$$+ \tilde{\Omega}_{i+2}\left(-\mu S \int_{x_{i+1}}^{x_{i+2}} \varphi_{i+2}\dot{\varphi}_{i+1}\right)$$

$$= -\left(\int_{x_i}^{x_{i+1}} f\varphi_{i+1} + \int_{x_{i+1}}^{x_{i+2}} f\varphi_{i+1}\right). \tag{6.162}$$

Likewise, for the other two minimisation conditions:

$$\frac{\partial J}{\partial \tilde{\omega}_i}(\tilde{v}, \tilde{\Omega}) = 0 = \frac{\partial J}{\partial \tilde{\omega}_{i+1}}(\tilde{v}, \tilde{\Omega}). \tag{6.163}$$

This time we have

$$\frac{\partial J}{\partial \tilde{\omega}_i}(\tilde{v}, \tilde{\Omega}) = 0 \iff \tilde{\Omega}_{i-1}\left(EI \int_{x_{i-1}}^{x_i} \dot{\varphi}_{i-1}\dot{\varphi}_i + \mu S \int_{x_{i-1}}^{x_i} \varphi_{i-1}\varphi_i\right)$$
$$+ \tilde{\Omega}_i\left(EI \int_{x_{i-1}}^{x_i} \dot{\varphi}_i^2 + \mathbf{EI} \int_{\mathbf{x_i}}^{\mathbf{x_{i+1}}} \dot{\varphi}_\mathbf{i}^2\right.$$
$$\left. + \mu S \int_{x_{i-1}}^{x_i} \varphi_i^2 + \mu\mathbf{S} \int_{\mathbf{x_i}}^{\mathbf{x_{i+1}}} \varphi_\mathbf{i}^2\right)$$
$$+ \tilde{\Omega}_{i+1}\left(\mathbf{EI} \int_{\mathbf{x_i}}^{\mathbf{x_{i+1}}} \dot{\varphi}_\mathbf{i}\dot{\varphi}_\mathbf{i+1} + \mu\mathbf{S} \int_{\mathbf{x_i}}^{\mathbf{x_{i+1}}} \varphi_\mathbf{i}\varphi_\mathbf{i+1}\right)$$
$$+ \tilde{v}_{i-1}\left(-\mu S \int_{x_{i-1}}^{x_i} \dot{\varphi}_{i-1}\varphi_i\right)$$
$$+ \tilde{v}_i\left(-\mu S \int_{x_{i-1}}^{x_i} \dot{\varphi}_i\varphi_i - \mu\mathbf{S} \int_{\mathbf{x_i}}^{\mathbf{x_{i+1}}} \dot{\varphi}_\mathbf{i}\varphi_\mathbf{i}\right)$$
$$+ \tilde{v}_{i+1}\left(-\mu\mathbf{S} \int_{\mathbf{x_i}}^{\mathbf{x_{i+1}}} \dot{\varphi}_\mathbf{i+1}\varphi_\mathbf{i}\right)$$
$$= 0 \qquad (6.164)$$

and

$$\frac{\partial J}{\partial \tilde{\omega}_{i+1}}(\tilde{v}, \tilde{\Omega}) = 0 \iff \tilde{\Omega}_i\left(\mathbf{EI} \int_{\mathbf{x_i}}^{\mathbf{x_{i+1}}} \dot{\varphi}_\mathbf{i}\dot{\varphi}_\mathbf{i+1} + \mu\mathbf{S} \int_{\mathbf{x_i}}^{\mathbf{x_{i+1}}} \varphi_\mathbf{i}\varphi_\mathbf{i+1}\right)$$
$$+ \tilde{\Omega}_{i+1}\left(\mathbf{EI} \int_{\mathbf{x_i}}^{\mathbf{x_{i+1}}} \dot{\varphi}_\mathbf{i+1}^2 + EI \int_{x_{i+1}}^{x_{i+2}} \dot{\varphi}_{i+1}^2\right.$$
$$\left. + \mu\mathbf{S} \int_{\mathbf{x_i}}^{\mathbf{x_{i+1}}} \varphi_\mathbf{i+1}^2 + \mu S \int_{x_{i+1}}^{x_{i+2}} \varphi_{i+1}^2\right)$$
$$+ \tilde{\Omega}_{i+2}\left(EI \int_{x_{i+1}}^{x_{i+2}} \dot{\varphi}_{i+1}\dot{\varphi}_{i+2} + \mu S \int_{x_{i+1}}^{x_{i+2}} \varphi_{i+1}\varphi_{i+2}\right)$$
$$+ \tilde{v}_i\left(-\mu\mathbf{S} \int_{\mathbf{x_i}}^{\mathbf{x_{i+1}}} \dot{\varphi}_\mathbf{i}\varphi_\mathbf{i+1}\right)$$
$$+ \tilde{v}_{i+1}\left(-\mu\mathbf{S} \int_{\mathbf{x_i}}^{\mathbf{x_{i+1}}} \varphi_\mathbf{i+1}\dot{\varphi}_\mathbf{i+1} - \mu S \int_{x_{i+1}}^{x_{i+2}} \varphi_{i+1}\dot{\varphi}_{i+1}\right)$$
$$+ \tilde{v}_{i+2}\left(-\mu S \int_{x_{i+1}}^{x_{i+2}} \dot{\varphi}_{i+2}\varphi_{i+1}\right)$$
$$= 0. \qquad (6.165)$$

Note that, for each group of Eqs. (6.161) and (6.162) and (6.164) and (6.165), the various contributions relating to the finite element $[x_i, x_{i+1}]$ are indicated in bold type.

The next step is to extract the weighting coefficients of the unknowns $\tilde{v}_i$, $\tilde{v}_{i+1}$, $\tilde{\Omega}_i$, and $\tilde{\Omega}_{i+1}$ from each of these four equations and to evaluate them using the trapezoidal rule. For example, the elementary matrix $A^{(i+1)}$ corresponds to

$$
A^{(i+1)} =
\begin{array}{cccc}
\tilde{v}_i & \tilde{\Omega}_i & \tilde{v}_{i+1} & \tilde{\Omega}_{i+1} \\
\downarrow & \downarrow & \downarrow & \downarrow
\end{array}
\left[
\begin{array}{cccc}
\dfrac{\mu S}{h} & \dfrac{\mu S}{2} & -\dfrac{\mu S}{h} & \dfrac{\mu S}{2} \\[2mm]
\dfrac{\mu S}{2} & \dfrac{EI}{h}+\dfrac{\mu Sh}{2} & -\dfrac{\mu S}{2} & -\dfrac{EI}{h} \\[2mm]
-\dfrac{\mu S}{h} & -\dfrac{\mu S}{2} & \dfrac{\mu S}{h} & -\dfrac{\mu S}{2} \\[2mm]
\dfrac{\mu S}{2} & -\dfrac{EI}{h} & -\dfrac{\mu S}{2} & \dfrac{EI}{h}+\dfrac{\mu Sh}{2}
\end{array}
\right]
\begin{array}{l}
\leftarrow \tilde{v}_i \\[2mm]
\leftarrow \tilde{\Omega}_i \\[2mm]
\leftarrow \tilde{v}_{i+1} \\[2mm]
\leftarrow \tilde{\Omega}_{i+1}
\end{array}
\tag{6.166}
$$

Note that the matrix (6.166) can be rewritten in the form (6.114) if we use the notation of (6.115).

Before initiating the assembly process and in order to distinguish the contribution of each mesh element $[x_i, x_{i+1}]$ to the global matrix, we adopt the following convention, rewriting the local elementary matrix $A^{(i+1)}$ in the form

$$
A^{(i+1)} =
\begin{array}{cccc}
\tilde{v}_i & \tilde{\Omega}_i & \tilde{v}_{i+1} & \tilde{\Omega}_{i+1} \\
\downarrow & \downarrow & \downarrow & \downarrow
\end{array}
\left[
\begin{array}{cccc}
\alpha^{(i+1)} & \beta^{(i+1)} & -\alpha^{(i+1)} & \beta^{(i+1)} \\[2mm]
\beta^{(i+1)} & \delta^{(i+1)} & -\beta^{(i+1)} & \gamma^{(i+1)} \\[2mm]
-\alpha^{(i+1)} & -\beta^{(i+1)} & \alpha^{(i+1)} & -\beta^{(i+1)} \\[2mm]
\beta^{(i+1)} & \gamma^{(i+1)} & -\beta^{(i+1)} & \delta^{(i+1)}
\end{array}
\right]
\begin{array}{l}
\leftarrow \tilde{v}_i \\[2mm]
\leftarrow \tilde{\Omega}_i \\[2mm]
\leftarrow \tilde{v}_{i+1} \\[2mm]
\leftarrow \tilde{\Omega}_{i+1}
\end{array}
\tag{6.167}
$$

To avoid any confusion concerning this notation, we stress that the exponents here serve only to identify the contribution of each mesh element $[x_i, x_{i+1}]$ in the global matrix.

However, regarding the values of the coefficients of the matrix $A^{(i+1)}$, these are constants, and do not depend on the mesh element.

Put another way,

$$
\forall i = 0, \ldots, N, \quad \alpha^{(i+1)} \equiv \alpha, \quad \beta^{(i+1)} \equiv \beta, \quad \gamma^{(i+1)} \equiv \gamma, \quad \delta^{(i+1)} \equiv \delta,
\tag{6.168}
$$

where α, β, γ, and δ are defined by (6.115).

13) In this question, we assemble the global matrix describing the linear system arising from the minimisation problem $\widehat{(\mathbf{MP})}$, in which the unknowns are $\tilde{v}_1$, $\tilde{\Omega}_1, \ldots, \tilde{v}_N, \tilde{\Omega}_N$, and $\tilde{\Omega}_{N+1}$.

We are careful to note for the node at x_{N+1} that the approximate solution $\tilde{v}$ is zero at this point, whereas the solution $\tilde{\Omega}$ introduces an unknown at this point, viz., $\tilde{\Omega}_{N+1}$.

In order to build the global matrix of the linear system (6.107) and (6.108), we process each element $[x_i, x_{i+1}]$, $(i = 0, \ldots, N)$, of the mesh and fill up the global matrix step by step with the local contributions. We initiate the assembly procedure by considering the mesh element $[x_0, x_1]$.

For this first element, only the degrees of freedom of the point x_1 need be considered, since the beam is clamped at $x_0 = 0$. In other words, for this first element, when we deal with the local matrix $A^{(1)}$, only the submatrix relating to the unknowns $(\tilde{v}_1, \tilde{\Omega}_1)$ need be taken into consideration.

This submatrix is shown in bold type in the elementary matrix $A^{(1)}$, as follows:

$$
A^{(1)} =
\begin{bmatrix}
\alpha^{(1)} & \beta^{(1)} & -\alpha^{(1)} & \beta^{(1)} \\
\beta^{(1)} & \delta^{(1)} & -\beta^{(1)} & \gamma^{(1)} \\
-\alpha^{(1)} & -\beta^{(1)} & \boldsymbol{\alpha^{(1)}} & \boldsymbol{-\beta^{(1)}} \\
\beta^{(1)} & \gamma^{(1)} & \boldsymbol{-\beta^{(1)}} & \boldsymbol{\delta^{(1)}}
\end{bmatrix}
\begin{matrix} \\ \\ \leftarrow \tilde{v}_1 \\ \leftarrow \tilde{\Omega}_1 \end{matrix}
\qquad (6.169)
$$

(columns headed by $\tilde{v}_1 \downarrow \quad \tilde{\Omega}_1 \downarrow$)

We continue now with the second mesh element $[x_1, x_2]$. In this case, the elementary matrix $A^{(2)}$ contributes fully, since two degrees of freedom $(\tilde{v}_1, \tilde{\Omega}_1)$ and two degrees of freedom $(\tilde{v}_2, \tilde{\Omega}_2)$ are associated with the nodes x_1 and x_2, respectively.

Maintaining the convention of using bold type for the coefficients of the matrix $A^{(2)}$ that should be taken into account in the assembly process, we obtain

$$
A^{(2)} =
\begin{bmatrix}
\boldsymbol{\alpha^{(2)}} & \boldsymbol{\beta^{(2)}} & \boldsymbol{-\alpha^{(2)}} & \boldsymbol{\beta^{(2)}} \\
\boldsymbol{\beta^{(2)}} & \boldsymbol{\delta^{(2)}} & \boldsymbol{-\beta^{(2)}} & \boldsymbol{\gamma^{(2)}} \\
\boldsymbol{-\alpha^{(2)}} & \boldsymbol{-\beta^{(2)}} & \boldsymbol{\alpha^{(2)}} & \boldsymbol{-\beta^{(2)}} \\
\boldsymbol{\beta^{(2)}} & \boldsymbol{\gamma^{(2)}} & \boldsymbol{-\beta^{(2)}} & \boldsymbol{\delta^{(2)}}
\end{bmatrix}
\begin{matrix} \leftarrow \tilde{v}_1 \\ \leftarrow \tilde{\Omega}_1 \\ \leftarrow \tilde{v}_2 \\ \leftarrow \tilde{\Omega}_2 \end{matrix}
\qquad (6.170)
$$

(columns headed by $\tilde{v}_1 \downarrow \quad \tilde{\Omega}_1 \downarrow \quad \tilde{v}_2 \downarrow \quad \tilde{\Omega}_2 \downarrow$)

This situation extends to all the other elements $[x_i, x_{i+1}]$ for all values of i from 2 to N. In other words, for all N of these elements, the elementary matrix $A^{(i+1)}$ contributes fully. Using the same notational conventions, we have

$$
\begin{array}{cccc}
\tilde{v}_i & \tilde{\Omega}_i & \tilde{v}_{i+1} & \tilde{\Omega}_{i+1} \\
\downarrow & \downarrow & \downarrow & \downarrow
\end{array}
$$

$$
A^{(i+1)} = \begin{bmatrix}
\alpha^{(i+1)} & \beta^{(i+1)} & -\alpha^{(i+1)} & \beta^{(i+1)} \\
\beta^{(i+1)} & \delta^{(i+1)} & -\beta^{(i+1)} & \gamma^{(i+1)} \\
-\alpha^{(i+1)} & -\beta^{(i+1)} & \alpha^{(i+1)} & -\beta^{(i+1)} \\
\beta^{(i+1)} & \gamma^{(i+1)} & -\beta^{(i+1)} & \delta^{(i+1)}
\end{bmatrix}
\begin{array}{l}
\leftarrow \tilde{v}_i \\
\leftarrow \tilde{\Omega}_i \\
\leftarrow \tilde{v}_{i+1} \\
\leftarrow \tilde{\Omega}_{i+1}
\end{array}
\qquad (6.171)
$$

We now proceed to incorporate all the elementary contributions from each finite element into the global matrix. For a comfortable visualisation of the global matrix A, we provide here three snapshots: top left, centre, and bottom right of the matrix.

Top Left Corner of the Rigidity Matrix

$$
\begin{array}{cccccc}
\tilde{v}_1 & \tilde{\Omega}_1 & \tilde{v}_2 & \tilde{\Omega}_2 & \cdots & \tilde{\Omega}_{N+1} \\
\downarrow & \downarrow & \downarrow & \downarrow & & \downarrow
\end{array}
$$

$$
A = \begin{bmatrix}
\alpha^{(1)}+\alpha^{(2)} & -\beta^{(1)}+\beta^{(2)} & -\alpha^{(2)} & \beta^{(2)} & \cdots & \cdots \\
-\beta^{(1)}+\beta^{(2)} & \delta^{(1)}+\delta^{(2)} & -\beta^{(2)} & \gamma^{(2)} & \cdots & \cdots \\
-\alpha^{(2)} & -\beta^{(2)} & \alpha^{(2)}+\alpha^{(3)} & -\beta^{(2)}+\beta^{(3)} & \cdots & \cdots \\
\beta^{(2)} & \gamma^{(2)} & -\beta^{(2)}+\beta^{(3)} & \delta^{(2)}+\delta^{(3)} & \cdots & \cdots \\
\cdots & \cdots & \cdots & \cdots & \cdots & \cdots \\
\cdots & \cdots & \cdots & \cdots & \cdots & \cdots
\end{bmatrix}
\begin{array}{l}
\leftarrow \tilde{v}_1 \\
\leftarrow \tilde{\Omega}_1 \\
\leftarrow \tilde{v}_2 \\
\leftarrow \tilde{\Omega}_2 \\
\\
\leftarrow \tilde{\Omega}_{N+1}
\end{array}
$$

$$(6.172)$$

Generic Centre of the Rigidity Matrix

$$
\begin{array}{cccccc}
\cdots & \tilde{v}_i & \tilde{\Omega}_i & \tilde{v}_{i+1} & \tilde{\Omega}_{i+1} & \cdots \\
& \downarrow & \downarrow & \downarrow & \downarrow &
\end{array}
$$

$$
A = \begin{bmatrix}
\cdots & \cdots & \cdots & \cdots & \cdots & \cdots \\
\cdots & \alpha^{(i)}+\alpha^{(i+1)} & -\beta^{(i)}+\beta^{(i+1)} & -\alpha^{(i+1)} & \beta^{(i+1)} & \cdots \\
\cdots & -\beta^{(i)}+\beta^{(i+1)} & \delta^{(i)}+\delta^{(i+1)} & -\beta^{(i+1)} & \gamma^{(i+1)} & \cdots \\
\cdots & -\alpha^{(i+1)} & -\beta^{(i+1)} & \alpha^{(i+1)}+\alpha^{(i+2)} & -\beta^{(i+1)}+\beta^{(i+2)} & \cdots \\
\cdots & \beta^{(i+1)} & \gamma^{(i+1)} & -\beta^{(i+1)}+\beta^{(i+2)} & \delta^{(i+1)}+\delta^{(i+2)} & \cdots \\
\cdots & \cdots & \cdots & \cdots & \cdots & \cdots
\end{bmatrix}
\begin{array}{l}
\cdots \\
\leftarrow \tilde{v}_i \\
\leftarrow \tilde{\Omega}_i \\
\leftarrow \tilde{v}_{i+1} \\
\leftarrow \tilde{\Omega}_{i+1} \\
\cdots
\end{array}
$$

$$(6.173)$$

Bottom Right Corner of the Rigidity Matrix

$$
\mathbf{A} =
\begin{array}{c}
\quad \tilde{v}_N \qquad\qquad \tilde{\Omega}_N \qquad\quad \tilde{\Omega}_{N+1} \\
\quad \downarrow \qquad\qquad \downarrow \qquad\qquad \downarrow \\
\left[
\begin{array}{cccc}
\cdots & \cdots & \cdots & \cdots \\
\cdots & \cdots & \cdots & \cdots \\
\cdots & \alpha^{(N)}+\alpha^{(N+1)} & -\beta^{(N)}+\beta^{(N+1)} & \beta^{(N+1)} \\
\cdots & -\beta^{(N)}+\beta^{(N+1)} & \delta^{(N)}+\delta^{(N+1)} & \gamma^{(N+1)} \\
\cdots & \beta^{(N+1)} & \gamma^{(N+1)} & \delta^{(N+1)}
\end{array}
\right]
\begin{array}{l}
\cdots \\
\cdots \\
\leftarrow \tilde{v}_N \\
\leftarrow \tilde{\Omega}_N \\
\leftarrow \tilde{\Omega}_{N+1}
\end{array}
\end{array}
\qquad (6.174)
$$

We end the assembly process by carrying out all the algebraic calculations, taking into account the sequences α_i, β_i, γ_i, and δ_i, as indicated in (6.168). We then obtain the rigidity matrix in its final form:

$$
\mathbf{A} =
\begin{array}{c}
\tilde{v}_1 \; \tilde{\Omega}_1 \; \tilde{v}_2 \; \tilde{\Omega}_2 \; \tilde{v}_3 \; \tilde{\Omega}_3 \; \tilde{v}_4 \; \tilde{\Omega}_4 \quad \cdots \quad \tilde{v}_N \; \tilde{\Omega}_N \; \tilde{\Omega}_{N+1} \\
\downarrow \; \downarrow \; \downarrow \; \downarrow \; \downarrow \; \downarrow \; \downarrow \; \downarrow \qquad\quad \downarrow \;\; \downarrow \;\; \downarrow \\
\left[
\begin{array}{ccccccccccc}
2\alpha & 0 & -\alpha & \beta & & & & & & & \\
0 & 2\delta & -\beta & \gamma & 0 & & & & \cdots & & \\
-\alpha & -\beta & 2\alpha & 0 & -\alpha & \beta & & & \cdots & & \\
\beta & \gamma & 0 & 2\delta & -\beta & \gamma & 0 & & \cdots & & \\
& 0 & -\alpha & -\beta & 2\alpha & 0 & -\alpha & \beta & \cdots & & \\
& \beta & \gamma & 0 & 2\delta & -\beta & \gamma & \cdots & & \\
& & 0 & -\alpha & -\beta & 2\alpha & 0 & \cdots & & \\
& & \beta & \gamma & 0 & 2\delta & \cdots & & \\
\cdots & \cdots & \cdots & \cdots & \cdots & \cdots & \cdots & \cdots & \cdots & \cdots & \cdots \\
& & & & & & & \cdots & 2\alpha & 0 & \beta \\
& & & & & & & \cdots & 0 & 2\delta & \gamma \\
& & & & & & & \cdots & \beta & \gamma & \delta
\end{array}
\right]
\begin{array}{l}
\leftarrow \tilde{v}_1 \\
\leftarrow \tilde{\Omega}_1 \\
\leftarrow \tilde{v}_2 \\
\leftarrow \tilde{\Omega}_2 \\
\leftarrow \tilde{v}_3 \\
\leftarrow \tilde{\Omega}_3 \\
\leftarrow \tilde{v}_4 \\
\leftarrow \tilde{\Omega}_4 \\
\cdots \\
\leftarrow \tilde{v}_N \\
\leftarrow \tilde{\Omega}_N \\
\leftarrow \tilde{\Omega}_{N+1}
\end{array}
\end{array}
$$

$$(6.175)$$

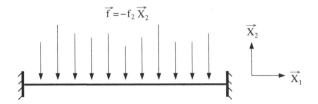

Fig. 6.7 Doubly clamped beam under bending load

6.3 Doubly Clamped Beam: Euler–Bernoulli Theory

6.3.1 Statement of the Problem

▶ **Theoretical Part: Variational Formulation of the Displacement Field and Principle of Virtual Work**

We consider a uniform beam Ω of length L, cross-section S, and density ρ, with isotropic elastic behaviour under small perturbations. The beam is made of a material with Young's modulus E and Poisson coefficient ν.

The axis of the beam is denoted by $(O; \mathbf{X}_1)$ and coordinatised by x. The cross-section of the beam is coordinatised by the geometric variables (y, z).

The beam is clamped at both ends, i.e., at $x = 0$ and $x = L$. In addition, a transverse force field $\mathbf{f} = -f_2 \mathbf{X}_2$ per unit volume is applied along the beam as shown in Fig. 6.7. The density f_2 is given and has all the necessary functional regularity to ensure that we may carry out any integrals arising in the first two questions of the theoretical part.

1) Here we use the Euler–Bernoulli kinematic theory to describe the real displacement field. We thus consider the virtual displacement fields $\mathbf{U}^*$ in the form

$$\mathbf{U}^* = -y\dot{u}^*(x)\mathbf{X}_1 + u^*(x)\mathbf{X}_2, \tag{6.176}$$

satisfying the boundary conditions

$$u^*(0) = u^*(L) = 0, \tag{6.177}$$

$$\dot{u}^*(0) = \dot{u}^*(L) = 0, \tag{6.178}$$

where we have used the notation $\dot{u}^* \equiv \mathrm{d}u^*/\mathrm{d}x$.

Furthermore, (6.177) and (6.178) express the clamping condition for all virtual fields $\mathbf{U}^*$. Let $\mathfrak{U}^*$ denote the set of all displacement fields satisfying the three conditions (6.176)–(6.178).

Determine the virtual work T^*_{int} done by all internal forces and the virtual work T^*_{ext} done by all external forces, these being defined by

$$T^*_{\text{int}} = -\int_{\Omega} \sigma_{ij} \varepsilon_{ij} (\mathbf{U}^*) \, d\Omega, \qquad T^*_{\text{ext}} = \int_{\Omega} \mathbf{f} . \mathbf{U}^* \, d\Omega, \qquad (6.179)$$

where σ is the stress tensor and $\varepsilon(\mathbf{U}^*)$ is the linear strain tensor associated with the virtual displacement field $\mathbf{U}^*$, defined by:

$$\varepsilon_{ij} (\mathbf{U}^*) = \frac{1}{2} \left(\frac{\partial U^*_i}{\partial x_j} + \frac{\partial U^*_j}{\partial x_i} \right). \qquad (6.180)$$

We define the bending moment M and the surface force density f by

$$M(x) = -\iint_{S(x)} y \sigma_{11} \, dS(x), \qquad f(x) = \iint_{S(x)} f_2 \, dS(x), \qquad (6.181)$$

where $S(x)$ is the cross-section at x.

2) Without trying to justify the convergence of the integrals, show that an application of the principle of virtual work leads to the following formal variational formulation **(EVP)**:

$$\textbf{(EVP)} \quad \begin{cases} \text{Find real-valued } M \text{ on } [0, L] \text{ solution of} \\[2mm] \displaystyle\int_0^L \left(\ddot{M}(x) + f \right) u^*(x) \, dx = 0, \quad \forall u^* \in \mathfrak{U}^*. \end{cases} \qquad (6.182)$$

3) Using the constitutive law of the material making up the beam, viz.,

$$M(x) = EI\ddot{u}(x),$$

where I is the moment of inertia of the beam about the axis $(O; \mathbf{X}_3)$, and assuming that the force density f is in $L^2(0, L)$, show that the continuous problem **(CP)** comprising the equilibrium equations of the beam can be expressed as follows:

Find $u \in H^4(0, L)$ solution of

$$\textbf{(CP)} \quad \begin{cases} -EI u^{(4)}(x) = f(x), \quad \forall x \in \,]0, 1[, \\[1mm] u(0) = u(L) = 0, \\[1mm] \dot{u}(0) = \dot{u}(L) = 0, \end{cases} \qquad (6.183)$$

where the Sobolev space $H^4(0, L)$ is defined by

$$H^4(0, L) = \left\{ v : [0, L] \to \mathbb{R}, \ v^{(k)} \in L^2(0, L), \ k = 0, \ldots, 4 \right\}. \qquad (6.184)$$

4) Let $V \equiv H_0^2(0, L)$ be the Sobolev space defined by

$$H_0^2(0, L) = \left\{ v^{(k)} \in L^2(0, L), \quad k = 0, 1, 2, \quad \text{s.t.} \quad v(0) = \dot{v}(0) = v(L) = \dot{v}(L) = 0 \right\}.$$
(6.185)

Show that the displacement variational formulation (**VP**) associated with the continuous problem (**CP**) can be expressed as follows:

$$(\mathbf{VP}) \quad \begin{cases} \text{Find } u \in V \text{ solution of} \\ EI \int_0^L \ddot{u}(x)\ddot{v}(x)\,dx + \int_0^L f(x)v(x)\,dx = 0, \quad \forall v \in V. \end{cases}$$
(6.186)

5) Show that there is a minimisation problem (**MP**) equivalent to the variational formulation (**VP**), defined by

$$(\mathbf{MP}) \quad \begin{cases} \text{Find } u \in V \text{ solution of } J(u) = \min_{v \in V} J(v), \\ \text{where } J(v) = \dfrac{EI}{2} \int_0^L \ddot{v}^2(x)\,dx + \int_0^L f(x)v(x)\,dx. \end{cases}$$
(6.187)

▶ Numerical Part

This part deals with the finite-element analysis of elastic beams in bending, as modelled by the Euler–Bernoulli theory. To this end, we shall implement a Hermitian finite-element analysis as follows.

We begin with a regular mesh on the interval $[0, L]$, with constant spacing h, such that

$$\begin{cases} x_0 = 0, \quad x_{N+1} = L, \\ x_{i+1} = x_i + h, \quad i = 0, \dots, N. \end{cases}$$
(6.188)

In addition, define the approximation space $\widetilde{W}$ by

$$\widetilde{W} = \left\{ \tilde{w} : [0, L] \longrightarrow \mathbb{R}, \text{ s.t. } \tilde{w} \in C^1([0, L]), \quad \tilde{w}|_{[x_i, x_{i+1}]} \in P_3([x_i, x_{i+1}]) \right\},$$
(6.189)

where $P_3([x_i, x_{i+1}])$ is the space of polynomials on $[x_i, x_{i+1}]$ of degree less than or equal to 3.

Likewise, we define the space $\widetilde{V}$ by

$$\widetilde{V} = \left\{ \tilde{v} \in \widetilde{W}, \text{ such that } \tilde{v}(0) = \dot{\tilde{v}}(0) = \tilde{v}(L) = \dot{\tilde{v}}(L) = 0 \right\}.$$
(6.190)

6) What are the dimensions of the spaces $\widetilde{W}$ and $\widetilde{V}$?

7) We consider the system of functions $(\varphi_i)_{i=1, 2N}$ in $\widetilde{V}$, divided into two groups in the following way:

$$\dot{\varphi}_{2k}(x_j) = \delta_{ij}, \qquad \varphi_{2k}(x_j) = 0, \quad \forall\, j = 1, \ldots, N, \tag{6.191}$$

$$\varphi_{2k+1}(x_j) = \delta_{ij}, \quad \dot{\varphi}_{2k+1}(x_j) = 0, \quad \forall\, j = 1, \ldots, N. \tag{6.192}$$

– What is the support of the functions $(\varphi_i)_{i=1, \ldots, 2N}$?

– Show that the $2N$ functions $(\varphi_i)_{i=1, 2N}$ form a basis for the space $\widetilde{V}$.

8) Define $\tilde{v} \in \widetilde{V}$ by

$$\tilde{v} = \sum_{k=1}^{N} \tilde{\alpha}_k \varphi_{2k} + \sum_{k=1}^{N} \tilde{\beta}_k \varphi_{2k-1}. \tag{6.193}$$

– Suggest a mechanical interpretation of the coefficients $\tilde{\alpha}_k$ and $\tilde{\beta}_k$.

– Write down the restriction of the approximation $\tilde{v}$ to a mesh element $[x_i, x_{i+1}]$.

– Express the 4 basis functions $\varphi_{2i-1}, \varphi_{2i}, \varphi_{2i+1}$, and φ_{2i+2} on the element $[x_i, x_{i+1}]$.

9) Let $\widetilde{(\mathbf{MP})}$ be the approximate minimisation problem associated with the problem $(\mathbf{MP})$, defined by

$$\widetilde{(\mathbf{MP})} \qquad \text{Find } \tilde{u} \in \widetilde{V} \text{ solution of } J(\tilde{u}) = \min_{\tilde{v} \in \widetilde{V}} J(\tilde{v}), \tag{6.194}$$

where J is the potential energy defined by (6.187).

Show that a necessary condition for $\tilde{u}$ to be a solution to the problem $\widetilde{(\mathbf{MP})}$ can be expressed as follows:

Determine the two sequences of components $(\tilde{u}_i)_{1, \ldots, N}$ and $(\dot{\tilde{u}}_i)_{1, \ldots, N}$ specifying the approximation $\tilde{u} \in \widetilde{V}$ which solves the global linear system:

$$EI \sum_{k=1}^{N} \left(\int_{S_{2i-1, 2k}} \ddot{\varphi}_{2i-1} \ddot{\varphi}_{2k} \, \mathrm{d}x \right) \dot{\tilde{u}}_k + \sum_{k=1}^{N} \left(\int_{S_{2i-1, 2k-1}} \ddot{\varphi}_{2i-1} \ddot{\varphi}_{2k-1} \, \mathrm{d}x \right) \tilde{u}_k$$

$$= -\int_{S_{2i-1}} f \varphi_{2i-1} \, \mathrm{d}x, \tag{6.195}$$

$$EI \sum_{k=1}^{N} \left(\int_{S_{2i, 2k}} \ddot{\varphi}_{2i} \ddot{\varphi}_{2k} \, \mathrm{d}x \right) \dot{\tilde{u}}_k + \sum_{k=1}^{N} \left(\int_{S_{2i, 2k-1}} \ddot{\varphi}_{2i} \ddot{\varphi}_{2k-1} \, \mathrm{d}x \right) \tilde{u}_k = -\int_{S_{2i}} f \varphi_{2i} \, \mathrm{d}x,$$

$$\tag{6.196}$$

for all $i = 1, \ldots, N$, where

$$S_{l,m} \equiv \text{supp } \varphi_l \cap \text{supp } \varphi_m, \qquad S_m \equiv \text{supp } \varphi_m. \tag{6.197}$$

10) We now introduce the elementary matrix $a^{(i+1)}$ which expresses the contribution of the segment $[x_i, x_{i+1}]$, i.e., the $(i+1)$ th element of the mesh on the interval $[0, L]$, to the global matrix for the linear system (6.195) and (6.196).

We have

$$a^{(i+1)} = \frac{2EI}{h^3} \begin{bmatrix} 6 & 3h & -6 & 3h \\ 3h & 2h^2 & -3h & h^2 \\ -6 & -3h & 6 & -3h \\ 3h & h^2 & -3h & 2h^2 \end{bmatrix}. \tag{6.198}$$

We also introduce the elementary vector $b^{(i+1)}$ defined by

$$b^{(i+1)} = \begin{bmatrix} b_1^{(i+1)} \\ b_2^{(i+1)} \\ b_3^{(i+1)} \\ b_4^{(i+1)} \end{bmatrix} \equiv \begin{bmatrix} -\int_{x_i}^{x_{i+1}} f\varphi_{2i-1}\, dx \\ -\int_{x_i}^{x_{i+1}} f\varphi_{2i}\, dx \\ -\int_{x_i}^{x_{i+1}} f\varphi_{2i+1}\, dx \\ -\int_{x_i}^{x_{i+1}} f\varphi_{2i+2}\, dx \end{bmatrix}. \tag{6.199}$$

Using Simpson's rule, suggest an approximation for the vector $b^{(i+1)}$. Recall that Simpson's rule is

$$\int_a^b f(x)dx \approx \frac{b-a}{6} \left[f(a) + 4f\left(\frac{a+b}{2}\right) + f(b) \right]. \tag{6.200}$$

Show that, in the special case of a uniformly distributed load p,

$$\mathbf{f} = -f_2 \mathbf{X}_2 \equiv -p\mathbf{X}_2,$$

we recover the standard result $b_*^{(i+1)}$ defined by

$${}^t b_*^{(i+1)} = \left[-\frac{ph}{2}, -\frac{ph^2}{12}, \frac{ph}{2}, \frac{ph^2}{12} \right]. \tag{6.201}$$

11) Consider the case where the mesh on the beam Ω comprises 3 elements of equal length h.

– Assemble the global matrix A with generic element A_{ij} and the vector b on the right-hand side, with component b_i, for the problem $\widetilde{\textbf{(MP)}}$.

– Deduce the form of the approximation of the displacement field at each point of the beam Ω.

– Do we recover the node equations obtained by the standard approximation technique applying Hermitian finite-element analysis to the variational formulation **(VP)**?

6.3.2 Solution

▶ **Theoretical Part: Variational Formulation of the Displacement Field and Principle of Virtual Work**

1) The structure of the virtual displacement field $\mathbf{U}^*$ defined by (6.176) implies that the strain tensor $\varepsilon(\mathbf{U}^*)$ [see (6.180)] associated with $\mathbf{U}^*$ can be written in the form

$$\varepsilon(\mathbf{U}^*) = \varepsilon_{11}^* \mathbf{X}_1 \otimes \mathbf{X}_1 = -y\ddot{u}^*(x)\mathbf{X}_1 \otimes \mathbf{X}_1, \tag{6.202}$$

where we use the tensor notation $\mathbf{X}_1 \otimes \mathbf{X}_1$, reflecting the fact that the tensor $\varepsilon(\mathbf{U}^*)$ has only one nonzero component, viz., ε_{11}^*.

The virtual work T_{int}^* done by internal forces can be evaluated as follows:

$$T_{\text{int}}^* = -\int_{\Omega} \sigma_{ij}\varepsilon_{ij}(\mathbf{U}^*)\,d\Omega = -\int_{\Omega} \sigma_{11}\varepsilon_{11}(\mathbf{U}^*)\,d\Omega \tag{6.203}$$

$$= -\int_0^L \left[-\iint_{S(x)} y\sigma_{11}\,dS(x) \right] \ddot{u}^*(x)\,dx = -\int_0^L M(x)\ddot{u}^*(x)\,dx,$$

where we used the definition (6.181) of the bending moment M.

The virtual work done by external forces can be obtained similarly:

$$T_{\text{ext}}^* = \int_{\Omega} \mathbf{f} \cdot \mathbf{U}^*\,d\Omega = -\int_{\Omega} f_2 u^*(x)\,d\Omega \tag{6.204}$$

$$= -\int_0^L \left[\iint_{S(x)} f_2\,dS(x) \right] u^*(x)\,dx = -\int_0^L f(x)u^*(x)\,dx.$$

2) We now apply the principle of virtual work in the context of an elastostatic problem:

$$T_{\text{int}}^* + T_{\text{ext}}^* = 0, \quad \forall u^* \in \mathfrak{U}^*. \tag{6.205}$$

Then using (6.203) and (6.204), we find that

$$\int_0^L M(x)\ddot{u}^*(x)\,dx + \int_0^L f(x)u^*(x)\,dx = 0, \quad \forall u^* \in \mathfrak{U}^*. \tag{6.206}$$

Integrating by parts twice, we transform the first integral of (6.206) to obtain

$$\int_0^L \left[\ddot{M}(x) + f(x) \right] u^*(x)\,dx - \left[\dot{M}u^* \right]_0^L + \left[M\dot{u}^* \right]_0^L = 0, \quad \forall u^* \in \mathfrak{U}^*. \tag{6.207}$$

With the boundary conditions (6.177) and (6.178) on the virtual fields in $\mathfrak{U}^*$, we obtain the following formal variational formulation **(EVP)**:

(EVP) $\begin{cases} \text{Find real-valued } M \text{ on } [0, L] \text{ solution of} \\ \displaystyle\int_0^L \left[\ddot{M}(x) + f(x)\right] u^*(x) dx = 0, \quad \forall u^* \in \mathfrak{U}^*. \end{cases}$ \hfill (6.208)

3) The continuous problem **(CP)** is obtained using density arguments of the kind already discussed on several occasions.

For this purpose, we return to the formulation (6.208) of the variational equation **(EVP)** and replace the bending moment M by its expression in terms of the solution displacement field u with the help of the constitutive law given in the statement of the problem.

We thereby obtain

$$\int_0^L \left[EI u^{(4)}(x) + f(x)\right] u^*(x) \, dx = 0, \quad \forall u^* \in \mathfrak{U}^*. \tag{6.209}$$

To guarantee that the integral in the variational formulation (6.209) makes mathematical sense, we seek the solution u in the space $H^4(0, L)$ defined by (6.184).

Indeed, given that the data f belongs to $L^2(0, L)$, if we require the fourth derivative $u^{(4)}$ and the virtual field u^* to belong to this same space $L^2(0, L)$, this ensures the existence of the integral in (6.209). In addition to these regularity conditions, we include the constraints imposed by the boundary conditions on u and u^*.

We thus consider solutions u and virtual fields u^* belonging to the Sobolev space $H_*^4([0; L])$ defined by

$$H_*^4(0, L) \equiv H^4(0, L) \cap \mathfrak{U}^*. \tag{6.210}$$

The variational formulation **(EVP)** then becomes

(EVP) $\begin{cases} \text{Find } u \in H_*^4(0, L) \text{ solution of} \\ \displaystyle\int_0^L \left[EI u^{(4)}(x) + f(x)\right] u^*(x) \, dx = 0, \quad \forall u^* \in H_*^4(0, L). \end{cases}$ \hfill (6.211)

The continuous problem **(CP)** is found by a density argument. Among all the functions of $H_*^4(0, L)$, we consider those belonging to $\mathscr{D}(0, L)$, which is a subset of $H_*^4(0, L)$, and we use the fact that $\mathscr{D}(0, L)$ is dense in $L^2(0, L)$.

The reader is referred to the demonstration of the density method discussed in Sect. 6.1.1 for the beam in traction. It is then a straightforward matter to show that

$$EI u^{(4)}(x) + f(x) = 0, \quad \forall x \in (0, 1). \tag{6.212}$$

Summary. We have considered the solution u of the variational equation (6.211) in the space $H_*^4(0, L)$ and we have established by a density argument that u satisfies the differential equation (6.212), while satisfying the homogeneous Dirichlet and Neumann boundary conditions at $x = 0$ and $x = L$, which are properties of the functions in $H_*^4(0, L)$ at the endpoints of the interval $[0, L]$.

In other words, u solves the continuous problem **(CP)** defined by (6.183).

Finally, we note that, when the force distribution f has greater regularity, let us say, at least continuity on the interval $[0, L]$, then the solution u of the continuous problem **(CP)** is the standard solution in $C^4([0, 1])$.

4) In order to establish the variational problem **(VP)** associated with the continuous problem **(CP)**, we consider a function v in a variational space V which remains to be defined.

We multiply the fourth order differential equation (6.183) by v and carry out two integrations by parts to yield the following formulation:

$$EI \int_0^L \left(\ddot{u}\ddot{v} + fv \right) dx + \left[u^{(3)}v \right]_0^L - \left[\ddot{u}\dot{v} \right]_0^L = 0, \quad \forall v \in V. \tag{6.213}$$

Let us now consider the function space V required for the formulation (6.213). We begin by observing that, following the two integrations by parts, the variational equation (6.213) does not explicitly involve the values of the solution u or its derivative $\dot{u}$ at $x = 0$ or $x = L$.

So in order to retain this information in the variational formulation **(VP)**, we require the functions v in V (the search space for the solution u) to satisfy the following boundary conditions:

$$v(0) = \dot{v}(0) = v(L) = \dot{v}(L) = 0. \tag{6.214}$$

Since the future solution u is one of the functions v in V, it will *ipso facto* satisfy the double clamping boundary conditions on the beam Ω.

As a consequence, (6.213) becomes

$$EI \int_0^L \left(\ddot{u}\ddot{v} + fv \right) dx = 0, \quad \forall v \in V. \tag{6.215}$$

Regarding the regularity of the functions v of V, we consider the Sobolev space $H^2(0, L)$ in order to ensure convergence of the two integrals in (6.215). We thus find the variational formulation **(VP)** specified in (6.186).

5) The variational formulation **(VP)** defined in (6.186) involves a bilinear form $a(., .)$ and a linear form $L(.)$:

$$a(u, v) = EI \int_0^L \ddot{u}\ddot{v}\, dx , \qquad L(v) = -\int_0^L fv\, dx, \tag{6.216}$$

where u and v range over $H_0^2(0, L)$. The variational formulation can thus be expressed in the generic form:

$$\text{Find } u \in V \text{ solution of } a(u, v) = L(v), \quad \forall v \in V. \tag{6.217}$$

It is also clear that the bilinear form a is symmetric and positive. These properties ensure the existence of an equivalent minimisation problem defined by (MP) (see Theorem 1.13).

▶ **Numerical Part**

6) The space $\widetilde{W}$ has dimension $2N + 4$ and the space $\widetilde{V}$ has dimension $2N$. Indeed, in order to determine the dimension of $\widetilde{W}$, we simply count the number of degrees of freedom characterising any function $\tilde{w}$ belonging to $\widetilde{W}$.

Now any such function is a third degree polynomial on each element $[x_i, x_{i+1}]$. We thus have 4 degrees of freedom per mesh element $[x_i, x_{i+1}]$, which makes a total of $4(N + 1)$ degrees of freedom over all $(N + 1)$ elements.

It remains to reduce this as a result of the continuity conditions at the junctions between adjacent mesh elements, not only for the function $\tilde{u}$, but also for its derivative $\dot{\tilde{u}}$, given that the functions in $\widetilde{W}$ are C^1 on the interval $[0, L]$.

So for the N junction points $x_1, \ldots, x_N$, there are a total of $2N$ continuity conditions for $\tilde{w}$ and $\dot{\tilde{w}}$. Finally, any function $\tilde{w}$ has $4(N + 1) - 2N$ degrees of freedom, making a total of $2N + 4$, whence dim $\widetilde{W} = 2N + 4$.

The dimension of $\widetilde{V}$ follows immediately, since we only need to take into account the zero values of $\tilde{v}$ and $\dot{\tilde{v}}$ at $x = 0$ and $x = L$, reducing the number of degrees of freedom of any function $\tilde{v}$ in $\widetilde{V}$ by 4. The dimension of $\widetilde{V}$ is therefore $2N$.

7) Let φ_i be one of the $2N$ functions of $\widetilde{V}$ defined by the conditions (6.191) and (6.192).

Supports of Functions φ_i. For any interval $[x_k, x_{k+1}]$ different from $[x_{i-1}, x_i]$ or $[x_i, x_{i+1}]$, φ_i is a function that will vanish along with its derivative $\dot{\varphi}_i$ at the points x_k and x_{k+1}.

Indeed, given that the function φ_i is a polynomial of degree less than or equal to 3 on the interval $[x_k, x_{k+1}]$, this polynomial is necessarily zero on the whole interval. To see this, assume that such a function could be nonzero somewhere on the interval $[x_k, x_{k+1}]$.

In this case, given the boundary conditions at x_k and x_{k+1}, this function would have to look like one of those illustrated in Fig. 6.8. We thus observe that the graph of φ_i must necessarily have two or three points of inflection, which is impossible for a third degree polynomial.

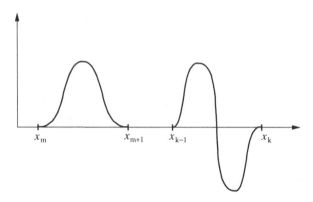

Fig. 6.8 Counterexamples for the argument concerning the supports of the functions φ_i

A more analytical demonstration would apply Rolle's theorem, using the fact that $\varphi_i(x_k) = \varphi_i(x_{k+1}) = 0$ and $\dot{\varphi}_i(x_k) = \dot{\varphi}_i(x_{k+1}) = 0$. In any case, it follows that the support of the function φ_i is the union of the intervals $[x_{i-1}, x_i]$ and $[x_i, x_{i+1}]$.

The System of $2N$ **Functions** $(\varphi_i)_{i=1, \ldots, 2N}$ **in** $\widetilde{V}$. To establish that the $2N$ functions $(\varphi_i)_{i=1, \ldots, 2N}$ form a basis for $\widetilde{V}$, we shall show that they form an independent set of $2N$ elements in the vector space of dimension $2N$.

To do this, we consider two sequences of real coefficients $(\xi_i)_{i=1, \ldots, N}$ and $(\eta_i)_{i=1, \ldots, N}$ and consider the linear combination

$$\sum_{k=1}^{N} \xi_k \varphi_{2k} + \sum_{k=1}^{N} \eta_k \varphi_{2k-1} = 0. \tag{6.218}$$

Evaluating (6.218) at the nodes x_i, we thus find that the coefficients η_i must all vanish. In addition, differentiating (6.218) and evaluating once again at the points x_i, we find that the coefficients ξ_i must also all vanish.

Note that this result follows essentially from the properties (6.191) and (6.192) of the functions φ_i and their derivatives $\dot{\varphi}_i$. The family of $2N$ functions φ_i is thus independent, and must therefore generate a vector space of dimension $2N$, and this must be the whole space $\widetilde{V}$.

8) Here we describe the Hermitian finite-element approximation. The aim is to interpolate the displacement field $\tilde{u}$ and its derivative $\dot{\tilde{u}}$, which represents the rotation of the cross-section in the Euler–Bernoulli theory, while in the natural theory, the rotation θ of the cross-section of the beam is an independent function of the displacement field **U**.

Mechanical Interpretation of the Coefficients $\tilde{\alpha}_k$ **and** $\tilde{\beta}_k$. Any function $\tilde{v}$ in $\widetilde{V}$ can be expanded in terms of the two families of basis functions $(\varphi_{2k})_{k=1, \ldots, N}$ and $(\varphi_{2k-1})_{k=1, \ldots, N}$ to give

$$\tilde{v} = \sum_{k=1}^{N} \tilde{\alpha}_k \varphi_{2k} + \sum_{k=1}^{N} \tilde{\beta}_k \varphi_{2k-1}. \tag{6.219}$$

If we now evaluate the decomposition (6.219) at the point x_i and take into account the properties (6.191) and (6.192) of the basis function, we find that

$$\tilde{v}(x_i) = \tilde{\beta}_i. \tag{6.220}$$

In the same spirit, if we differentiate (6.219) and evaluate the derivative of $\tilde{v}$ at the point x_i, we obtain

$$\dot{\tilde{v}}(x_i) = \tilde{\alpha}_i. \tag{6.221}$$

For this reason, the coefficients α_i and β_i are interpreted as the approximation of the derivative $\dot{\tilde{v}}_i$ (the rotation of the cross-section) and the displacement $\tilde{v}_i$ at the node x_i.

Finally, the approximation $\tilde{v}$ can be written in the form

$$\tilde{v} = \sum_{k=1}^{N} \dot{\tilde{v}}_k \varphi_{2k} + \sum_{k=1}^{N} \tilde{v}_k \varphi_{2k-1}, \tag{6.222}$$

where

$$\tilde{v}_k \equiv \tilde{v}(x_k), \qquad \dot{\tilde{v}}_k \equiv \dot{\tilde{v}}(x_k). \tag{6.223}$$

Restriction of the Approximation $\tilde{v}$ to One Mesh Element $[x_i, x_{i+1}]$. In the last question, we showed that the supports of the functions $(\varphi_i)_{i=1,\dots,2N}$ are just the union of the relevant intervals $[x_{i-1}, x_i]$ and $[x_i, x_{i+1}]$.

So on the element $[x_i, x_{i+1}]$, the restriction of a function $\tilde{v}$ in $\tilde{V}$ can be written in the form

$$\tilde{v} = \tilde{v}_i \varphi_{2i-1} + \dot{\tilde{v}}_i \varphi_{2i} + \tilde{v}_{i+1} \varphi_{2i+1} + \dot{\tilde{v}}_{i+1} \varphi_{2i+2}. \tag{6.224}$$

Expression for the Basis Functions on the Element $[x_i, x_{i+1}]$. The only basis functions having part of their support on the interval $[x_i, x_{i+1}]$ are $\varphi_{2i-1}, \varphi_{2i}, \varphi_{2i+1}$, and φ_{2i+2}. These four functions are third degree polynomials and must satisfy (6.191) and (6.192).

The function $\varphi_{2i-1}(x)$ is defined by

$$\varphi_{2i-1}(x_i) = 1, \qquad \dot{\varphi}_{2i-1}(x_i) = 0, \tag{6.225}$$

$$\varphi_{2i-1}(x_{i+1}) = 0, \qquad \dot{\varphi}_{2i-1}(x_{i+1}) = 0. \tag{6.226}$$

Given the cubic polynomial structure of φ_{2i-1}, the conditions (6.226) imply that φ_{2i-1} can be written in the form

$$\varphi_{2i-1}(x) = (x - x_{i+1})^2(Ax + B). \tag{6.227}$$

Furthermore, the boundary conditions (6.225) can be used to evaluate the coefficients A and B, and the function φ_{2i-1} thus takes its final form

$$\varphi_{2i-1}(x) = \frac{1}{h^3}(x - x_{i+1})^2[2(x - x_i) + h]. \tag{6.228}$$

Likewise, by similar reasoning, we obtain the other basis functions as follows:

$$\varphi_{2i}(x) = \frac{1}{h^2}(x - x_{i+1})^2(x - x_i), \tag{6.229}$$

$$\varphi_{2i+1}(x) = \frac{1}{h^3}(x - x_i)^2[2(x_{i+1} - x) + h], \tag{6.230}$$

$$\varphi_{2i+2}(x) = \frac{1}{h^2}(x - x_i)^2(x - x_{i+1}). \tag{6.231}$$

9) To obtain the approximate formulation of the minimisation problem (**MP**), we evaluate the functional J defined in (6.187) at the point $\tilde{v}$ given by the expansion (6.193).

To simplify the notation, we use the repeated index convention, whence

$$J(\tilde{v}) = \frac{EI}{2} \int_0^L \left(\tilde{v}_k \ddot{\varphi}_{2k} + \tilde{v}_k \ddot{\varphi}_{2k-1} \right)^2 dx + \int_0^L f\left(\tilde{v}_k \varphi_{2k} + \tilde{v}_k \varphi_{2k-1} \right) dx. \tag{6.232}$$

A necessary condition for minimisation of the functional J is then

$$\forall k = 1, \dots, N, \quad \frac{\partial J}{\partial \tilde{v}_k}(\tilde{u}) = 0 = \frac{\partial J}{\partial \tilde{v}_k}(\tilde{u}). \tag{6.233}$$

The $2N$ conditions (6.233) can be written

$$EI \int_0^L \left(\dot{\tilde{u}}_k \ddot{\varphi}_{2k} + \tilde{u}_k \ddot{\varphi}_{2k-1} \right) \ddot{\varphi}_{2k-1} \, dx = - \int_0^L f\varphi_{2k-1} \, dx, \quad \forall k = 1, \dots, N, \tag{6.234}$$

$$EI \int_0^L \left(\dot{\tilde{u}}_k \ddot{\varphi}_{2k} + \tilde{u}_{k+1} \ddot{\varphi}_{2k+1} \right) \ddot{\varphi}_{2k} \, dx = - \int_0^L f\varphi_{2k} \, dx, \quad \forall k = 1, \dots, N. \tag{6.235}$$

Note finally that the integrands occurring in (6.234) and (6.235) involve only the intersection of the supports of the basis functions $(\varphi_i)_{i=1,\dots,2N}$, whence the formulation (6.195) and (6.196).

10) The elementary matrix $a^{(i+1)}$ defined by (6.198) expresses the contribution from the mesh element $[x_i, x_{i+1}]$ to the global matrix A of the system (6.195) and (6.196). In a similar way, the contribution of this same element to the right-hand side b is denoted by $b^{(i+1)}$.

Approximation of the Elementary Vector $b^{(i+1)}$ by Simpson's Rule. Each of the 4 components of the vector $b^{(i+1)}$ is approximately evaluated using Simpson's rule (6.200) for numerical integration:

$$b_1^{(i+1)} = -\int_{x_i}^{x_{i+1}} f\varphi_{2i-1}\,dx = -\int_{x_i}^{x_{i+1}} \frac{1}{h^3}(x-x_{i+1})^2[2(x-x_i)+h]f(x)\,dx$$

$$\approx -\frac{h}{6h^3}\left\{ f_i h^3 + 4(x_{i+1/2}-x_{i+1})^2[2(x_{i+1/2}-x_i)+h]f_{i+1/2}\right\}$$

$$\approx -\frac{h}{6}(f_i + 2f_{i+1/2}). \tag{6.236}$$

Similarly,

$$b_2^{(i+1)} = -\int_{x_i}^{x_{i+1}} f\varphi_{2i}\,dx = -\int_{x_i}^{x_{i+1}} \frac{1}{h^2}(x-x_{i+1})^2(x-x_i)f(x)\,dx$$

$$\approx -\frac{h}{6h^2}\left[0\times f_i + 4(x_{i+1/2}-x_{i+1})^2(x_{i+1/2}-x_i)f_{i+1/2}+0\times f_{i+1}\right]$$

$$\approx -\frac{h^2}{12}f_{i+1/2}. \tag{6.237}$$

Then,

$$b_3^{(i+1)} = -\int_{x_i}^{x_{i+1}} f\varphi_{2i+1}\,dx = -\int_{x_i}^{x_{i+1}} \frac{1}{h^3}(x-x_i)^2[2(x_{i+1}-x)+h]f(x)\,dx$$

$$\approx -\frac{h}{6h^3}\left\{ +4(x_{i+1/2}-x_i)^2[2(x_{i+1}-x_{i+1/2})+h]f_{i+1/2}+h^3 f_{i+1}\right\}$$

$$\approx -\frac{h}{6}(f_{i+1} + 2f_{i+1/2}). \tag{6.238}$$

The last component $b_4^{(i+1)}$ is

$$b_4^{(i+1)} = -\int_{x_i}^{x_{i+1}} f\varphi_{2i+2}\,dx = -\int_{x_i}^{x_{i+1}} \frac{1}{h^2}(x-x_i)^2(x-x_{i+1})f(x)\,dx$$

$$\approx -\frac{h}{6h^2}\left[0\times f_i + 4(x_{i+1/2}-x_i)^2(x_{i+1/2}-x_{i+1})f_{i+1/2}+0\times f_{i+1}\right]$$

$$\approx \frac{h^2}{12}f_{i+1/2}. \tag{6.239}$$

In conclusion, the approximation of the elementary vector $b^{(i+1)}$ can be written in the form

$$
{}^{\mathrm{t}}b^{(i+1)} \approx \left[-\frac{h}{6}(f_i + 2f_{i+1/2}), \;\; -\frac{h^2}{12}f_{i+1/2}, \;\; -\frac{h}{6}(f_{i+1} + 2f_{i+1/2}), \;\; \frac{h^2}{12}f_{i+1/2} \right].
$$
$$(6.240)$$

Special Case of Uniform Load $f_2 \equiv p$. When the force density f_2 is constant along the beam Ω, the elementary vector $b^{(i+1)}$ becomes

$$
{}^{\mathrm{t}}b_*^{(i+1)} = -p\left[\int_{x_i}^{x_{i+1}} \varphi_{2i-1}\,\mathrm{d}x, \;\; \int_{x_i}^{x_{i+1}} \varphi_{2i}\,\mathrm{d}x, \;\; \int_{x_i}^{x_{i+1}} \varphi_{2i+1}\,\mathrm{d}x, \;\; \int_{x_i}^{x_{i+1}} \varphi_{2i+2}\,\mathrm{d}x \right].
$$
$$(6.241)$$

Therefore, since the basis functions $\widetilde{W}$ are third degree polynomials, Simpson's rule is exact on such polynomials and the approximation put forward in (6.240) becomes exact if we replace f_i, $f_{i+1/2}$, and f_{i+1} by the constant load p.

Finally, we have

$$
{}^{\mathrm{t}}b_*^{(i+1)} = \left[-\frac{ph}{2}, \;\; -\frac{ph^2}{12}, \;\; -\frac{ph}{2}, \;\; \frac{ph^2}{12} \right].
$$
$$(6.242)$$

11) We now subdivide the beam into three equal elements of length h. The four nodes of the mesh resulting from this discretisation are x_0, x_1, x_2, and x_3.

Before assembling the matrix A and the vector b representing the linear system of the minimisation problem $\widetilde{(\mathrm{MP})}$, let us consider the degrees of freedom corresponding to the Hermitian finite-element approximation.

To do this, note first that, given the double clamping of the beam at $x_0 = 0$ and $x_3 = L$, only the nodes at x_1 and x_2 have degrees of freedom, which we denote by $\tilde{u}_1$, $\dot{\tilde{u}}_1$, $\tilde{u}_2$, and $\dot{\tilde{u}}_2$.

The approximation of the displacement field $\tilde{u}$ along the beam Ω is thus

$$
\tilde{u} = \tilde{u}_1\varphi_1 + \dot{\tilde{u}}_1\varphi_2 + \tilde{u}_2\varphi_3 + \dot{\tilde{u}}_2\varphi_4.
$$
$$(6.243)$$

Assembling the Global Matrix A and Vector b for the Problem $\widetilde{(\mathrm{MP})}$. We use the assembly procedure already implemented in Sect. 6.1.1, starting with the mesh element $[x_0, x_1]$.

In this case, only the node at x_1 contributes to the global matrix A. That is, if we consider the elementary matrix $a^{(1)}$ relating to the element $[x_0, x_1]$, it is the submatrix corresponding to the degrees of freedom at the node x_1, viz., $\tilde{u}_1$ and $\dot{\tilde{u}}_1$, which must be taken into consideration when assembling the matrix A.

The elements of the matrix $a^{(1)}$ defined in (6.198) are shown in bold type here:

$$
a^{(1)} = \frac{2EI}{h^3}
\begin{array}{cccc}
 & \tilde{u}_1 & \dot{\tilde{u}}_1 & \\
 & \downarrow & \downarrow & \\
\left[\begin{array}{cccc}
6 & 3h & -6 & 3h \\
3h & 2h^2 & -3h & h^2 \\
\mathbf{-6} & \mathbf{-3h} & \mathbf{6} & \mathbf{-3h} \\
\mathbf{3h} & \mathbf{h^2} & \mathbf{-3h} & \mathbf{2h^2}
\end{array}\right] &
\begin{array}{c}
\\ \\ \leftarrow \tilde{u}_1 \\ \leftarrow \dot{\tilde{u}}_1
\end{array}
\end{array}
\qquad (6.244)
$$

Having taken into account the contribution from the element $[x_0, x_1]$, the global matrix A takes the form

$$
A = \frac{2EI}{h^3}
\begin{array}{cccc}
\tilde{u}_1 & \dot{\tilde{u}}_1 & \tilde{u}_2 & \dot{\tilde{u}}_2 \\
\downarrow & \downarrow & & \\
\left[\begin{array}{cccc}
6 & -3h & 0 & 0 \\
-3h & 2h^2 & 0 & 0 \\
0 & 0 & 0 & 0 \\
0 & 0 & 0 & 0
\end{array}\right]
\begin{array}{c}
\leftarrow \tilde{u}_1 \\ \leftarrow \dot{\tilde{u}}_1 \\ \tilde{u}_2 \\ \dot{\tilde{u}}_2
\end{array}
\end{array}
\qquad (6.245)
$$

Likewise, according to the general definition (6.199), the contribution of the elementary vector $b^{(1)}$ corresponds solely to the two components $b_3^{(1)}$ and $b_4^{(1)}$ relating to the degrees of freedom $\tilde{u}_1$ and $\dot{\tilde{u}}_1$.

So after incorporating the contribution from the first mesh element $[x_0, x_1]$, the vector b has the form

$$
{}^t b = \left[b_3^{(1)}, \ b_4^{(1)}, \ 0, \ 0 \right]. \qquad (6.246)
$$

We now consider the second element $[x_1, x_2]$ of the beam Ω. In this case, the nodes at x_1 and x_2 have the degrees of freedom $(\tilde{u}_1, \dot{\tilde{u}}_1)$ and $(\tilde{u}_2, \dot{\tilde{u}}_2)$, respectively, so the elementary matrix $a^{(2)}$ relating to the element $[x_1, x_2]$ is full. According to the definition (6.199),

$$
a^{(2)} = \frac{2EI}{h^3}
\begin{array}{cccc}
\tilde{u}_1 & \dot{\tilde{u}}_1 & \tilde{u}_2 & \dot{\tilde{u}}_2 \\
\downarrow & \downarrow & \downarrow & \downarrow \\
\left[\begin{array}{cccc}
6 & 3h & -6 & 3h \\
3h & 2h^2 & -3h & h^2 \\
-6 & -3h & 6 & -3h \\
3h & h^2 & -3h & 2h^2
\end{array}\right]
\begin{array}{c}
\leftarrow \tilde{u}_1 \\ \leftarrow \dot{\tilde{u}}_1 \\ \leftarrow \tilde{u}_2 \\ \leftarrow \dot{\tilde{u}}_2
\end{array}
\end{array}
\qquad (6.247)
$$

We now assemble the global matrix A, incorporating the matrix $a^{(2)}$. This contribution is shown in bold type in A, while the terms from the elementary matrix $a^{(1)}$ are in normal type:

$$A = \frac{2EI}{h^3} \begin{array}{cccc} \tilde{u}_1 & \dot{\tilde{u}}_1 & \tilde{u}_2 & \dot{\tilde{u}}_2 \\ \downarrow & \downarrow & \downarrow & \downarrow \end{array} \left[\begin{array}{cccc} 6+6 & -3h+3h & -6 & 3h \\ -3h+3h & 2h^2+2h^2 & 3h & h^2 \\ -6 & -3h & 6 & -3h \\ 3h & h^2 & -3h & 2h^2 \end{array} \right] \begin{array}{c} \leftarrow \tilde{u}_1 \\ \leftarrow \dot{\tilde{u}}_1 \\ \leftarrow \tilde{u}_2 \\ \leftarrow \dot{\tilde{u}}_2 \end{array} \qquad (6.248)$$

After carrying out the elementary calculations in the coefficients of A, we have

$$A = \frac{2EI}{h^3} \left[\begin{array}{cccc} 12 & 0 & -6 & 3h \\ 0 & 4h^2 & -3h & h^2 \\ -6 & -3h & 6 & -3h \\ 3h & h^2 & -3h & 2h^2 \end{array} \right]. \qquad (6.249)$$

Regarding the contribution of the element $[x_1, x_2]$ to the vector b, the elementary vector $b^{(2)}$ is full, once again, and we have

$$^t b^{(2)} = \left[b_1^{(2)}, \ b_2^{(2)}, \ b_3^{(2)}, \ b_4^{(2)} \right]. \qquad (6.250)$$

Assembling the vector b, we obtain

$$b = \left[\begin{array}{c} b_3^{(1)} + b_1^{(2)} \\ b_4^{(1)} + b_2^{(2)} \\ b_3^{(2)} \\ b_4^{(2)} \end{array} \right]. \qquad (6.251)$$

We end the assembly of the matrix A by taking into account the contribution from the last mesh element $[x_2, x_3]$. This case is perfectly analogous to the one encountered for the mesh element $[x_0, x_1]$.

This time it is the node at x_3 which is clamped and we only need to take into account the degrees of freedom of the node at x_2, viz., $\tilde{u}_2$ and $\dot{\tilde{u}}_2$. The coefficients of the elementary matrix $a^{(3)}$ to be taken into account are those of the submatrix indicated here:

$$
a^{(3)} = \frac{2EI}{h^3}
\begin{array}{cc}
& \begin{array}{cccc} \tilde{u}_2 & \dot{\tilde{u}}_2 \\ \downarrow & \downarrow \end{array} \\
\left[\begin{array}{cccc}
6 & 3h & -6 & 3h \\
3h & 2h^2 & -3h & h^2 \\
-6 & -3h & 6 & -3h \\
3h & h^2 & -3h & 2h^2
\end{array}\right]
&
\begin{array}{c}
\leftarrow \tilde{u}_2 \\
\leftarrow \dot{\tilde{u}}_2 \\
\\
\\
\end{array}
\end{array}
\tag{6.252}
$$

Assembling into the global matrix, we obtain

$$
A = \frac{2EI}{h^3}
\begin{array}{cc}
& \begin{array}{cccc} \tilde{u}_1 & \dot{\tilde{u}}_1 & \tilde{u}_2 & \dot{\tilde{u}}_2 \\ & & \downarrow & \downarrow \end{array} \\
\left[\begin{array}{cccc}
12 & 0 & -6 & 3h \\
0 & 4h^2 & -3h & h^2 \\
-6 & -3h & 6+6 & -3h+3h \\
3h & h^2 & -3h+3h & 2h^2+2h^2
\end{array}\right]
&
\begin{array}{c}
\tilde{u}_1 \\
\dot{\tilde{u}}_1 \\
\leftarrow \tilde{u}_2 \\
\leftarrow \dot{\tilde{u}}_2
\end{array}
\end{array}
\tag{6.253}
$$

Finally, the matrix A corresponding to the three-element mesh on the beam Ω is

$$
A = \frac{2EI}{h^3}
\left[\begin{array}{cccc}
12 & 0 & -6 & 3h \\
0 & 4h^2 & -3h & h^2 \\
-6 & -3h & 12 & 0 \\
3h & h^2 & 0 & 4h^2
\end{array}\right].
\tag{6.254}
$$

The contribution of the element $[x_2, x_3]$ on the right-hand side is represented *via* the elementary vector $b^{(3)}$, and this only through its components $b_1^{(3)}$ and $b_2^{(3)}$, given the clamp at $x_3 = L$.

The final assembly of the vector b thus leads to

$$
b =
\left[\begin{array}{c}
b_3^{(1)} + b_1^{(2)} \\
b_4^{(1)} + b_2^{(2)} \\
b_3^{(2)} + b_1^{(3)} \\
b_4^{(2)} + b_2^{(3)}
\end{array}\right]
\approx
\left[\begin{array}{c}
-\dfrac{h}{6}(2 f_{1/2} + 2 f_1 + 2 f_{3/2}) \\
\dfrac{h^2}{12}(f_{1/2} - f_{3/2}) \\
-\dfrac{h}{6}(2 f_{3/2} + 2 f_2 + 2 f_{5/2}) \\
\dfrac{h^2}{12}(f_{3/2} - f_{5/2})
\end{array}\right],
\tag{6.255}
$$

where we have used the generic result of the approximation (6.240) for the elementary vector $b^{(i+1)}$.

Approximation of the Displacement Field at All Points of the Beam Ω. To establish an approximation of the displacement field $\tilde{u}$ at all points of the beam Ω, we simply solve the linear system specified by the matrix A in (6.254) and the right-hand side b given in (6.255).

The corresponding analytic solution, produced by computer, is:

$$\tilde{u}_1 = \frac{1}{171h}\Big[(11b_3 + 24b_1)h - 17b_4 - 4b_2\Big], \tag{6.256}$$

$$\dot{\tilde{u}}_1 = \frac{1}{57h^2}\Big[(5b_3 + 4b_1)h - 6b_4 + 12b_2\Big], \tag{6.257}$$

$$\tilde{u}_2 = \frac{1}{171h}\Big[(16b_3 + 9b_1)h - 4b_4 - 11b_2\Big], \tag{6.258}$$

$$\dot{\tilde{u}}_2 = -\frac{1}{57h^2}\Big[(4b_3 + 7b_1)h - 20b_4 + 2b_2\Big], \tag{6.259}$$

where b_1, b_2, b_3 and b_4 are the components of the vector b as approximated in (6.255). The displacement field $\tilde{u}$ is then specified at all points x along the beam Ω by the formula (6.219), after adapting it to the three-element mesh, viz.,

$$\tilde{u}(x) = \tilde{u}_1\varphi_1(x) + \dot{\tilde{u}}_1\varphi_2(x) + \tilde{u}_2\varphi_3(x) + \dot{\tilde{u}}_2\varphi_4(x), \tag{6.260}$$

where the coefficients $\tilde{u}_1$, $\tilde{u}_2$, $\tilde{u}_3$, and $\tilde{u}_4$ are given by (6.256)–(6.259) and the functions $(\varphi_i)_{i=1,2,3,4}$ are the basis functions corresponding to the expressions (6.228)–(6.231).

Application of the Hermitian Finite-Element Method to the Variational Formulation (VP). The variational formulation (**VP**) is given in (6.186). To obtain the approximate variational formulation $(\widetilde{\textbf{VP}})$, we make the usual substitutions:

$$u(x) \longrightarrow \tilde{u}(x) = \sum_{k=1}^{N} \dot{\tilde{u}}_k\varphi_{2k}(x) + \sum_{k=1}^{N} \tilde{u}_k\varphi_{2k-1}(x), \tag{6.261}$$

$$v(x) \longrightarrow \tilde{v}(x) = \varphi_i(x). \tag{6.262}$$

The approximate formulation $(\widetilde{\textbf{VP}})$ then becomes the following:

$$\left\{ \begin{array}{l} \text{Find } \tilde{u} \text{ solution of} \\[2mm] EI \sum_{k=1}^{N} \int_0^L \Big(\dot{\tilde{u}}_k\varphi_{2k} + \tilde{u}_k\varphi_{2k-1}\Big)\, \varphi_i\, dx = -\int_0^L f\varphi_i\, dx\,, \quad i = 1, \ldots, 2N. \end{array} \right. \tag{6.263}$$

We then note that the $2N$ equations labelled by i in the formulation (6.263) can be divided into two groups: those corresponding to even values of i and those corresponding to odd values of i.

This distinction then delivers precisely the same formulation as the system of Eqs. (6.234) and (6.235) for the minimisation problem $\widetilde{(\mathbf{MP})}$.

We deduce that the node equations of the approximate variational formulation $\widetilde{(\mathbf{VP})}$ will lead to the same linear system, with the matrix A defined by (6.254) and vector b defined by (6.255), for the case of the three-element mesh.

This result is a consequence of the equivalence between the variational problem $(\mathbf{VP})$ and the minimisation problem $(\mathbf{MP})$, given the properties of the bilinear form $a(., .)$ and the linear form $L(.)$ defined by (6.216).

Reference

1. H. Brézis, *Analyse Fonctionnelle, Théorie et Applications* (Masson, Paris, 1983)

Chapter 7
Nonlinear Problems

7.1 Burgers' Equation with Viscosity

We begin this section with a warning. We discuss here the viscous Burgers equation as an approximation to the Navier–Stokes equation in one space dimension.

In order to focus on a mixed formulation considering a finite-element analysis in space and a finite-difference analysis in time that will be accessible to students at the beginning graduate level who have not yet mastered the functional analysis required for the resulting treatment, the following presentation contains no consideration of the necessary functional framework.

In other words, we discuss only formal features of the variational formulations and numerical implementation of the finite-element analysis.

7.1.1 Statement of the Problem

1) We shall be concerned with a scalar function u of the variables (x, t) that solves the following system of partial differential equations:

$$
\text{(CP)} \quad
\begin{cases}
\dfrac{\partial u}{\partial t} + u \dfrac{\partial u}{\partial x} = v \dfrac{\partial^2 u}{\partial x^2}, & \forall (x, t) \in (0, L) \times (0, \infty), \\[2mm]
u(0, t) = 0, \ \dfrac{\partial u}{\partial x}(L, t) = f(t), \ \forall t \in (0, \infty), \\[2mm]
u(x, 0) = u_0(x), & \forall x \in (0, L),
\end{cases}
\tag{7.1}
$$

where v is the kinematic viscosity of the fluid, L a given characteristic length for the flow, and f a sufficiently regular function, also given.

– What is the key property of the partial differential equation of the continuous problem?

J. Chaskalovic, *Mathematical and Numerical Methods for Partial Differential Equations*, Mathematical Engineering, DOI: 10.1007/978-3-319-03563-5_7, © Springer International Publishing Switzerland 2014

2) Let v be a test function of the variable x alone. Show that the continuous problem (**CP**) can be given the following variational formulation (**VP**):

$$\begin{cases} \text{Find } u \in V \text{ solution of} \\ \dfrac{d}{dt} \int_0^L uv\,dx + v \int_0^L \dfrac{\partial u}{\partial x}\dfrac{dv}{dx}\,dx + \int_0^L u\dfrac{\partial u}{\partial x}v\,dx = vf(t)v(L), \quad \forall v \in V. \end{cases}$$
$$(7.2)$$

– Specify the boundary conditions that must be satisfied by the functions v of V, but without considering issues of functional regularity.

3) The variational problem (**VP**) is approximated using a P_1 Lagrange finite-element analysis. To do this, we introduce a regular mesh on the interval $[0, L]$, with constant spacing h, such that

$$\begin{cases} x_0 = 0 ,\ x_{N+1} = L , \\ x_{i+1} = x_i + h ,\ i = 0, \ldots, N. \end{cases}$$
$$(7.3)$$

In addition, we define the approximation space $\widetilde{V}$ by

$$\widetilde{V} = \left\{ \tilde{v} : [0, L] \longrightarrow \mathbb{R}, \ \tilde{v} \in C^0([0, 1]), \ \tilde{v}|_{[x_i, x_{i+1}]} \in P_1([x_i, x_{i+1}]), \ \tilde{v}(0) = 0 \right\},$$
$$(7.4)$$

where $P_1([x_i, x_{i+1}])$ is the space of polynomials on $[x_i, x_{i+1}]$ with degree less than or equal to 1.

– What is the dimension of $\widetilde{V}$?

4) Let φ_i, $(i = 1, \ldots, \dim \widetilde{V})$, be the basis of $\widetilde{V}$ satisfying $\varphi_i(x_j) = \delta_{ij}$. Write down the approximate variational formulation with solution $\tilde{u}$ associated with the problem (**VP**) and show that if we choose

$$\tilde{v}(x) = \varphi_i(x), \qquad \tilde{u}(x, t) = \sum_{j=1,\ldots,\dim \widetilde{V}} \tilde{u}_j(t)\varphi_j(x),$$
$$(7.5)$$

we obtain the time-domain differential system (**DS**) specified by

$$A_{ij}\tilde{u}'_j(t) + B_{ij}\tilde{u}_j(t) + C_{ijk}\tilde{u}_j(t)\tilde{u}_k(t) = \tilde{F}_i(t), \quad \forall i \in \{1, \ldots, \dim \widetilde{V}\}, \quad \forall t \geq 0,$$
$$(7.6)$$

where

$$\tilde{u}'_j(t) = \frac{d}{dt}\tilde{u}_j(t), \quad A_{ij} = \int_0^L \varphi_i\varphi_j\,dx, \quad B_{ij} = v\int_0^L \frac{d\varphi_i}{dx}\frac{d\varphi_j}{dx}\,dx,$$
$$(7.7)$$

$$C_{ijk} = \int_0^L \varphi_i\varphi_j\frac{d\varphi_k}{dx}\,dx, \quad \tilde{F}_i(t) = vf(t)\varphi_i(L).$$
$$(7.8)$$

Here we have used the repeated index convention as usual, e.g.,

$$X_j Y_j \equiv \sum_{j=1,\ldots,\dim \tilde{V}} X_j Y_j. \tag{7.9}$$

– What is the main feature of the differential system (**DS**)?

▶ **Basis Function φ_i Characterizing a Node Strictly in the Interior of $[0, L]$**

5) Given the regularity of the mesh, the generic equation of the system (**DS**) associated with any basis function φ_i characterizing a node strictly in the interior of $[0, L]$ can be written in the form

$$(\mathbf{DS}_1) \quad \begin{cases} A_{i,i-1}\tilde{u}'_{i-1}(t) + A_{i,i}\tilde{u}'_i(t) + A_{i,i+1}\tilde{u}'_{i+1}(t) \\[4pt] + B_{i,i-1}\tilde{u}_{i-1}(t) + B_{i,i}\tilde{u}_i(t) + B_{i,i+1}\tilde{u}'_{i+1}(t) \\[4pt] + C_{i,i-1,i-1}\tilde{u}^2_{i-1}(t) + C_{i,i-1,i}\tilde{u}_{i-1}(t)\tilde{u}_i(t) + C_{i,i,i-1}\tilde{u}_i(t)\tilde{u}_{i-1}(t) \\[4pt] + C_{i,i,i}\tilde{u}^2_i(t) + C_{i,i,i+1}\tilde{u}_i(t)\tilde{u}_{i+1}(t) + C_{i,i+1,i}\tilde{u}_{i+1}(t)\tilde{u}_i(t) \\[4pt] + C_{i,i+1,i+1}\tilde{u}^2_{i+1}(t) = 0. \end{cases} \tag{7.10}$$

– Using the trapezoidal rule, calculate the 13 coefficients A_{ij}, B_{ij}, and C_{ijk} of the system (**DS$_1$**) and write down the corresponding node equation. Recall that the trapezoidal rule is

$$\int_a^b \xi(s)\mathrm{d}s \approx \frac{b-a}{2}\big[\xi(a) + \xi(b)\big].$$

6) Show that we recover the centered finite-difference scheme associated with the partial differential equation of the problem (**CP**). Of what order is it?

▶ **Basis Function φ_{N+1} for the Node at $x_{N+1} = L$**

7) We proceed in the same manner for the basis function φ_{N+1} characterizing the end node at x_{N+1}. The corresponding equation of the system (**DS**) is thus

$$(\mathbf{DS}_2) \quad \begin{cases} A_{N+1,N}\,\tilde{u}'_N(t) + A_{N+1,N+1}\,\tilde{u}'_{N+1}(t) \\[4pt] + B_{N+1,N}\,\tilde{u}_N(t) + B_{N+1,N+1}\,\tilde{u}_{N+1}(t) \\[4pt] + C_{N+1,N+1,N+1}\tilde{u}^2_{N+1}(t) + C_{N+1,N,N+1}\tilde{u}_N(t)\tilde{u}_{N+1}(t) \\[4pt] + C_{N+1,N,N}\tilde{u}^2_N(t) + C_{N+1,N+1,N}\tilde{u}_N(t)\tilde{u}_{N+1}(t) = vf(t). \end{cases} \tag{7.11}$$

Using the trapezoidal rule, calculate the eight coefficients A_{ij}, B_{ij}, and C_{ijk} of the equation (**DS$_2$**) and write down the corresponding node equation.

8) Show that we recover the second order finite-difference scheme by discretising the Neumann boundary condition of the problem (**CP**).

▶ **Time Domain Finite-Difference Discretization**

9) Suggest a time-domain finite-difference discretization of the differential system (**DS$_1$**)–(**DS$_2$**).

7.1.2 Solution

1) Naturally, the main feature of the Burgers equation in the continuous problem (**CP**) is its nonlinearity, due to the convection–advection coupling term $u \partial u / \partial x$. A great deal of research has focused on this equation, which is a special case of a nonlinear hyperbolic partial differential equation.

The interested reader is referred to the elementary presentation by Euvrard [1], aimed at beginning graduate students in physics or mechanics, or the more detailed study by Godlewski and Raviart [2], which requires a broader knowledge of functional analysis.

2) The aim here is to establish a variational formulation that can be handled by a P_1 Lagrange finite-element analysis on the space dimension. To this end, we consider real-valued test functions v on $[0, L]$, i.e., test functions of a *single* spatial variable x.

The partial differential equation of the continuous problem (**CP**), specified in (7.1), is multiplied by v and integrated over the interval $[0, L]$, which yields

$$\int_0^L \frac{\partial u}{\partial t} v(x) \mathrm{d}x + \int_0^L u \frac{\partial u}{\partial x} v(x) \mathrm{d}x = v \int_0^L \frac{\partial^2 u}{\partial x^2} v(x) \mathrm{d}x. \tag{7.12}$$

We integrate by parts on the right-hand side of (7.12) to obtain

$$\int_0^L \frac{\partial u}{\partial t} v \, \mathrm{d}x + \int_0^L u \frac{\partial u}{\partial x} v \, \mathrm{d}x = -v \int_0^L \frac{\partial u}{\partial x} \frac{\mathrm{d}v}{\mathrm{d}x} \mathrm{d}x + v \left[\frac{\partial u}{\partial x}(L, t) v(L) - \frac{\partial u}{\partial x}(0, t) v(0) \right]. \tag{7.13}$$

Using the boundary condition at $x = L$ in the problem (**CP**) and requiring the test functions v to vanish at $x = 0$, in order to conserve all the spatial information contained in the continuous problem, we obtain the following variational formulation:

$$(\mathbf{VP}) \quad \begin{cases} \text{Find } u \in V \text{ solution of} \\[2mm] \displaystyle \int_0^L \frac{\partial u}{\partial t} v \, \mathrm{d}x + \int_0^L u \frac{\partial u}{\partial x} v \, \mathrm{d}x = -v \int_0^L \frac{\partial u}{\partial x} \frac{\mathrm{d}v}{\mathrm{d}x} \mathrm{d}x + v f(t) v(L), \\[4mm] \forall v \text{ such that } v(0) = 0. \end{cases}$$

$$\tag{7.14}$$

Note that this space-domain variational formulation (**VP**) is purely formal, since we have deliberately neglected to specify the function space V in which this formulation would make mathematical sense.

For more information about this, the reader who has mastered the basic techniques of functional analysis, and in particular those relating to Sobolev spaces $H^m[0, T; L^p(\mathbb{R})]$, is referred once again to the work by Godlewski and Raviart [2].

3) The dimension of the approximation space $\widetilde{V}$ is estimated as follows. Functions in the space $\widetilde{V}$ are specified by the $N + 1$ values they take at the nodes $(x_i)_{i=1,\dots,N+1}$ of the mesh (7.3). Note that the node at $x_0 = 0$ contributes no degree of freedom, because the value of every function $\tilde{v}$ in $\widetilde{V}$ is zero at this point.

To see this, consider a graphic representation of such functions. Every function $\tilde{v}$ in the approximation space $\widetilde{V}$ is a continuous piecewise affine function, being affine on each mesh element $[x_i, x_{i+1}]$, and such that the internal nodes $x_i, (i = 1, \dots, N)$, are points of continuity between adjacent elements.

For this reason, the values at the $N + 1$ nodes $x_1, x_2, \dots, x_{N+1}$ each express one degree of freedom for a function $\tilde{v}$ in $\widetilde{V}$. Altering one of these $N + 1$ values would immediately change the element $\tilde{v}$ of $\widetilde{V}$ under consideration into a different function $\tilde{w}$ of $\widetilde{V}$.

We thus see, without the need for formal demonstration, that specifying a function $\tilde{v}$ in $\widetilde{V}$ is equivalent to specifying a vector in $\mathbb{R}^{N+1}$ whose components are the $N + 1$ values $\tilde{v}_1, \dots, \tilde{v}_{N+1}$ of the function at the nodes of the mesh.

In other words, the dimension of $\widetilde{V}$ is equal to $N+1$, since this space is isomorphic to $\mathbb{R}^{N+1}$.

4) The approximate variational formulation is obtained by replacing the functions (u, v) of the function space V by their corresponding approximations $(\tilde{u}, \tilde{v})$ in the space $\widetilde{V}$.

The approximate variational formulation $\widetilde{(\textbf{VP})}$ can thus be stated as follows:

$$
\widetilde{(\textbf{VP})} \begin{cases} \text{Find } \tilde{u} \in \widetilde{V} \text{ solution of} \\ \dfrac{d}{dt} \displaystyle\int_0^L \tilde{u}\tilde{v}\, dx + \nu \int_0^L \frac{\partial \tilde{u}}{\partial x}\frac{d\tilde{v}}{dx}\, dx + \int_0^L \tilde{u}\frac{\partial \tilde{u}}{\partial x}\tilde{v}\, dx = \nu f(t)\tilde{v}(L), \quad \forall \tilde{v} \in \widetilde{V}. \end{cases}
$$

$$(7.15)$$

We now consider the special case in which the functions $\tilde{v}$ are basis functions φ_i in the space $\widetilde{V}$, expanding the approximate solution $\tilde{u}$ relative to this basis:

$$\tilde{v}(x) = \varphi_i(x), \qquad \tilde{u}(x,t) = \sum_{j=1,\dots,N+1} \tilde{u}_j(t)\varphi_j(x). \qquad (7.16)$$

Given the inherent nonlinearity of the convection–advection term

$$\tilde{u}\frac{\partial \tilde{u}}{\partial x}\tilde{v},$$

we must introduce two summation indices j and k in the expansion (7.16) of the approximation $\tilde{u}$.

The approximate variational formulation $\widetilde{(\textbf{VP})}$ in (7.15) can then be rewritten in the form

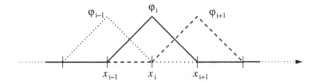

Fig. 7.1 Basis functions φ_{i-1}, φ_i, and φ_i

$$\tilde{u}'_j(t) \int_0^L \varphi_i \varphi_j \, dx + \left(\nu \int_0^L \frac{d\varphi_i}{dx} \frac{d\varphi_j}{dx} \, dx \right) \tilde{u}_j(t) + \left(\int_0^L \varphi_i \varphi_j \frac{d\varphi_k}{dx} \, dx \right) \tilde{u}_j(t) \tilde{u}_k(t)$$

$$= \nu f(t) \varphi_i(L), \quad \forall i = 1, \ldots, N+1, \tag{7.17}$$

where we have used the repeated index convention (7.9).

Then introducing the notation of (7.7) and (7.8), we see that the variational approximation (7.17) produces the time-domain differential system (**DS**) defined by

$$A_{ij} \tilde{u}'_j(t) + B_{ij} \tilde{u}_j(t) + C_{ijk} \tilde{u}_j(t) \tilde{u}_k(t) = \tilde{F}_i(t), \quad \forall i = 1, \ldots, N+1, \quad \forall t \geq 0. \tag{7.18}$$

Naturally, since the Burgers equation of the continuous problem (**CP**) defined by (7.1) is nonlinear, this feature is omnipresent, appearing both in the variational formulation (7.2) and in its approximate form (7.15), and then consequently also in the nonlinear system (**DS**).

This nonlinearity would require appropriate numerical methods, such as Newton's method [3], to produce an approximation to the differential system (**DS**).

5) We now consider a basis function φ_i, $(i = 1, \ldots, N)$, characterizing a node strictly in the interior of the mesh defined by (7.3).

For each of these basis functions φ_i, since its support is a union of two intervals $[x_{i-1}, x_i]$ and $[x_i, x_{i+1}]$, as shown in Fig. 7.1, the coefficients of the matrices A_{ij} and B_{ij} and also the coefficients C_{ijk} of (7.7) and (7.8) will produce nonzero terms if and only if the supports of the functions φ_j and φ_k have a nonempty intersection with the support of the function φ_i under consideration.

In order to determine which terms should be considered for the construction of the differential system (**DS**), we fix the index $i \in \{1, \ldots, N\}$. In this case, regarding the matrices A_{ij} and B_{ij}, it follows immediately that only the basis functions φ_{i-1}, φ_i, and φ_{i+1} can produce a nonzero integration when multiplied by the basis function φ_i (see Fig. 7.1).

For this reason, only the coefficients A_{ij} and B_{ij} for $j = i - 1, i, i + 1$ need be retained in the statement of the differential system (**DS**).

What about the term C_{ijk}? A more detailed examination is required here. Having fixed $i \in \{1, \ldots, N\}$, given the structure of the coefficient C_{ijk}, we need to consider only the following three values of j: $i - 1, i$, and $i + 1$.

We must then deal with the following pairs of indices: $(i, i-1), (i, i)$, and $(i, i+1)$. For each of these pairs, let us therefore determine the values of the index k likely to produce nonzero terms in the nonlinear system (**DS**).

Regarding the pair $(i, i - 1)$, only those values of k equal to $i - 1$ and i need be retained. Indeed, the value $k = i+1$ would lead to a zero coefficient $C_{i, i-1, i+1}$, since the support of the basis function φ_{i-1} is disjoint from the support of the function φ_{i+1}. We thus retain the triplets $(i, i - 1, i - 1)$ and $(i, i - 1, i)$.

In the same way, the pair (i, i) leads us to consider the following triplets: $(i, i, i - 1), (i, i, i)$, and $(i, i, i + 1)$.

Finally, the pair $(i, i + 1)$ is the mirror image of the pair $(i, i - 1)$, and we retain the following triplets: $(i, i + 1, i)$ and $(i, i + 1, i + 1)$.

Let us therefore list all the coefficients likely to produce nonzero terms in the nonlinear system (**DS**):

$$C_{i,i-1,i-1}, \quad C_{i,i-1,i}, \quad C_{i,i,i-1}, \quad C_{i,i,i}, \quad C_{i,i,i+1}, \quad C_{i,i+1,i}, \quad C_{i,i+1,i+1}. \qquad (7.19)$$

So if we consider the coefficients $A_{i,j}, B_{i,j}$, and $C_{i,j,k}$ just identified as being relevant to the statement of the generic equation in the nonlinear differential system (**DS**), we obtain the N equations (**DS$_1$**) specified in (7.10).

Note also that the right-hand side $\tilde{F}_i(t) \equiv \nu f(t)\varphi_i(L)$ is identically zero for these basis functions $\varphi_i, (i = 1, \ldots, N)$, given that

$$\varphi_i(L) = \varphi_i(x_{N+1}) = 0, \quad \forall i = 1, \ldots, N. \qquad (7.20)$$

Furthermore, all the basis functions φ_i that we consider here satisfy the property

$$\varphi_i(x_j) = \delta_{ij}, \quad \forall (i, j) \in \{1, \ldots, N + 1\}. \qquad (7.21)$$

Approximation of the Coefficients $A_{ij}, j = i - 1, i, i + 1$, by the Trapezoidal Rule.

Given the relative locations of the supports of the basis functions φ_{i-1}, φ_i, and φ_{i+1}, we have

$$A_{i,i-1} \equiv \int_0^L \varphi_i \varphi_{i-1} \, dx = \int_{x_{i-1}}^{x_i} \varphi_i \varphi_{i-1} \, dx$$

$$\approx \frac{h}{2}\left[\varphi_i(x_{i-1})\varphi_{i-1}(x_{i-1}) + \varphi_i(x_i)\varphi_{i-1}(x_i)\right] = 0, \qquad (7.22)$$

where we have used the property (7.21) of the basis functions φ_i. Likewise,

$$A_{i,i+1} \equiv \int_0^L \varphi_i \varphi_{i+1} \, dx = \int_{x_i}^{x_{i+1}} \varphi_i \varphi_{i+1} \, dx$$

$$\approx \frac{h}{2} \left[\varphi_i(x_i)\varphi_{i+1}(x_i) + \varphi_i(x_{i+1})\varphi_{i+1}(x_{i+1}) \right] = 0. \qquad (7.23)$$

The coefficient $A_{i,i}$ is evaluated as follows:

$$A_{i,i} = \int_0^L \varphi_i^2 \, dx = \int_{x_{i-1}}^{x_i} \varphi_i^2 \, dx + \int_{x_i}^{x_{i+1}} \varphi_i^2 \, dx$$

$$\approx \frac{h}{2}(1+0) + \frac{h}{2}(0+1) = h. \qquad (7.24)$$

Estimation of the Coefficients B_{ij}, $j = i - 1,\ i,\ i + 1$.

The three coefficients B_{ij} can be found either exactly, or approximately using the trapezoidal rule, where one must bear in mind that the latter is exact for constant functions.

Indeed, since the functions φ_i are piecewise affine, it follows that they have constant derivative on each mesh element $[x_i, x_{i+1}]$.

We thus have

$$B_{i,i-1} = v \int_0^L \frac{d\varphi_i}{dx} \frac{d\varphi_{i-1}}{dx} \, dx = v \int_{x_{i-1}}^{x_i} \frac{d\varphi_i}{dx} \frac{d\varphi_{i-1}}{dx} \, dx$$

$$= vh \times \left(\frac{1}{h}\right)\left(-\frac{1}{h}\right) = -\frac{v}{h}. \qquad (7.25)$$

Furthermore, the symmetry of the matrix B_{ij} and the translational invariance of the mesh with constant spacing h mean that we can write

$$\underset{\text{symmetry}}{B_{i,i-1}} \quad = \quad \underset{\text{invariance}}{B_{i-1,i}} \quad = \quad B_{i,i+1} = -\frac{v}{h}. \qquad (7.26)$$

Finally, the coefficient $B_{i,i}$ is found as follows:

$$B_{i,i} = v \int_0^L \left(\frac{d\varphi_i}{dx}\right)^2 dx = v \int_{x_{i-1}}^{x_i} \left(\frac{d\varphi_i}{dx}\right)^2 dx + v \int_{x_i}^{x_{i+1}} \left(\frac{d\varphi_i}{dx}\right)^2 dx$$

$$= vh \left[\frac{1}{h^2} + \left(-\frac{1}{h}\right)^2 \right] = \frac{2v}{h}. \qquad (7.27)$$

Estimation of the Coefficients C_{ijk}.

We begin with a qualitative remark. Given that the coefficients C_{ijk} involve the double product $\varphi_i \times \varphi_j \times \varphi_k'$, use of the trapezoidal rule to calculate the coefficients will lead only to terms of the form

$$\pm \frac{1}{h} \varphi_i(x_l) \times \varphi_j(x_m).$$

Since all the basis functions φ_i have the property

$$\varphi_i(x_j) = \delta_{ij},$$

when the index k is fixed (see the above analysis), the only case we need to deal with for this value of k is the one with $i = j$.

Put another way, the coefficients $C_{i,i,i-1}$, $C_{i,i,i}$, and $C_{i,i,i+1}$ are worth calculating, while the other four in the set (7.19) are all identically zero. Indeed, for the other coefficients C_{ijk}, when, for example, $\varphi_i(x_i)$ is equal to 1, $\varphi_{i-1}(x_i)$ and $\varphi_{i+1}(x_i)$ would be identically zero.

This implies that

$$C_{i,i,i-1} = \int_0^L (\varphi_i)^2 \frac{d\varphi_{i-1}}{dx} \, dx = \int_{x_{i-1}}^{x_i} (\varphi_i)^2 \frac{d\varphi_{i-1}}{dx} \, dx$$

$$\approx -\frac{1}{h} \times \left[\frac{h}{2}(1 + 0) \right] = -\frac{1}{2}. \tag{7.28}$$

Likewise,

$$C_{i,i,i+1} = \int_0^L (\varphi_i)^2 \frac{d\varphi_{i+1}}{dx} \, dx = \int_{x_i}^{x_{i+1}} (\varphi_i)^2 \frac{d\varphi_{i+1}}{dx} \, dx$$

$$\approx \frac{1}{h} \times \left[\frac{h}{2}(1 + 0) \right] = \frac{1}{2}. \tag{7.29}$$

Finally, the coefficient $C_{i,i,i}$ is obtained as follows:

$$C_{i,i,i} = \int_0^L (\varphi_i)^2 \frac{d\varphi_i}{dx} \, dx = \int_{x_{i-1}}^{x_i} (\varphi_i)^2 \frac{d\varphi_i}{dx} \, dx + \int_{x_i}^{x_{i+1}} (\varphi_i)^2 \frac{d\varphi_i}{dx} \, dx$$

$$\approx \frac{h}{2} \left[\frac{1}{h}(1 + 0) \right] + \frac{h}{2} \left[-\frac{1}{h}(0 + 1) \right] = 0. \tag{7.30}$$

The node equation (7.18) associated with the basis function φ_i characterizing a node x_i, $(i = 1, \ldots, N)$, is then obtained by gathering the results (7.19) and (7.30):

$$h\tilde{u}_i'(t) - \frac{\nu}{h}\left[\tilde{u}_{i-1}(t) + \tilde{u}_{i+1}(t)\right] + \frac{2\nu}{h}\tilde{u}_i(t) - \frac{1}{2}\tilde{u}_i(t)\tilde{u}_{i-1}(t) + \frac{1}{2}\tilde{u}_i(t)\tilde{u}_{i+1}(t) = 0.$$

$$(7.31)$$

6) The space-domain finite-difference scheme associated with the viscous Burgers equation is obtained by rewriting the node equation (7.31) in the form

$$\tilde{u}_i'(t) + \tilde{u}_i(t)\frac{\tilde{u}_{i+1}(t) - \tilde{u}_{i-1}(t)}{2h} = \nu\frac{\tilde{u}_{i-1}(t) - 2\tilde{u}_i(t) + \tilde{u}_{i+1}(t)}{h^2}. \qquad (7.32)$$

Indeed, we easily recognize the second-order finite-difference approximation to the second partial derivative $\partial^2 u/\partial x^2$ at the point (x_i, t):

$$\frac{\partial^2 u}{\partial x^2}(x_i, t) = \frac{u(x_{i+1}, t) - 2u(x_i, t) + u(x_{i-1}, t)}{h^2} + O(h^2). \qquad (7.33)$$

Moreover, Taylor expanding to order 3, both forward and backward from the point x_i, we have

$$u(x_{i+1}, t) = u(x_i, t) + h\frac{\partial u}{\partial x}u(x_i, t) + \frac{h^2}{2}\frac{\partial^2 u}{\partial x^2}(x_i, t) + O(h^3), \qquad (7.34)$$

$$u(x_{i-1}, t) = u(x_i, t) - h\frac{\partial u}{\partial x}u(x_i, t) + \frac{h^2}{2}\frac{\partial^2 u}{\partial x^2}(x_i, t) + O(h^3). \qquad (7.35)$$

Subtracting (7.35) from (7.34), we find that

$$\frac{\partial u}{\partial x}(x_i, t) = \frac{u(x_{i+1}, t) - u(x_{i-1}, t)}{2h} + O(h^2). \qquad (7.36)$$

We can now show that the node equation (7.32) is nothing but the second-order finite-difference approximation to the viscous Burgers equation of the continuous problem **(CP)** defined by (7.1). To see this, we simply substitute the finite differences (7.33) and (7.36) into the Burgers equation of the problem (7.1) to obtain

$$\frac{\partial u}{\partial t}(x_i, t) + \frac{u(x_{i+1}, t) - u(x_{i-1}, t)}{2h} = \nu\frac{u(x_{i+1}, t) - 2u(x_i, t) + u(x_{i-1}, t)}{h^2} + O(h^2).$$

$$(7.37)$$

Then neglecting the terms $O(h^2)$ in (7.37) and introducing the approximations $\tilde{u}$ to maintain equality between the two sides of the equation, we obtain precisely the node Eq. (7.32), which is thus approximated to second order.

7) To obtain the equation **(DS$_2$)**, we return to the approximate variational formulation (7.18) and consider the particular case of a generic basis function φ_{N+1} characterizing the last node x_{N+1} in the mesh on the interval $[0, L]$.

Here we keep the expansion (7.16) of the approximation $\tilde{u}$ in the canonical basis $\varphi_i, (i = 1, \ldots, N + 1)$.

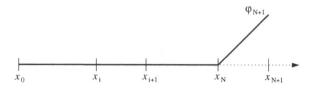

Fig. 7.2 Basis function φ_{N+1}

The approximate variational Eq. (7.18) can then be written

$$A_{N+1,j}\,\tilde{u}'_j(t) + B_{N+1,j}\,\tilde{u}_j(t) + C_{N+1,j,k}\,\tilde{u}_j(t)\tilde{u}_k(t) = \tilde{F}_{N+1}(t),\, \forall\, t \ge 0, \quad (7.38)$$

using the repeated index convention of (7.9) in the usual way.

We now ask which coefficients in (7.38) might make a nonzero contribution. Given that the basis function φ_{N+1} looks like half of a hat, as shown in Fig. 7.2, with support reduced to the interval $[x_N, x_{N+1}]$, the only values of j we need to consider are $j = N$ and $j = N + 1$, corresponding to functions whose supports intersect the support of the function φ_{N+1}.

Indeed, any other value of the index j would lead us to consider a basis function φ_j whose support would have an empty intersection with the support of φ_{N+1}.

As a result, the coefficients of the matrices A_{ij} and B_{ij} that we need to retain in this evaluation are

$$A_{N+1,N}, \quad A_{N+1,N+1}, \quad B_{N+1,N}, \quad B_{N+1,N+1}. \quad (7.39)$$

In the same way, for these two values of the index j, the only values of k we need to consider when estimating the nonzero coefficients C_{ijk} in Eq. (7.38) of the nonlinear system **(DS)** are $k = N$ and $k = N + 1$.

The coefficients C_{ijk} that we need to evaluate are therefore

$$C_{N+1,N,N}, \quad C_{N+1,N,N+1}, \quad C_{N+1,N+1,N}, \quad C_{N+1,N+1,N+1}. \quad (7.40)$$

Finally, we note that the right-hand side $\tilde{F}_{N+1}(t)$ of (7.38) is given precisely by

$$\tilde{F}_{N+1}(t) = \nu f(t)\varphi_{N+1}(x_{N+1}) = \nu f(t).$$

Approximation of the Coefficients $A_{N+1,N}$, $A_{N+1,N+1}$, $B_{N+1,N}$, and $B_{N+1,N+1}$ Using the Trapezoidal Rule.

The coefficients $A_{N+1,N}$ and $A_{N+1,N+1}$, then $B_{N+1,N}$ and $B_{N+1,N+1}$, are calculated by the same kind of argument as in Question 5.

We thus have

$$A_{N+1,\,N} \equiv \int_0^L \varphi_N \varphi_{N+1}\, dx = \int_{x_N}^{x_{N+1}} \varphi_N \varphi_{N+1}\, dx$$

$$\approx \frac{h}{2}\Big[\varphi_N(x_N)\varphi_{N+1}(x_N) + \varphi_N(x_{N+1})\varphi_{N+1}(x_{N+1})\Big] = 0. \quad (7.41)$$

Likewise,

$$A_{N+1,\,N+1} = \int_0^L (\varphi_{N+1})^2\, dx = \int_{x_N}^{x_{N+1}} (\varphi_{N+1})^2\, dx \approx \frac{h}{2}(1+0) = \frac{h}{2}. \quad (7.42)$$

Then,

$$B_{N+1,\,N} = \nu \int_0^L \frac{d\varphi_{N+1}}{dx}\frac{d\varphi_N}{dx}\, dx = \nu \int_{x_N}^{x_{N+1}} \frac{d\varphi_{N+1}}{dx}\frac{d\varphi_N}{dx}\, dx$$

$$= \nu h \times \left(\frac{1}{h}\right)\left(-\frac{1}{h}\right) = -\frac{\nu}{h}. \quad (7.43)$$

Finally,

$$B_{N+1,\,N+1} = \nu \int_0^L \left(\frac{d\varphi_{N+1}}{dx}\right)^2 dx = \nu \int_{x_N}^{x_{N+1}} \left(\frac{d\varphi_{N+1}}{dx}\right)^2 dx$$

$$= \nu h \times \left(\frac{1}{h}\right)^2 = \frac{\nu}{h}. \quad (7.44)$$

Approximation of $C_{N+1,\,N,\,N}$, $C_{N+1,\,N,\,N+1}$, $C_{N+1,\,N+1,\,N}$, **and** $C_{N+1,\,N+1,\,N+1}$
Using the Trapezoidal Rule.

The structural remarks that we made when estimating the coefficients C_{ijk} in Question 5 remain equally valid in the present case. For this reason, we estimate only the coefficients $C_{N+1,\,N+1,\,N}$ and $C_{N+1,\,N+1,\,N+1}$, since the others are trivially zero. We thus have

$$C_{N+1,\,N+1,\,N} = \int_0^L (\varphi_{N+1})^2\frac{d\varphi_N}{dx}\, dx = \int_{x_N}^{x_{N+1}} (\varphi_{N+1})^2\frac{d\varphi_N}{dx}\, dx$$

$$\approx -\frac{1}{h}\left[\frac{h}{2}(1+0)\right] = -\frac{1}{2}. \quad (7.45)$$

Likewise,

$$C_{N+1,\,N+1,\,N+1} = \int_0^L (\varphi_{N+1})^2 \frac{d\varphi_{N+1}}{dx}\, dx = \int_{x_N}^{x_{N+1}} (\varphi_{N+1})^2 \frac{d\varphi_{N+1}}{dx}\, dx$$

$$\approx \frac{1}{h}\left[\frac{h}{2}(1+0)\right] = \frac{1}{2}. \tag{7.46}$$

The node equation (7.38) associated with the basis function φ_{N+1} characterizing the node at x_{N+1} is obtained as usual by gathering together the results (7.41–7.46), whence we find that

$$\frac{h}{2}\tilde{u}'_{N+1}(t) + \frac{\nu}{h}\tilde{u}_{N+1}(t) - \frac{\nu}{h}\tilde{u}_N(t) + \frac{1}{2}\tilde{u}^2_{N+1}(t) - \frac{1}{2}\tilde{u}_N(t)\tilde{u}_{N+1}(t) = \nu f(t), \tag{7.47}$$

or again, after a little rearrangement,

$$\frac{h}{2}\tilde{u}'_{N+1}(t) + \frac{\nu}{h}\left[\tilde{u}_{N+1}(t) - \tilde{u}_N(t)\right] + \frac{1}{2}\tilde{u}_{N+1}(t)\left[\tilde{u}_{N+1}(t) - \tilde{u}_N(t)\right] = \nu f(t). \tag{7.48}$$

8) We now obtain the space-domain finite-difference scheme associated with the Neumann condition of the continuous problem **(CP)** specified in (7.1) by

$$\frac{\partial u}{\partial x}(L, t) \equiv \frac{\partial u}{\partial x}(x_{N+1}, t) = 0, \quad \forall t \geq 0. \tag{7.49}$$

For this purpose, we write down the backward Taylor expansion at the point x_{N+1} of the solution u to the problem **(CP)**, assumed sufficiently regular in the vicinity of this point:

$$u(x_N, t) = u(x_{N+1}, t) - h\frac{\partial u}{\partial x}u(x_{N+1}, t) + \frac{h^2}{2}\frac{\partial^2 u}{\partial x^2}(x_{N+1}, t) + O(h^3). \tag{7.50}$$

Since this brings in the second partial derivative with respect to x at the point (x_{N+1}, t) in order to evaluate the first partial derivative with respect to x at the same point, this might not seem to be a useful step to take.

However, we were forced to consider such an expansion in order to obtain a finite-difference approximation to the same, i.e., second, order as the one established for the approximation to the viscous Burgers equation (7.37).

In order to make use of the expansion (7.50) while getting rid of the second partial derivative of u at the point (x_{N+1}, t), we write the Burgers equation at this point in order to express the second partial derivative with respect to x in the following way (a strong hypothesis that should eventually be justified):

$$\frac{\partial^2 u}{\partial x^2}(x_{N+1}, t) = \frac{1}{\nu}\left[\frac{\partial u}{\partial t}(x_{N+1}, t) + u\frac{\partial u}{\partial x}(x_{N+1}, t)\right]. \tag{7.51}$$

The Taylor expansion (7.50) then becomes

$$u(x_N, t) = u(x_{N+1}, t) - h\frac{\partial u}{\partial x}(x_{N+1}, t)$$

$$+ \frac{h^2}{2\nu}\left[\frac{\partial u}{\partial t}(x_{N+1}, t) + u(x_{N+1}, t)\frac{\partial u}{\partial x}(x_{N+1}, t)\right] + O(h^3). \quad (7.52)$$

Hence the first partial derivative with respect to x of the solution u at the point (x_{N+1}, t) appears at two levels in (7.52): a first time with a weighting coefficient that is a multiple of h and a second time with a weighting going as h^2.

Since we hope to produce an approximation to second order (after dividing by h), we replace the first appearance of the first partial derivative by its value (7.49) and make a first-order approximation for the second appearance of the first partial derivative in (7.52).

The first-order approximation of the partial derivative with respect to x at the point (x_{N+1}, t) is the standard one given by

$$\frac{\partial u}{\partial x}(x_{N+1}, t) = \frac{u(x_{N+1}, t) - u(x_N, t)}{h} + O(h), \quad (7.53)$$

and (7.52) can then be written in the form

$$u(x_N, t) = u(x_{N+1}, t) - hf(t)$$

$$+ \frac{h^2}{2\nu}\left[\frac{\partial u}{\partial t}(x_{N+1}, t) + u(x_{N+1}, t)\left(\frac{u(x_{N+1}, t) - u(x_N, t)}{h} + O(h)\right)\right]$$

$$+ O(h^3). \quad (7.54)$$

We now turn to the approximations:

$$\tilde{u}_N(t) = \tilde{u}_{N+1}(t) - hf(t) + \frac{h^2}{2\nu}\left[\tilde{u}'_{N+1}(t) + \tilde{u}_{N+1}(t)\frac{\tilde{u}_{N+1}(t) - \tilde{u}_N(t)}{h}\right]. \quad (7.55)$$

Rearranging (7.55), we find that

$$\frac{h}{2}\tilde{u}'_{N+1}(t) + \frac{\nu}{h}\left[\tilde{u}_{N+1}(t) - \tilde{u}_N(t)\right] + \frac{1}{2}\tilde{u}_{N+1}(t)\left[\tilde{u}_{N+1}(t) - \tilde{u}_N(t)\right] = \nu f(t),$$

$$(7.56)$$

which leads precisely to the node equation (7.48).

9) Since the differential system (DS_1)–(DS_2) is of first order in time, if we wish to obtain an unconditionally stable finite-difference approximation, we must apply a θ finite-difference scheme for values of θ greater than or equal to $1/2$ [1].

To do this, we consider a time step $k \equiv \Delta t$ and the discrete sequence of times $t^{(n)}$ defined by $t^{(n)} = kn$.

We also introduce the approximation sequence $\bar{u}_i^{(n)}$ defined by

$$\bar{u}_i^{(n)} \approx \tilde{u}_i(t^{(n)}) \approx u(x_i, t^{(n)}), \tag{7.57}$$

where the sequence $\tilde{u}_i(t^{(n)})$ solves the differential system $(\mathbf{DS_1})$–$(\mathbf{DS_2})$. At time $t^{(n)}$, (7.31) and (7.48) can be written formally in the form

$$\tilde{u}_i'(t^{(n)}) = \tilde{\Phi}_i(t^{(n)}), \quad \forall i = 1, \dots, N+1, \tag{7.58}$$

where the functional $\tilde{\Phi}$ is defined by

$$\tilde{\Phi}_i(t^{(n)}) = -\tilde{u}_i(t^{(n)}) \frac{\tilde{u}_{i+1}(t^{(n)}) - \tilde{u}_{i-1}(t^{(n)})}{2h} \tag{7.59}$$

$$+ v \frac{\tilde{u}_{i-1}(t^{(n)}) - 2\tilde{u}_i(t^{(n)}) + \tilde{u}_{i+1}(t^{(n)})}{h^2}, \quad \forall i \neq N+1,$$

$$\tilde{\Phi}_{N+1}(t^{(n)}) = \frac{2v}{h} f(t^{(n)}) - \frac{2v}{h^2} \left[\tilde{u}_{N+1}(t^{(n)}) - \tilde{u}_N(t^{(n)}) \right] \tag{7.60}$$

$$- \frac{\tilde{u}_{N+1}}{h} \left[\tilde{u}_{N+1}(t^{(n)}) - \tilde{u}_N(t^{(n)}) \right].$$

We now apply a θ-scheme to the functional Eq. (7.58) in the following way:

$$\bar{u}_i^{(n+1)} = \bar{u}_i^{(n)} + k \left[\theta \bar{\Phi}(t^{(n+1)}) + (1-\theta)\bar{\Phi}(t^{(n)}) \right], \tag{7.61}$$

where the approximation $\bar{\Phi}$ has the same definition as $\tilde{\Phi}$ in (7.59) and (7.60), once we have replaced the quantities $\tilde{u}_i(t^{(n)})$ by the new approximations $\bar{u}_i^{(n)}$:

$$\bar{\Phi}_i(t^{(n)}) = -\bar{u}_i^{(n)} \frac{\bar{u}_{i+1}^{(n)} - \bar{u}_{i-1}^{(n)}}{2h} + v \frac{\bar{u}_{i-1}^{(n)} - 2\bar{u}_i^{(n)} + \bar{u}_{i+1}^{(n)}}{h^2}, \quad \forall i \neq N+1, \tag{7.62}$$

$$\bar{\Phi}_{N+1}(t^{(n)}) = \frac{2v}{h} f(t^{(n)}) - \frac{2v}{h^2} \left[\bar{u}_{N+1}^{(n)} - \bar{u}_N^{(n)} \right] - \frac{\bar{u}_{N+1}}{h} \left[\bar{u}_{N+1}^{(n)} - \bar{u}_N^{(n)} \right]. \tag{7.63}$$

Note. The approximation scheme (7.61–7.63) results from a mixed space-domain P_1 finite-element–time-domain finite-difference approximation.

Given that the spatial mesh (7.3) on the interval $[0, L]$ employs a constant discretization spacing h, we have shown that the node equations (7.32) and (7.47) associated with each of the basis functions φ_i, $(i = 1, \dots, N+1)$, coincide with a spatial discretization by finite difference.

In other words, in the precise case of a uniform spatial mesh, the global approximation system (7.61–7.63) is exactly what would be obtained by a finite-difference scheme in the pair of variables (x, t).

It is then legitimate to consider the stability of such a numerical scheme according to the usual methods for dynamical equations, solved by finite-difference methods. However, as discussed at the beginning of this question, the choice of a θ scheme for time-domain discretization is motivated precisely in order to ensure the stability of the resulting method.

We then consider the values of the parameter θ that guarantee the stability of the scheme (7.61–7.63), viz., $\theta \geq 1/2$. Furthermore, we note that apart from the specific value $\theta = 1/2$, the scheme is of first order in time and of second order in space.

Finally, when $\theta = 1/2$, the θ scheme coincides with the Crank–Nicolson method [1], and the approximation to the system (7.61–7.63) is of second order in both time and space.

7.2 Nonlinear Integrodifferential Equation

7.2.1 Statement of the Problem

The aim here is to use a finite-element method to solve a second-order nonlinear integrodifferential equation. More precisely, we shall seek solutions to the following continuous problem:

Find $u \in H^2$ (0,1) solution of

$$(\text{CP}) \quad \begin{cases} -u''(x) + u(x) \displaystyle\int_0^1 u(t)\, dt = f(x), & 0 \le x \le 1, \\ u(0) = 0, \quad u'(1) = \alpha, \end{cases} \tag{7.64}$$

where f is a given function in $L^2(0, 1)$, and α a given parameter.

1) Prove that if u belongs to $H^2(0, 1)$, then the integral of u in the continuous problem (**CP**) is convergent.

▶ Variational Formulation

2) Let v be a real-valued test function defined on $[0, 1]$ and belonging to a space V. Show that the continuous problem (**CP**) can be cast into a variational form (**VP**) defined as follows:

$$(\text{VP}) \quad \begin{cases} \text{Find } u \in V \text{ solution of} \\ a(u, v) = L(v), \quad \forall v \in V. \end{cases} \tag{7.65}$$

Specify the *nonlinear* form $a(., .)$, the *linear* form $L(.)$, and the function space V.

▶ P_1 Lagrange Finite-Element Analysis

3) The variational problem (**VP**) is approximated using a P_1 Lagrange finite-element method. To do this, we introduce a regular mesh on the interval $[0, 1]$, with constant spacing h, such that

$$\begin{cases} x_0 = 0, \quad x_{N+1} = 1, \\ x_{i+1} = x_i + h, i = 0, \dots, N. \end{cases} \tag{7.66}$$

We now define the approximation space $\widetilde{V}$ by

$$\tilde{V} = \left\{\tilde{v} : [0, 1] \longrightarrow \mathbb{R}, \quad \tilde{v} \in C^0([0, 1]), \quad \tilde{v}|_{[x_i, x_{i+1}]} \in P_1([x_i, x_{i+1}]), \quad \tilde{v}(0) = 0\right\},$$
$$(7.67)$$

where $P_1([x_i, x_{i+1}])$ is the space of polynomials on $[x_i, x_{i+1}]$ with degree less than or equal to 1.

− What is the dimension of $\tilde{V}$?

▶ **Approximate Variational Formulation**

4) Let φ_i, $(i = 1, \ldots, \dim \tilde{V})$, be the canonical basis of $\tilde{V}$ satisfying $\varphi_i(x_j) = \delta_{ij}$. Write down the approximate variational formulation $\widetilde{(VP)}$, with solution $\tilde{u}$, associated with the variational problem **(VP)**, and show that by choosing

$$\tilde{v}(x) = \varphi_i(x), \quad i = 1, \ldots, \dim \tilde{V}, \quad \tilde{u}(x) = \sum_{j=1,\ldots,\dim \tilde{V}} \tilde{u}_j \varphi_j, \quad (7.68)$$

we obtain the following system $\widetilde{(VP)}$:

$$\widetilde{(VP)} \quad \sum_{j=1}^{\dim \tilde{V}} A_{ij}\tilde{u}_j + \sum_{j=1}^{\dim \tilde{V}} \sum_{k=1}^{\dim \tilde{V}} B_{ijk}\tilde{u}_j\tilde{u}_k = C_i, \quad \forall i = 1, \ldots, \dim \tilde{V}, \quad (7.69)$$

where

$$A_{ij} = \int_0^1 \varphi_i' \varphi_j' \, dx, \quad B_{ijk} = \left(\int_0^1 \varphi_i \varphi_j \, dx\right)\left(\int_0^1 \varphi_k \, dx\right), \quad (7.70)$$

and

$$C_i = \int_0^1 f\varphi_i \, dx + \alpha\varphi_i(1). \quad (7.71)$$

5) Using the trapezoidal rule, show that B_{ijk} can be estimated by

$$B_{ijk} \approx \begin{cases} h D_{ij}, & \forall k = 1, \ldots, N, \\ \dfrac{h}{2} D_{ij}, & k = N + 1, \end{cases} \quad (7.72)$$

where D_{ij} is defined by

$$D_{ij} \equiv \int_0^1 \varphi_i \varphi_j \, dx. \quad (7.73)$$

Deduce that the approximate variational formulation $\widetilde{(VP)}$ can be written

$$\widetilde{(\text{VP})} \quad \sum_{j=1}^{\dim \tilde{V}} A_{ij}\tilde{u}_j + h\left(\frac{\tilde{u}_{N+1}}{2} + \sum_{k=1}^{N} \tilde{u}_k\right)\left(\sum_{j=1}^{\dim \tilde{V}} D_{ij}\tilde{u}_j\right) = C_i. \quad (7.74)$$

Recall that the *composite* trapezoidal rule is

$$\int_a^b \xi(s)\,ds \approx \frac{h}{2}\left[\xi(a) + \xi(b) + 2\sum_{i=1}^{N}\xi(x_i)\right].$$

▶ **Basis Function φ_i Characterizing a Node in the Interior of [0, 1].**

Given the regularity of the mesh, the generic node equation of the system $\widetilde{(\text{VP})}$ associated with any basis function φ_i, $(i = 1, \ldots, \dim \tilde{V} - 1)$, characterizing a node strictly in the interior of [0, 1] can be written in the following form:

$$\widetilde{(\text{VP}_{\text{int}})} \quad \begin{cases} \forall i = 1, \ldots, \dim \tilde{V} - 1, \\[1ex] A_{i,i-1}\tilde{u}_{i-1} + A_{i,i}\tilde{u}_i + A_{i,i+1}\tilde{u}_{i+1} \\[1ex] +h\left(\dfrac{\tilde{u}_{N+1}}{2} + \displaystyle\sum_{k=1}^{N}\tilde{u}_k\right)\left(D_{i,i-1}\tilde{u}_{i-1} + D_{i,i}\tilde{u}_i + D_{i,i+1}\tilde{u}_{i+1}\right) = C_i. \end{cases}$$

$$(7.75)$$

Use the trapezoidal rule to calculate the seven coefficients A_{ij}, D_{ij}, and C_i.

6) Combine these results to write down the corresponding node equation.

7) Show that we recover the centered finite-difference scheme associated with the differential equation of the continuous problem **(CP)**.

– What is the order of accuracy?

▶ **Basis function φ_{N+1} Characterizing the Node at x_{N+1}**

8) We proceed in the same way for the basis function φ_{N+1} characterizing the last node at x_{N+1}. The corresponding equation of the system $\widetilde{(\text{VP})}$ is

$$\widetilde{(\text{VP}_{N+1})} \quad \begin{cases} A_{N+1,N}\tilde{u}_N + A_{N+1,N+1}\tilde{u}_{N+1} \\[1ex] +h\left(\dfrac{\tilde{u}_{N+1}}{2} + \displaystyle\sum_{k=1}^{N}\tilde{u}_k\right)\left(D_{N+1,N}\tilde{u}_N + D_{N+1,N+1}\tilde{u}_{N+1}\right) = C_{N+1}. \end{cases}$$

$$(7.76)$$

Use the trapezoidal rule to calculate the five coefficients $A_{N+1,N}$, $A_{N+1,N+1}$, $D_{N+1,N}$, $D_{N+1,N+1}$, and C_{N+1}.

9) Combine these results to write down the corresponding node equation.

10) Recover this node equation by discretizing to second order the Neumann boundary condition of the continuous problem (**CP**) at $x_{N+1} = 1$ using the finite-difference method.

7.2.2 Solution

1) The integrodifferential equation of the continuous problem **(CP)** exhibits nonlinearity through the coupling term between u and the integral $\int_0^1 u(x)\,dx$.

We observe that the convergence of this integral is ensured by the function space in which the continuous problem is posed, namely, $H^2(0, 1)$.

Indeed, using the Cauchy–Schwarz inequality, we have

$$\left| \int_0^1 1 \times u(x)\,dx \right| \leq \left(\int_0^1 |1|^2\,dx \right)^{1/2} \left[\int_0^1 |u(x)|^2\,dx \right]^{1/2} \leq \left[\int_0^1 |u(x)|^2\,dx \right]^{1/2}. \tag{7.77}$$

Put another way, if we seek the solution u of the continuous problem **(CP)** in the Sobolev space $H^2(0, 1)$, then u automatically belongs to $L^2(0, 1)$, hence also $L^1(0, 1)$, according to the inequality (7.77), and this implies the integrability of u over the interval $[0, 1]$.

2) Let v be a real-valued test function on $[0, 1]$. We multiply the integrodifferential equation of the continuous problem **(CP)** by v and integrate the resulting equation from 0 to 1.

This yields

$$- \int_0^1 u''v\,dx + \int_0^1 \left[\int_0^1 u(s)\,ds \right] uv\,dx = \int_0^1 fv\,dx. \tag{7.78}$$

As usual, we specify the function space V once we have firmly established the variational formulation.

Integrating by parts and using the Neumann condition

$$u'(1) = \alpha,$$

we then have

$$\int_0^1 \left\{ u'v' + \left[\int_0^1 u(s)\,ds \right] uv \right\} dx + u'(0)v(0) = \int_0^1 fv\,dx + \alpha v(1). \tag{7.79}$$

Since this formulation does not allow us to take into account the homogeneous Dirichlet condition $u(0) = 0$, we require the functions v to satisfy this condition, i.e., $v(0) = 0$.

The variational formulation **(VP)** thus becomes

$$
\textbf{(VP)} \quad \begin{cases} \text{Find } u \in V \text{ solution of } a(u, v) = L(v), \quad \forall, v \in V, \\[2mm] a(u, v) \equiv \displaystyle\int_0^1 \left\{ u'(x)v'(x) + \left[\int_0^1 u(s)\, ds \right] u(x)v(x) \right\} dx, \\[4mm] L(v) \equiv \displaystyle\int_0^1 f(x)v(x)\, dx + \alpha v(1). \end{cases} \quad (7.80)
$$

We can now consider the nature of the function space V, the aim being to ensure that the variational formulation **(VP)** makes mathematical sense.

Regarding the integrals of $u'v'$ and fv, we have seen on several occasions, e.g., the Dirichlet problem in Sect. 4.1 and the Neumann problem in Sect. 4.2, that the Cauchy–Schwarz inequality can be used to ensure their existence.

Regarding the integral of the nonlinear term, we simply note that

$$
\int_0^1 \left\{ \left[\int_0^1 u(s)\, ds \right] u(x)v(x) \right\} dx = \left[\int_0^1 u(s)\, ds \right] \left[\int_0^1 u(x)v(x)\, dx \right]. \quad (7.81)
$$

Convergence of the integral is once again ensured by applying the Cauchy–Schwarz inequality to the integral of uv.

The function space V that gives mathematical sense to the variational formulation **(VP)** is therefore defined by

$$
V \equiv H^1(0, 1) \cap \left\{ v : (0, 1) \to \mathbb{R}, \text{ such that } v(0) = 0 \right\}. \quad (7.82)
$$

3) The dimension of the approximation space $\tilde{V}$ can be determined in several ways. The simplest is to note that the functions $\tilde{v}$ in $\tilde{V}$ are basically made up of straight line segments.

In fact, they are affine on each mesh element $[x_i, x_{i+1}]$ and vanish at $x = 0$. Since the whole mesh on the interval $[0, 1]$ contains $N + 2$ discretization points, two functions of $\tilde{V}$ can be distinguished by the differences between their values at the $N + 1$ points $x_1, \ldots, x_{N+1}$, where we note, of course, that any function $\tilde{v}$ in V must satisfy $\tilde{v}_0 = 0$.

Put another way, a function $\tilde{v}$ in $\tilde{V}$ is fully determined by the $(N + 1)$-tuplet $(\tilde{v}_1, \ldots, \tilde{v}_{N+1})$. This implies that the space $\tilde{V}$ is isomorphic to $\mathbb{R}^{N+1}$.

In conclusion, we deduce that the dimension of $\tilde{V}$ is $N + 1$.

4) The approximate variational formulation $\widetilde{\textbf{(VP)}}$ is obtained by substituting the approximation functions $\tilde{u}$ and $\tilde{v}$ for the functions u and v in the variational formulation **(VP)**.

We use the expressions given in (7.68) to obtain

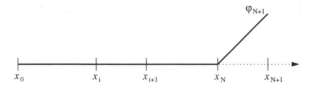

Fig. 7.3 Basis function φ_{N+1}

$$
\left|
\begin{array}{l}
\text{Find } (\tilde{u}_j) \ j = 1, \ldots, N+1, \quad \text{solution of} \\[2mm]
\displaystyle\sum_{j=1,\ldots,N+1} \left(\int_0^1 \left\{ \varphi_j' \varphi_i'(x) + \left[\sum_{k=1,\ldots,N+1} \int_0^1 \tilde{u}_k \varphi_k(s)\, ds \right] \varphi_j(x)\varphi_i(x) \right\} dx \right) \tilde{u}_j \\[6mm]
\displaystyle = \int_0^1 f(x)\varphi_i(x)\, dx + \alpha\varphi_i(1).
\end{array}
\right.
$$

(7.83)

We then note that the integral of $\tilde{u}_k \varphi_k(s)$ is independent of the variable x in the main integral of (7.83).

By comparison, we therefore obtain the expressions (7.70) and (7.71).

5) To estimate the quantities B_{ijk}, we use the trapezoidal rule to approximate the integral of the function φ_k :

$$
\int_0^1 \varphi_k(s)\, ds = \int_{x_{k-1}}^{x_{k+1}} \varphi_k(s)\, ds = \int_{x_{k-1}}^{x_k} \varphi_k(s)\, ds + \int_{x_k}^{x_{k+1}} \varphi_k(s)\, ds
$$

$$
\approx \frac{h}{2}(1+0) + \frac{h}{2}(0+1) = h, \quad \forall k = 1, \ldots, N. \qquad (7.84)
$$

The basis function φ_{N+1} must be treated separately since its support contains only the interval $[x_N, x_{N+1}]$ (see Fig. 7.3).

We deduce that

$$
\int_0^1 \varphi_{N+1}(s)\, ds = \int_{x_N}^{x_{N+1}} \varphi_{N+1}(s)\, ds \approx \frac{h}{2}(1+0) = \frac{h}{2}. \qquad (7.85)
$$

We then insert the evaluations from (7.84)–(7.85) into the generic equation of the approximate variational problem (7.83), which leads to (7.74).

6) We now consider the basis functions φ_i, $i = 1, \ldots, N$, characterizing the nodes strictly in the interior of the integration interval $[0, 1]$.

A priori, the only terms in the generic equation of the system (7.83) that could be nonzero are those corresponding to functions φ_j whose supports intersect the support of the chosen function φ_i (see Fig. 7.4), i.e., the basis functions φ_{i-1}, φ_i, and φ_{i+1}.

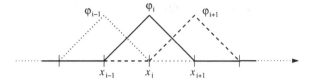

Fig. 7.4 Basis functions φ_{i-1}, φ_i, and φ_i

For this reason, the equation $(\widetilde{\mathbf{VP}}_{\mathbf{int}})$ contains only the terms $A_{i,i-1}$, $A_{i,i}$, $A_{i,i+1}$, on the one hand, and $D_{i,i-1}$, $D_{i,i}$, and $D_{i,i+1}$, on the other.

Exact Calculation of the Coefficients A_{ij}, $j = i - 1, i, i + 1$.

1. *Calculation of the coefficient A_{ii}.*

   Here we have

$$A_{ii} = \int_0^1 (\varphi_i')^2 \, dx = \int_{\text{supp}\,\varphi_i'} (\varphi_i')^2 \, dx = \int_{x_{i-1}}^{x_i} (\varphi_i')^2 \, dx + \int_{x_i}^{x_{i+1}} (\varphi_i')^2 \, dx.$$
(7.86)

Since the basis functions φ_i in $\tilde{V}$ are piecewise affine, the derivatives φ_i' are constant on each mesh element of the form $[x_i, x_{i+1}]$.

We may thus evaluate each integral in (7.86) exactly, or apply the trapezoidal rule, which is exact for constant functions:

$$A_{ii} = h \times \left(\frac{1}{h}\right)^2 + h \times \left(-\frac{1}{h}\right)^2 = \frac{2}{h}.$$
(7.87)

2. *Calculation of the coefficient $A_{i,i-1}$.*

   This time we have

$$A_{i,i-1} = \int_0^1 \varphi_i' \varphi_{i-1}' \, dx = \int_{\text{supp}\,\varphi_i' \,\cap\, \text{supp}\,\varphi_{i-1}'} \varphi_i' \varphi_{i-1}' \, dx = \int_{x_{i-1}}^{x_i} \varphi_i' \varphi_{i-1}' \, dx$$

$$= h \times \left(\frac{1}{h} \times -\frac{1}{h}\right) = -\frac{1}{h}.$$
(7.88)

3. *Calculation of the coefficient $A_{i,i+1}$.*

   This is direct, since we simply note that

$$A_{i,i+1} = A_{i+1,i} = A_{i,i-1}.$$
(7.89)

Here we used the symmetry of the matrix $A_{i,j}$ and the horizontal translational invariance along the mesh, a consequence of its uniformity.

Approximate Calculation of the Coefficients D_{ij}, $j = i - 1, i, i + 1$.

1. *Calculation of the coefficient D_{ii}.*

   Here we have

$$D_{ii} = \int_0^1 \varphi_i^2 \, dx = \int_{\text{supp}\,\varphi_i} \varphi_i^2 \, dx = \int_{x_{i-1}}^{x_i} \varphi_i^2 \, dx + \int_{x_i}^{x_{i+1}} \varphi_i^2 \, dx$$

$$\approx \frac{h}{2}(0 + 1) + \frac{h}{2}(1 + 0) = h. \tag{7.90}$$

2. *Calculation of the coefficient $D_{i,i-1}$.*

   In this case,

$$D_{i,i-1} = \int_0^1 \varphi_i \varphi_{i-1} \, dx = \int_{\text{supp}\,\varphi_i \,\cap\, \text{supp}\,\varphi_{i-1}} \varphi_i \varphi_{i-1} \, dx = \int_{x_{i-1}}^{x_i} \varphi_i \varphi_{i-1} \, dx$$

$$\approx \frac{h}{2}(0 \times 1 + 1 \times 0) = 0. \tag{7.91}$$

3. *Calculation of the coefficient $D_{i,i+1}$.*

   For reasons of symmetry, as in the calculation of the coefficients $A_{i,i+1}$, we obtain

$$D_{i,i+1} = D_{i+1,i} = D_{i,i-1} \equiv 0. \tag{7.92}$$

Calculation of C_i.

Given the property of the basis functions that $\forall i = 1, \ldots, N$, $\varphi_i(1) = 0$, the right-hand side C_i can be estimated as follows:

$$C_i = \int_0^1 f \varphi_i \, dx = \int_{x_{i-1}}^{x_i} f \varphi_i \, dx + \int_{x_i}^{x_{i+1}} f \varphi_i \, dx$$

$$\approx \frac{h}{2}(0 + f_i) + \frac{h}{2}(f_i + 0), \tag{7.93}$$

$$C_i \approx h f_i. \tag{7.94}$$

7) The node equation associated with a basis function φ_i, $(i = 1, \ldots, N)$, is obtained by gathering the results from the last question:

$$\forall i = 1, \ldots, N, \quad -\frac{\tilde{u}_{i-1} - 2\tilde{u}_i + \tilde{u}_{i+1}}{h^2} + h\left(\frac{\tilde{u}_{N+1}}{2} + \sum_{k=1}^N \tilde{u}_k\right)\tilde{u}_i = f_i. \tag{7.95}$$

8) We now recover the node equation (7.95) associated with any basis function φ_i, $i = 1, \ldots, N$, using the finite-difference method.

To do this, we write down the integrodifferential equation of the continuous problem **(CP)** at the point x_i, then approximate the second derivative of u and the integral of the solution u over the interval $[0, 1]$.

Concerning the second derivative, we use the forward and backward Taylor expansions:

$$u(x_{i+1}) = u(x_i) + hu'(x_i) + \frac{h^2}{2}u''(x_i) + \frac{h^3}{3!}u^3(x_i) + O(h^4), \qquad (7.96)$$

$$u(x_{i-1}) = u(x_i) - hu'(x_i) + \frac{h^2}{2}u''(x_i) - \frac{h^3}{3!}u^3(x_i) + O(h^4). \qquad (7.97)$$

Subtracting one from the other, we then have

$$u''(x_i) = \frac{u(x_{i+1}) - 2u(x_i) + u(x_{i-1})}{h^2} + O(h^2). \qquad (7.98)$$

Regarding the integral of u from 0 to 1, we use the composite trapezoidal rule, noting that the solution u satisfies a homogeneous Dirichlet condition at $x = 0$:

$$\int_0^1 u(s)\,ds = \frac{h}{2}\left[u(x_{N+1}) + 2\sum_{i=1}^N u(x_i)\right] + O(h^2). \qquad (7.99)$$

The integrodifferential equation of the continuous problem **(CP)** at the point x_i is then rewritten, replacing $u''(x_i)$ by (7.98) and $\int_0^1 u(s)\,ds$ by (7.99):

$$-\frac{u(x_{i+1}) - 2u(x_i) + u(x_{i-1})}{h^2} + u(x_i)\frac{h}{2}\left[u(x_{N+1}) + 2\sum_{i=1}^N u(x_i)\right] = f(x_i) + O(h^2). \qquad (7.100)$$

We now apply the finite-difference method by substituting the sequence of approximations $\tilde{u}_i$ for the true values $u(x_i)$.

We can thus neglect the second-order remainder $O(h^2)$ in (7.100), whereupon we obtain precisely the node equation (7.95) found by the finite-element method.

In addition, the finite-difference scheme (7.95) is of second order, since we neglected the term going as $O(h^2)$.

A General Remark

As usual when applying the Taylor expansion in such a context, we note that we have assumed a certain level of regularity for the solution u of the continuous problem **(CP)**: we assume in our case that u is at least C^4 on $[0, 1]$ in order to be able to Taylor expand to order 4, whereas it would seem at first glance that it should rather be C^2 on $[0, 1]$.

In fact, even though in some situations it would be possible to establish that the solution u has greater regularity, most cases deserve some kind of explanation. Indeed, the regularity of the solution u depends on the regularity of the right-hand side, but also on the regularity intrinsic to the structure of the differential operator.

So how can we explain the freedom to choose the regularity appropriate to a solution u of the continuous problem, in order to obtain a suitable Taylor expansion, when in most cases the differential equations would not actually allow it?

This solution does not seem totally satisfactory, but it is often the kind of compromise that must be made in numerical analysis (as in many other fields, scientific or otherwise). This is why the Taylor expansions used are valid only for the class of differential equations that have "sufficiently regular" solutions.

And even when a differential equation has a solution that can be shown to fall short of the regularity required for the Taylor expansion, we nevertheless retain that expansion, knowing full well that in the end, the objective is to construct a sequence of plausible approximations that really will approach the exact solution to the problem.

So having assumed more regularity than was really justified, we will be able to exploit an algebraic process to suggest an approximation of the differential operator of suitable quality.

This quality, known in the jargon of numerical analysis as the order of the finite-difference scheme, allows us at the end of the day to measure and assess, for given and sufficient regularity of the solutions of the differential equations, the performance of the various finite-difference schemes.

▶ Basis Function φ_{N+1} Characterizing the Node at x_{N+1}

9) We now obtain the equation of the system $\widetilde{(\mathbf{VP})}$ defined by (7.74) for the particular basis function φ_{N+1}, which leads to (7.76). The five coefficients of this node equation are found using similar calculations to thoaw presented in answer to the last question.

Exact Calculation of the Coefficients $A_{N+1, N+1}$. and $A_{N+1, N}$.

1. *Calculation of the coefficient $A_{N+1, N+1}$.*

We have

$$A_{N+1, N+1} = \int_0^1 (\varphi'_{N+1})^2 \, dx = \int_{\text{supp}\,\varphi'_{N+1}} (\varphi'_{N+1})^2 \, dx = \int_{x_N}^{x_{N+1}} (\varphi'_{N+1})^2 \, dx$$

$$= h \times \left(\frac{1}{h}\right)^2 = \frac{1}{h}. \tag{7.101}$$

2. *Calculation of the coefficient $A_{N+1,\,N}$.*

In this case,

$$
\begin{aligned}
A_{N+1,\,N} &= \int_0^1 \varphi'_{N+1}\varphi'_N \,dx = \int_{\operatorname{supp}\varphi'_{N+1}\cap\operatorname{supp}\varphi'_N} \varphi'_{N+1}\varphi'_N \,dx \\
&= \int_{x_N}^{x_{N+1}} \varphi'_{N+1}\varphi'_N \,dx = h \times \left(\frac{1}{h}\times -\frac{1}{h}\right) = -\frac{1}{h}. \quad (7.102)
\end{aligned}
$$

Approximate Calculation of the Coefficients $D_{N+1,\,N+1}$ and $D_{N+1,\,N}$.

1. *Calculation of the coefficient $D_{N+1,\,N+1}$.*

We have

$$
\begin{aligned}
D_{N+1,\,N+1} &= \int_0^1 \varphi^2_{N+1}\,dx = \int_{\operatorname{supp}\varphi_{N+1}} \varphi^2_{N+1}\,dx = \int_{x_N}^{x_{N+1}} \varphi^2_{N+1}\,dx \\
&\approx \frac{h}{2}(0+1) = \frac{h}{2}. \quad (7.103)
\end{aligned}
$$

2. *Calculation of the coefficient $D_{N+1,\,N}$.*

In this case,

$$
\begin{aligned}
D_{N+1,\,N} &= \int_0^1 \varphi_{N+1}\varphi_N \,dx = \int_{\operatorname{supp}\varphi_{N+1}\cap\operatorname{supp}\varphi_N} \varphi_{N+1}\varphi_N \,dx \\
&= \int_{x_N}^{x_{N+1}} \varphi_{N+1}\varphi_N \,dx \approx \frac{h}{2}(0\times 1 + 1\times 0) = 0. \quad (7.104)
\end{aligned}
$$

Calculation of C_{N+1}.

We have

$$
\begin{aligned}
b_{N+1} &= \int_0^1 f\varphi_{N+1}\,dx = \int_{x_N}^{x_{N+1}} f\varphi_{N+1}\,dx + \alpha \\
&\approx \frac{h}{2}(0 + f_{N+1}) + \alpha, \quad (7.105)
\end{aligned}
$$

$$
b_{N+1} \approx \frac{h}{2}f_{N+1} + \alpha.
$$

10) We gather the results of the last question and construct the node equation corresponding to the basis function φ_{N+1}, whence

$$
-\frac{1}{h}\tilde{u}^N + \frac{1}{h}\tilde{u}^{N+1} + h\left(\frac{\tilde{u}_{N+1}}{2} + \sum_{k=1}^N \tilde{u}_k\right)\frac{h}{2}\tilde{u}_{N+1} = \frac{h}{2}f_{N+1} + \alpha. \quad (7.106)
$$

Rearranging, we obtain

$$\frac{2}{h^3}(\tilde{u}_{N+1} - \tilde{u}_N) + \frac{\tilde{u}_{N+1}^2}{2} + \tilde{u}_{N+1} \sum_{k=1}^{N} \tilde{u}_k = \frac{f_{N+1}}{h} + \frac{2\alpha}{h^2}. \tag{7.107}$$

11) We now carry out the finite-difference discretization of the Neumann condition $u'(1) = \alpha$.

To do this, we do a backward Taylor expansion to order 3, the point being to attain order 2 for the finite-difference approximation to the continuous problem **(CP)**:

$$u(x_N) = u(x_{N+1}) - hu'(x_{N+1}) + \frac{h^2}{2}u''(x_{N+1}) + O(h^3). \tag{7.108}$$

We also assume that the integrodifferential equation is valid at $x = 1$. This presupposes once again that the solution u of the continuous problem **(CP)** has sufficient regularity.

We can now replace the second derivative of u at x_{N+1} by its expression in terms of the values of u at the other nodes of the mesh:

$$u(x_N) = u(x_{N+1}) - \alpha h + \frac{h^2}{2}\left[u(x_{N+1}) \int_0^1 u(s)\,ds - f(x_{N+1}) \right] + O(h^3). \tag{7.109}$$

Using the composite trapezoidal rule, we obtain finally

$$u(x_N) = u(x_{N+1}) - \alpha h - \frac{h^2}{2}f(x_{N+1}) + \frac{h^3}{2}u(x_{N+1})\left[\frac{u(x_{N+1})}{2} + \sum_{k=1}^{N} \tilde{u}_k \right] + O(h^3). \tag{7.110}$$

Finally, we multiply (7.110) by $2/h^3$ and rearrange the terms to obtain precisely the node equation (7.107).

————————————

7.3 Riccati Differential Equation

7.3.1 Statement of the Problem

In this section, we describe the numerical solution of the nonlinear Riccati equation using a P_1 finite-element analysis. In other words, we will consider the scalar function u of the variable x that solves the continuous problem **(CP)** specified as follows:
Find $u \in H^1(0, 1)$ solution of

$$\textbf{(CP)} \quad \begin{cases} u'(x) + u^2(x) = f(x) \text{ for } 0 \le x \le 1, \\ u(0) = 0, \end{cases} \tag{7.111}$$

where f is a given function in $L^2(0, 1)$.

1) Let v be a real-valued test function on $[0, 1]$ belonging to the variational space V. Show that the continuous problem **(CP)** can be cast into a variational form **(VP)** to be specified.

– What are the properties that the functions v in V must satisfy?

►**Lagrange P_1 Finite-Element Analysis**

2) The variational problem **(VP)** is approximated by a Lagrange P_1 finite-element analysis. To do this, we introduce a regular mesh on the interval $[0, 1]$, with constant spacing h, such that

$$\begin{cases} x_0 = 0, \ x_{N+1} = 1, \\ x_{i+1} = x_i + h, \ i = 0, \dots, N. \end{cases} \tag{7.112}$$

We now define the approximation space $\tilde{V}$ by

$$\tilde{V} = \left\{ \tilde{v} : [0, 1] \longrightarrow \mathbb{R} \text{ s.t. } \tilde{v} \in C^0([0, 1]), \ \tilde{v}|_{[x_i, x_{i+1}]} \in P_1([x_i, x_{i+1}]), \ \tilde{v}(0) = 0 \right\}, \tag{7.113}$$

where $P_1([x_i, x_{i+1}])$ is the space of polynomials on $[x_i, x_{i+1}]$ of degree less than or equal to 1.

– What is the dimension of $\tilde{V}$?

3) Let φ_i, $(i = 1, \dots, \dim \tilde{V})$, be the basis of $\tilde{V}$ satisfying $\varphi_i(x_j) = \delta_{ij}$. Write down the approximate variational formulation with solution $\tilde{u}$ associated with the variational problem **(VP)**, and show that by choosing

$$\tilde{v}(x) = \varphi_i(x), \qquad \tilde{u}(x) = \sum_{j=1,\dots,\dim \tilde{V}} \tilde{u}_j \varphi_j, \qquad (7.114)$$

we obtain the following system $\widetilde{(\mathbf{VP})}$:

$$\widetilde{(\mathbf{VP})} \qquad \sum_{j=1,\dots,\dim \tilde{V}} A_{ij}\tilde{u}_j + \sum_{(j,k)\in\{1,\dots,\dim \tilde{V}\}} B_{ijk}\tilde{u}_j\tilde{u}_k = C_i, \quad \forall i \in \{1,\dots,\dim \tilde{V}\},$$

$$(7.115)$$

where

$$A_{ij} = \int_0^1 \varphi_i \varphi_j' \, dx, \quad B_{ijk} = \int_0^1 \varphi_i \varphi_j \varphi_k \, dx, \quad C_i = \int_0^1 f \varphi_i \, dx. \qquad (7.116)$$

What is the key feature of the system $\widetilde{(\mathbf{VP})}$?

▶ Basis Function φ_i Characterizing a Node Strictly in the Interior of [0, 1]

4) Due to the regularity of the mesh, the generic node equation of the system $\widetilde{(\mathbf{VP})}$ associated with any basis function φ_i characterizing a node strictly in the interior of [0, 1] can be written in the form

$$\widetilde{(\mathbf{VP}_{\text{int}})} \begin{cases} \forall i = 1,\dots,\dim \tilde{V} - 1, \\[4pt] A_{i,i-1}\,\tilde{u}_{i-1} + A_{i,i}\tilde{u}_i + A_{i,i+1}\tilde{u}_{i+1} \\[4pt] +(B_{i,i-1,i} + B_{i,i,i-1})\tilde{u}_i\tilde{u}_{i-1} + B_{i,i-1,i-1}\tilde{u}_{i-1}^2 + B_{i,i,i}\tilde{u}_i^2 \\[4pt] +(B_{i,i,i+1} + B_{i,i+1,i})\tilde{u}_i\tilde{u}_{i+1} + B_{i,i+1,i+1}\tilde{u}_{i+1}^2 = C_i. \end{cases} \qquad (7.117)$$

Using the trapezoidal rule, calculate the 11 coefficients $A_{i,j}$, $B_{i,j,k}$, and C_i. Recall that the trapezoidal rule is

$$\int_a^b \xi(s)\,ds \approx \frac{b-a}{2}\big[\xi(a) + \xi(b)\big].$$

5) Gather the above results to write down the corresponding node equation.

6) Show that we recover the centered finite-difference scheme associated with the Riccati equation of the continuous problem **(CP)**.

– What is the order of accuracy?

► Basis Function φ_{N+1} Characterizing the Node at x_{N+1}

7) We proceed in the same way for the basis function φ_{N+1} characterizing the final node x_{N+1}. The corresponding equation of the system $\widetilde{(VP)}$ is then

$$\widetilde{(PV_{ext})} \quad \begin{cases} A_{N+1,N}\tilde{u}_N + A_{N+1,N+1}\,\tilde{u}_{N+1} + B_{N+1,N,N}\tilde{u}_N^2 \\ +\left(B_{N+1,N,N+1} + B_{N+1,N+1,N}\right)\tilde{u}_N\tilde{u}_{N+1} + B_{N+1,N+1,N+1}\tilde{u}_{N+1}^2 = C_{N+1}. \end{cases}$$
$$(7.118)$$

Using the trapezoidal rule, calculate the seven coefficients $A_{i,j}$, $B_{i,j,k}$, and C_i of equation $\widetilde{(VP_{ext})}$.

8) Combine these results to write down the corresponding node equation.

9) Show that we recover the node equation $\widetilde{(VP_{ext})}$ by the finite-difference method. What is the order of approximation in the resulting scheme?

7.3.2 Solution

▶ **Theoretical Part**

1) Let v be a real-valued function on $[0, 1]$ belonging to the variational space V. This space will be characterized with hindsight once the variational formulation has been established formally.

Multiplying the Riccati equation of the continuous problem **(CP)** specified by (7.111) by v and integrating from 0 to 1, we obtain a variational formulation **(VP)** of the following form:

(VP)
$$\begin{cases} \text{Find } u \in V \text{ solution of} \\ \displaystyle\int_0^1 u'(x)v(x)\,dx + \int_0^1 u^2(x)v(x)\,dx = \int_0^1 f(x)v(x)\,dx, \quad \forall v \in V. \end{cases}$$
$$(7.119)$$

Before specifying the space V, note that given the structure of the two integrands, no integration by parts would lead to a more tractable variational formulation.

For this reason, the variational Eq. (7.119) will be retained throughout the rest of the problem.

We now consider the regularity of the functions v in V in order to establish sufficient conditions for the existence of the integrals in (7.119). We bound the first integral of (7.119) using the Cauchy–Schwarz inequality as follows:

$$\left| \int_0^1 u'(x)v(x)\,dx \right| \le \int_0^1 |u'(x)v(x)|\,dx$$
$$\le \left[\int_0^1 |u'(x)|^2\,dx \right]^{1/2} \left[\int_0^1 |v(x)|^2\,dx \right]^{1/2}. \quad (7.120)$$

So if u' and v are two functions that we require to have the regularity of the functions in $L^2(0, 1)$, then the integral $\displaystyle\int_0^1 u'(x)v(x)\,dx$ will converge.

Regarding the second integral, we proceed as follows:

$$\left| \int_0^1 u^2(x)v(x)\,dx \right| \le \int_0^1 |u^2(x)v(x)|\,dx$$
$$\le C \int_0^1 |v(x)|\,dx \le C \left[\int_0^1 |v(x)|^2\,dx \right]^{1/2}, \quad (7.121)$$

where

$$C = \max_{x \in [0, 1]} [u^2(x)].$$

In other words, if we take u and v to run over the space $H^1(0, 1)$, we can exploit the continuous injection of the Sobolev space

$$H^1(0, 1) \subset C^0([0, 1]), \tag{7.122}$$

whereupon we may justify introducing the finite constant C as the maximum of the function u^2. Furthermore, the integral $\int_0^1 u'(x)v(x)\,\mathrm{d}x$ will converge as a consequence.

Finally, in order to conserve all the information contained in the formulation of the continuous problem (CP), i.e., the Dirichlet boundary condition $u(0) = 0$, we require the functions v in V to satisfy this same boundary condition, i.e., $v(0) = 0$, so that the solution u of the variational problem (VP) retains this homogeneity property at $x = 0$, being a specific function v in V.

We thus end up with

$$V \equiv H_0^1(0, 1) = H^1(0, 1) \cap \left\{v : v(0) = 0\right\}.$$

▶ Lagrange P_1 Finite-Element Analysis

2) The dimension of the approximation space $\tilde{V}$ can be determined in several ways. The simplest is to note that the functions $\tilde{v}$ of $\tilde{V}$ are essentially composed of straight line segments, affine on each element $[x_i, x_{i+1}]$ of the mesh and vanishing at $x = 0$.

For this reason, since there are $N + 2$ discretization points for the whole mesh on the interval $[0, 1]$, two functions in $\tilde{V}$ can be distinguished by the difference between their values at the $N + 1$ points $x_1, \ldots, x_{N+1}$, where we bear in mind that every function $\tilde{v}$ of V must satisfy $\tilde{v}_0 = 0$.

Put another way, a function $\tilde{v}$ in $\tilde{V}$ is fully determined by the $(N + 1)$-tuplet $(\tilde{v}_1, \ldots, \tilde{v}_{N+1})$ specifying its values at the discretization nodes $x_1, \ldots, x_{N+1}$.

This, in turn, implies that the space $\tilde{V}$ is isomorphic to $\mathbb{R}^{N+1}$. In conclusion, we deduce that the dimension of $\tilde{V}$ is $N + 1$.

3) The approximate variational formulation $\widetilde{(VP)}$ is obtained from the variational formulation (VP) by replacing the pair of functions $(u, v) \in V \times V$ by the functions $(\tilde{u}, \tilde{v})$ ranging over $\tilde{V} \times \tilde{V}$, whence we obtain

$$\int_0^1 \tilde{u}'(x)\tilde{v}(x)\,\mathrm{d}x + \int_0^1 \tilde{u}^2(x)\tilde{v}(x)\,\mathrm{d}x = \int_0^1 f(x)\tilde{v}(x)\,\mathrm{d}x, \quad \forall \tilde{v} \in \tilde{V}. \tag{7.123}$$

We now choose $\tilde{v} = \varphi_i$ and expand $\tilde{u}$ in terms of the basis functions $(\varphi_j)_{j=1,\ldots,\dim \tilde{V}}$ of $\tilde{V}$. This leads immediately to the variational formulation $\widetilde{(VP)}$ as specified in (7.115) and (7.116).

The key feature of this algebraic system (7.115) and (7.116) in the unknown quantities $\tilde{u}_1, \ldots, \tilde{u}_{N+1}$ is that it is nonlinear in those quantities. Naturally, this nonlinearity has been inherited from the same property of the Riccati equation, having survived the discretization process.

And it is not totally unproblematic, since methods of solution such as those of Gauss, Jacobi, Gauss–Seidel, successive relaxation, or again, simple or conjugate gradient methods, will be of no assistance in such an instance. We thus turn to Newton-type methods, as discussed in [3], for example.

▶ Basis Function φ_i Characterizing a Node Strictly in the Interior of [0, 1]

4) In the system (7.115) and (7.116), we now consider the equations corresponding to basis functions φ_i characterizing nodes in the interior of the mesh, i.e., equal to 1 at the given node and 0 at all the others.

In other words, in the present case, we are concerned with the values of i between 1 and N.

In addition, given that the support of a function φ_i comprises solely the segment $[x_{i-1}, x_{i+1}]$, only the basis functions φ_{i-1}, φ_i, and φ_{i+1} can produce nonzero contributions when we calculate the coefficients A_{ij} and B_{ijk}. Indeed, we have

$$A_{ij} = \int_0^1 \varphi_i \varphi_j' \, dx = \int_{\text{supp}\,\varphi_i \,\cap\, \text{supp}\,\varphi_j'} \varphi_i \varphi_j' \, dx = \int_{\text{supp}\,\varphi_i \,\cap\, \text{supp}\,\varphi_j} \varphi_i \varphi_j' \, dx \,,$$

$$B_{ijk} = \int_0^1 \varphi_i \varphi_j \varphi_k \, dx = \int_{\text{supp}\,\varphi_i \,\cap\, \text{supp}\,\varphi_j \,\cap\, \text{supp}\,\varphi_k} \varphi_i \varphi_j \varphi_k \, dx.$$

$$\text{(7.124)}$$

So when i is fixed (between 1 and N, since we are concerned only with internal nodes), the corresponding equation of the system (7.115) and (7.116) can be expressed by (7.117).

Approximate Calculation of the Coefficients A_{ij}, $j = i - 1, i, i + 1$.

We now estimate the coefficients A_{ij} and B_{ijk} using the trapezoidal rule:

1. *Calculation of the coefficient A_{ii}.*

Here we have

$$A_{ii} = \int_0^1 \varphi_i \varphi_i' \, dx = \int_{\text{supp}\,\varphi_i} \varphi_i \varphi_i' \, dx = \int_{x_{i-1}}^{x_i} \varphi_i \varphi_i' \, dx + \int_{x_i}^{x_{i+1}} \varphi_i \varphi_i' \, dx \,,$$

$$\approx \frac{1}{h} \times \frac{h}{2} \Big[\varphi_i(x_{i-1}) + \varphi_i(x_i) \Big] - \frac{1}{h} \times \frac{h}{2} \Big[\varphi_i(x_i) + \varphi_i(x_{i+1}) \Big]$$

$$= \frac{1}{2}(0 + 1) - \frac{1}{2}(1 + 0) = 0. \tag{7.125}$$

2. *Calculation of the coefficient $A_{i,i-1}$.*

   In this case,

$$A_{i,i-1} = \int_0^1 \varphi_i \varphi'_{i-1} \, dx = \int_{\text{supp } \varphi_i \, \cap \, \text{supp } \varphi'_{i-1}} \varphi_i \varphi'_{i-1} \, dx = \int_{x_{i-1}}^{x_i} \varphi_i \varphi'_{i-1} \, dx$$

$$\approx -\frac{1}{h} \times \frac{h}{2} \left[\varphi_i(x_{i-1}) + \varphi_i(x_i) \right]$$

$$= -\frac{1}{2}(0+1) = -\frac{1}{2} . \tag{7.126}$$

3. *Calculation of the coefficient $A_{i,i+1}$.*

   Here, we have

$$A_{i,i+1} = \int_0^1 \varphi_i \varphi'_{i+1} \, dx = \int_{\text{supp } \varphi_i \, \cap \, \text{supp } \varphi'_{i+1}} \varphi_i \varphi'_{i+1} \, dx = \int_{x_i}^{x_{i+1}} \varphi_i \varphi'_{i+1} \, dx$$

$$\approx +\frac{1}{h} \times \frac{h}{2} \left[\varphi_i(x_i) + \varphi_i(x_{i+1}) \right]$$

$$= \frac{1}{2}(1+0) = +\frac{1}{2} . \tag{7.127}$$

Approximate Calculation of the Coefficients B_{ijk}, $(j,k) \in \{i-1, i, i+1\}^2$.

From the general expression (7.124) for the coefficients B_{ijk}, and given the characteristic property of each of the basis functions φ_i, viz., $\varphi_i(x_j) = \delta_{ij}$, only the coefficient $B_{i,i,i}$ can be nonzero.

Indeed, considering the application of the trapezoidal rule to evaluate the integrals in the expression for the coefficient B_{ijk}, we systematically obtain terms of the form $\varphi_i(x_l)\varphi_j(x_l)\varphi_k(x_l)$.

This is why all the coefficients B_{ijk} will be zero in the approximation by the trapezoidal rule, with the exception of $B_{i,i,i}$, for which we obtain

$$B_{i,i,i} = \int_{x_{i-1}}^{x_i} \varphi_i^3(x) \, dx + \int_{x_i}^{x_{i+1}} \varphi_i^3(x) \, dx \approx \frac{h}{2}(0+1) + \frac{h}{2}(1+0) = h. \tag{7.128}$$

Approximate Calculation of the Coefficient C_i.

The right-hand side C_i is found similarly, whence

$$C_i = \int_0^1 f(x)\varphi_i(x) \, dx = \int_{x_{i-1}}^{x_i} f(x)\varphi_i(x) \, dx + \int_{x_i}^{x_{i+1}} f(x)\varphi_i(x) \, dx$$

$$\approx \frac{h}{2}(0 + f_i) + \frac{h}{2}(f_i + 0) = hf_i , \tag{7.129}$$

where f_i is the value of the function f at the node x_i.

5) With the results of the above calculations, we can write down the corresponding node equation for each basis function φ_i characterizing a node in the interior of the mesh:

$$\frac{1}{2}\left(\tilde{u}_{i+1} - \tilde{u}_{i-1}\right) + h\tilde{u}_i^2 = hf_i, \quad i = 1, \ldots, N. \tag{7.130}$$

6) The node equation (7.130) corresponds exactly to the centered finite-difference scheme associated with the continuous problem (CP). Indeed, this is immediate, since it consists in writing the Riccati equation (7.111) at the point x_i, then replacing the first derivative of u by a second-order approximation:

$$u'(x_i) + u^2(x_i) = f(x_i), \quad i = 1, \ldots, N. \tag{7.131}$$

We now carry out forward and backward Taylor expansions, i.e.,

$$u(x_{i+1}) = u(x_i) + hu'(x_i) + \frac{h^2}{2}u''(x_i) + O(h^3), \tag{7.132}$$

$$u(x_{i-1}) = u(x_i) - hu'(x_i) + \frac{h^2}{2}u''(x_i) + O(h^3), \tag{7.133}$$

and subtract (7.133) from (7.132) to obtain

$$u'(x_i) = \frac{1}{2h}\left[u(x_{i+1}) - u(x_{i-1})\right] + O(h^2). \tag{7.134}$$

Finally, inserting the last result in (7.132), it follows that

$$\frac{1}{2h}\left[u(x_{i+1}) - u(x_{i-1})\right] + O(h^2) + u(x_i)^2 = f(x_i), \quad i = 1, \ldots, N. \tag{7.135}$$

As usual in a finite-difference discretization, we neglect the infinitesimal, in this case $O(h^2)$, which turns the Eq. (7.135) into an approximation:

$$\frac{1}{2h}\left[u(x_{i+1}) - u(x_{i-1})\right] + u(x_i)^2 \approx f(x_i), \quad i = 1, \ldots, N. \tag{7.136}$$

In order to obtain a genuine equality, we replace the sequence of unknown quantities $(u_i)_{i=1,\ldots,N}$ by the numerical sequence of approximations $(\tilde{u}_i)_{i=1,\ldots,N}$ defined by the recurrence relation

$$\frac{1}{2h}\left(\tilde{u}_{i+1} - \tilde{u}_{i-1}\right) + h\tilde{u}_i^2 = hf_i, \quad i = 1, \ldots, N. \tag{7.137}$$

Clearly, there is every reason to expect the sequence of approximations $(\tilde{u}_i)_{i=1,\ldots,N}$ to produce a satisfactory approximation to the values of $(u_i)_{i=1,\ldots,N}$, given that the process for constructing it was directly motivated by the one for the exact values $(u_i)_{i=1,\ldots,N}$, up to an error of $O(h^2)$.

In conclusion, the finite-difference scheme (7.137) corresponds exactly to the node equation (7.130) found for any function φ_i characterizing a node strictly in the interior of the integration interval $[0, 1]$.

▶ **Basis Function φ_{N+1} Characterizing the Node at x_{N+1}**

7) We now consider the basis function φ_{N+1} characterizing the node at $x = 1$. In this case, and by similar arguments to those developed in the last question for the nodes strictly in the interior of $[0, 1]$, the approximate variational formulation $\widetilde{(\mathbf{VP})}$ implies the equation $(\widetilde{\mathbf{VP}_{\text{ext}}})$ expressed as follows:

$$(\widetilde{\mathbf{VP}_{\text{ext}}}) \begin{cases} A_{N+1,N}\tilde{u}_N + A_{N+1,N+1}\tilde{u}_{N+1} + B_{N+1,N,N}\tilde{u}_N^2 + \cdots \\ (B_{N+1,N,N+1} + B_{N+1,N+1,N})\tilde{u}_N\tilde{u}_{N+1} + B_{N+1,N+1,N+1}\tilde{u}_{N+1}^2 = C_{N+1}. \end{cases}$$
$$(7.138)$$

Once again, we have used the property of the supports of the basis functions φ_N and φ_{N+1}.

Approximate Calculation of the Coefficients $A_{N+1,N}$ and $A_{N+1,N+1}$. We can now calculate the coefficients:

1. *Calculation of $A_{N+1,N}$.*

   We have

   $$A_{N+1,N} = \int_0^1 \varphi_{N+1}\varphi_N'\, dx = \int_{\text{supp }\varphi_{N+1} \cap \text{supp }\varphi_N'} \varphi_{N+1}\varphi_N'\, dx$$

   $$= \int_{x_N}^{x_{N+1}} \varphi_{N+1}\varphi_N'\, dx$$

   $$\approx -\frac{1}{h} \times \frac{h}{2}\left[\varphi_{N+1}(x_N) + \varphi_{N+1}(x_{N+1})\right]$$

   $$= -\frac{1}{2}(0+1) = -\frac{1}{2}. \tag{7.139}$$

2. *Calculation of $A_{N+1,N+1}$.*

This time, we have

$$A_{N+1,N+1} = \int_0^1 \varphi_{N+1}\varphi'_{N+1}\,dx = \int_{\text{supp } \varphi_{N+1}} \varphi_{N+1}\varphi'_{N+1}dx$$

$$= \int_{x_N}^{x_{N+1}} \varphi_{N+1}\varphi'_{N+1}\,dx$$

$$\approx \frac{1}{h} \times \frac{h}{2}\big[\varphi_{N+1}(x_N) + \varphi_{N+1}(x_{N+1})\big]$$

$$= \frac{1}{2}(0+1) = \frac{1}{2}. \tag{7.140}$$

Approximate Calculation of $B_{N+1,N,N+1}$, $B_{N+1,N+1,N}$, and $B_{N+1,N+1,N+1}$.

For the same reasons as those discussed in the answer to Question 4, only the coefficient $B_{N+1,N+1,N+1}$ can be nonzero, given that we are using the trapezoidal rule.

We thus have

$$B_{N+1,N+1,N+1} = \int_{x_N}^{x_{N+1}} \varphi^3_{N+1}(x)\,dx \approx \frac{h}{2}(1+0) = \frac{h}{2}. \tag{7.141}$$

Approximate Calculation of the Coefficients C_{N+1}.

Here we obtain

$$C_{N+1} = \int_0^1 f(x)\varphi_{N+1}(x)\,dx = \int_{x_N}^{x_{N+1}} f(x)\varphi_{N+1}(x)\,dx$$

$$\approx \frac{h}{2}(0 + f_{N+1}) = \frac{h}{2}f_{N+1}, \tag{7.142}$$

where f_{N+1} is the value of the function f at the node $x_{N+1} = 1$.

8) We now bring together the results of these calculations to write down the corresponding node equation:

$$\frac{1}{h}\big(\tilde{u}_{N+1} - \tilde{u}_N\big) + \tilde{u}^2_{N+1} = f_{N+1}. \tag{7.143}$$

9) The node equation (7.143) corresponds exactly to the equation that would be obtained by a finite-difference discretization of the Riccati equation (7.111). However, there is a major difference with the finite-difference scheme established in Question 4 for the points strictly in the interior of the mesh.

Indeed, in the present case, it is quite clear that the finite-difference scheme (7.143) is no longer of second order, but of only first order.

On the other hand, it is only a local reduction in the order of the method, since the finite-difference scheme is of order 2 as far as the node at x_N, while only the last node x_{N+1} suffers a local degradation in the order of approximation.

Such a result should come as no surprise if we recall the Bramble–Hilbert lemma (see Lemma 2.7), which tells us that the finite-element method will be of order k if the variational approximation space $\tilde{V}$ contains the space P_k, i.e., the set of polynomials of degree less than or equal to k.

In our case, it is clear that k is equal to 1, and the second order that we obtain here is just a bonus due to the uniformity of the mesh we have used.

References

1. D. Euvrard, *Résolution des équations aux dérivées partielles de la physique, de la mécanique et des sciences de l'ingénieur* (Masson, Paris, 1994)
2. P.A. Raviart, E. Godlewski, *Numerical Approximation of Hyperbolic Systems of Conservation Laws*. Appl. Math. Sci., **118** (Springer, New York, 1996)
3. M. Crouzeix, A.L. Mignot, *Analyse numérique des équations différentielles* (Masson, Paris, 1983)

Index

Printed in the United States
By Bookmasters